We are indebted to our wives, Debbie and Tina, and our daughters, Christa, Kary, Katy, Kelly, and Kimberly, whose patience, support, and understanding helped bring this effort to fruition.

PREFACE

The 1980s have brought an increased emphasis on market planning and decision-making, particularly as it relates to the challenge of developing new products and services. With the continued pressures faced by companies from economic, environmental, social, political, and cultural issues, it has become necessary for management to more carefully appraise the existing marketing program(s) to identify opportunities for innovation and develop marketing strategies to enhance the objectives of the organization with existing products and services. The risk associated with the development of innovations has brought new challenges to executives in all types of industries. Yet companies, large and small, continue to innovate under these demanding circumstances.

In the opinion of the authors, a systematic approach to planning, developing, and controlling a product or service throughout its life cycle can minimize the risk associated with the innovative process. Since the cost of developing new products and services has risen significantly in the past decade, managers have adopted a more cautious approach to introducing innovations. It is our intent to present the reader with methods and applications that can minimize the risk of new product development and provide managers with the techniques and understanding to plan and control products and services.

The basic theme of the book is indicated in three important managerial concepts, which are clearly designated as sections of the text:

1. *Planning*—The planning process is important in the pre-venture stage, as well as in each stage of the product life cycle. Two levels of planning are discussed in the text: strategic planning, which describes the long-term corporate planning process and its relationship to product development and strategy; and market/product planning, which serves as the annual framework for short-term marketing decision-making to help management achieve marketing objectives within the long-range plan of the organization.

2. *Development*—Given the continued pressures of the market environment, it is increasingly necessary for a firm to establish a system whereby new products can be developed and introduced to the marketplace with a minimum amount of risk. One important mechanism for realizing this objective is to establish an effective organization for the product development process, which allows for periodic evaluation and review of new ideas and concepts at appropriate stages until they are introduced into the marketplace. An organizational structure that operates efficiently, creatively, and effectively in the organization's internal environment will ensure that the new product is prepared to meet the needs of the market.

3. *Control*—Effective introduction and marketing of new products and services require that management carefully monitor the product or service over its life cycle. Performance standards are needed for periodic review to determine if the product or service is meeting the

established market objectives. Appropriate strategy modification during this review and monitoring process allows for effective control of the product and enhances the probability of success. As the product reaches maturity, decisions relating to stimulating sales through spin-offs and finding new market segments or new uses for the product are considered. This procedure also ensures effective control of the product in the later stage of its life cycle. If the product is no longer able to meet market objectives, abandonment may be considered.

The book is divided into seven major sections. Part 1, "The Role of New Products and Their Environment," introduces the product development process and discusses reasons for the eventual success or failure of new products. Social and environmental issues and their relationship to the product planning process are also presented, in addition to the factors relating to legal issues, competition, government pressures and constraints, and consumer preferences.

Part 2, "Strategic Market Planning," provides the framework for effective planning at the corporate level and specific product level. Assessment of the product mix, using product life cycle analysis, product portfolio analysis, or perceptual mapping, is presented in terms of the strategic plan and the market/product plan.

"Organizing and Planning New Products," the topics of Part 3, introduce alternative organizational structures used for developing new products and strategies throughout the life cycle.

Part 4, "New Product Ideas: Obtainment and Evaluation," provides the framework for effectively evaluating a new idea prior to committing extensive resources to its commercialization, and discusses the initial stages of the evaluation and development process. This is followed by the fifth section, "The New Product: Acceptance and Demand Determination," which focuses on the techniques and processes related to analyzing the target markets and forecasting market demand in these market segments.

Part 6, "The New Product's Marketing Program," introduces the marketing mix for the new product. An in-depth discussion of the basic elements of the new product's marketing plan—including branding, packaging, distribution, pricing, and promotion—is presented.

Finally, Part 7, "Controlling and Managing the Product Line," focuses on the control of the new product introduction and management decision-making at the mature and decline stage of the product life cycle. Factors to be considered at these stages include spin-offs, new market segments, new uses, product/marketing modifications, and abandonment.

Many people—business executives, professors, publishers, and students—have contributed to this textbook and we personally want to express our sincere gratitude to all of them. Of great assistance were the detailed and thoughtful comments of the reviewers: Cathie H. Michlitsch, Saint Louis University; Malcolm L. Morris, The University of Oklahoma; and Bruce M. Smackey, Lehigh University. Special thanks are given to Anne Shenkman, without whom this manuscript would not have been typed on time, and to Patricia Phelar for her unselfish support in providing research material and editorial assistance.

CONTENTS

I

The Role of
New Products and
Their Environments

New Products:
Past and Future

The 1980s are expected to bring important challenges to management of both large and small companies. With constant changes occurring in the economic, social, and political environment, managers will be required to be more systematic and effective in their decisions to develop, plan, and market new and existing products and services.

As government encourages increased technological development through research, industry cries for tax incentives, less government intervention, more resources, more import quotas, and preferred treatment for United States firms. While these restrictions or requirements may benefit one industry, they often negatively affect some other industry, firm, or customer group. For example, increased quotas on footwear or automobiles have raised prices for consumers. Increased needs of consumers and companies for natural resources have resulted in shortages and consequently higher prices. Tax incentives for some industries generate resentment from those who are not given government incentives. Who makes the tax incentive decisions? Does everyone benefit from these decisions? What criteria are used in making these decisions? The solutions are not easy, nor are they likely to satisfy everyone. However one thing is clear—the realities of the economic, social, and political environment dictate that, in order to survive, a firm will have to continuously change its product line to maintain a competitive balance within the changing environment.

In the past firms spent millions of dollars on research and development, which led to a proliferation of new products and services in our society. However current economic realities necessitate finding alternative methods to enhance the product line. Modifications, reformulations, and extensions of the product line will likely dominate the marketing strategy in the next decade. This does not preclude the development of new products through research and development, however. Although the number of new products is generally declining, we still continue to see an abundance of products introduced each year. In fact economic challenge often provides the stimulus for new product development. Scarcity of natural resources and increased environmental controls, for example, have led firms to develop alternative energy sources and energy-efficient products. The increased consciousness of industries to find more efficient means to communicate, analyze, and adjust their marketing strategy in a complex environment has led to remarkable improvements in data communication and processing. These improvements in data communication and information processing in the business environment have led to the development of computers for home use.

Companies must continually seek new products through whatever means are available. New product development must be a necessary element in the planning process since a status quo strategy may lead to the eventual decline and death of the firm.

The cutback in the number of new products is already evident in some industries. In 1972 more than 9,000 new items were introduced in supermarkets. The following year this number dropped to 7,000 and more recently has leveled off to 5,000.

Firms that fail to develop new products can become more dependent on existing successful products, which in the long run can lead to a "milking strategy" of cutting marketing budgets, cheapening formulas, and raising prices for short-term earnings growth. Standard Brands, for example, relied on slow growth and a "milking strategy" from 1964 to 1976. It became dependent on its highly cyclical corn syrup refining business, which accounted for about 15 percent of sales but almost 50 percent of earnings by 1976. When sugar prices crashed in that year, Standard Brands was left with extensive losses. By 1978, however, the firm made a complete turnaround by developing and marketing new products in such categories as candy, frozen foods, liquors, margarine, and corn syrup. These new products have contributed significantly to the company's recent success.

Currently Procter & Gamble, the largest marketer of consumer packaged goods, is reportedly developing, testing, and expanding distribution of more new products than at any other time in its history.[1] In addition the firm is entering institutional and commercial fields, such as medical products, drugs, and synthetic food products.

Standard Brands and Procter & Gamble are only two examples of firms using new product development to maintain a strong market position. New product development is the means to achieve a successful market position over the long term.

SIGNIFICANCE OF NEW PRODUCT DEVELOPMENT

As stated earlier, new and improved products are necessary to the survival and growth of any firm in a complex, changing environment. The following sections outline common reasons for new product development.

Growth   Probably the most important reason for developing new products is to stimulate growth in the firm. Growth conveys an image of success to investors and the financial community. Growth or an increase in market share is the ultimate goal of many organizations since it often means higher profitability and cash flow. The relationship between growth and profits is discussed in Chapter 4.

Response to competition   Since not everyone can be the first to introduce a new product, firms are generally responding to the innovator in the marketplace. Expanding the product line in response to innovations in the field protects the firm's competitive position. Improvements in the competitor's product can sometimes even lead to a dominant market position.

Excess capacity   The development of new products is also stimulated by excess capacity in the firm. Machinery, plant space, and labor all require certain fixed expenditures whether they are used 50 or 100 percent of the time. A firm can enhance cost and resource effectiveness by developing alternative products that use existing excess capacity and job skills.

---

[1]"P & G's New New-Product Onslaught," *Business Week*, 1 October 1979, 76–82.

Increase in cash flow and profitability   Firms involved in markets where competition intensifies in response to dwindling resources or demand must be able to support the downturn in their particular industry. Mergers, acquisitions, and licensing provide alternative means to market new products that may enhance cash flow and profitability during economic downturns. For example, a large copper manufacturer recently purchased a coal mining company to enhance cash flow and profitability to offset the weak short-term copper market.

Stimulation of sales of other items in the line   In the industrial market, buyers often "trade up" as their needs change. This is particularly evident in the computer market where the development of a small computer system by Honeywell, for example, has enhanced sales of its larger systems. By marketing the new system to the first-time user, Honeywell enhances its chances of selling a larger system to this same customer once its data processing needs increase. With distributive networks in the computer industry, the opposite is also occurring. Manufacturers of small computer systems, such as Digital Equipment and Data General, are developing larger systems to integrate with the smaller computers in a data communication network. Besides stimulating sales of other products, this rounding out of the product line can also enhance market share and make more efficient use of any excess capacity.

Response to changing environment   To continue successfully, a firm must seek alternative products when consumer preferences change or environmental conditions dictate a bleak future for an existing product. For example, in response to changes in consumer tastes toward alcohol, distillers have introduced many new, lighter products, such as 80-proof whiskey. In a similar way, industry has responded to the growing consumer awareness of increased energy costs by developing many new energy-efficient products. Even new legislation can force firms to develop new products or modify existing ones to accommodate new regulations (e.g., the automobile industry's development of pollution controls).

Cost reduction   Many innovations may be designed to reduce the cost of an existing product. For example, synthetics in shoes and clothing helped to reduce the costs of these products as well as enhance their durability. Plastic as a substitute for glass reduced the cost of shipping and breakage for many soft drink companies, hence reducing the total cost to the manufacturer. Substitution of aluminum for steel in automobiles not only made them lighter (providing better gas mileage) but also lowered the cost of materials.

Creativity of management   The innovativeness or creativity of top management cannot be overlooked as an important stimulus to new product development. Edwin Land is recognized as one of the most creative and innovative people in industry today. His resourcefulness and interest in creating new products contributed significantly to the growth of Polaroid Corporation.

Thus for many different reasons firms are modifying their product lines. A balance between new and mature products is necessary to provide a positive cash flow, as well as to generate funds for future product development. This

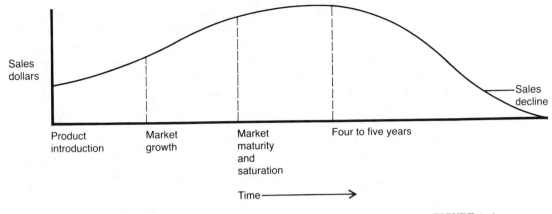

FIGURE 1–1
The product life cycle

balance is necessary because products have different life spans. In some industries this life span may be less than a year (style of clothing) and in others
it may be many years (Ivory Snow). Because of this varying life cycle, management needs to maintain a balance of products at different stages.

PRODUCT LIFE CYCLE

Figure 1–1 shows that the life cycle of a new product is divided into four major
stages: product introduction, market growth, market maturity and saturation,
and sales decline. Time, represented on the horizontal axis, can vary from
days to years, depending on the nature of the market. The product's marketing
mix[2] also varies during these stages as a result of consumer tastes and attitudes, cultural changes, competition, government, and economic factors. The
most profitable stage for the seller is the growth stage. Sales of the total industry rise rapidly in this stage as more and more customers enter the market. To
take advantage of this profitability, manufacturers try to extend the growth
stage by introducing a stream of new products or modifying them to bring
them back through the life cycle. Mature products tend to generate more cash,
while growth products require large amounts of cash to sustain growth. Products being developed through research also require significant funding. With
these factors in mind, the importance of balancing products in various stages
of their life cycle becomes clear. This concept will be discussed further in
Chapter 4.

The proliferation of new products has continued even after the shock of
an energy crisis and severe inflation. For example, toy manufacturers are continuing to develop new products to maintain market shares in highly competitive markets. Although caution prevails, one firm's manager recently stated
that his company had no intention of letting the economy affect its new

---

[2]This concept will be referred to throughout the text in relation to the market mix elements,
(i.e., pricing, distribution, and promotion).

product development. New household products have also continued to enter the market at a rapid pace. New window cleaners, oven cleaners, floor wax removers, laundry care products, rug cleaners, insect sprays, and bowl cleaners have all recently been introduced with some new chemical, packaging innovation, or other modification. Most consumer product markets continue to grow because of the increased competition and intense pressure to develop that one outstanding new product. Generally research and development departments are focusing on specific market segments. Eastman Kodak, Raytheon, and RCA are among those companies that emphasize specific market segments and their needs. This careful focusing of research and development is intended to enhance the successful growth of these firms.[3]

The industrial market, although a bit more cautious, continues its heavy spending on research and development. The computer industry continues to initiate innovations that reduce the average product life cycle of a computer to about four or five years. Although firms such as RCA, General Electric, and Xerox have left the market, their business has been absorbed by others, thus increasing the competition and reinforcing the need for new product development.

The most recent trend in the computer industry is in the mini or personal computer market. This field has grown significantly in the past few years as markets for small businesses, hobbyists, and home users have developed. Many spin-offs from these personal computers (e.g., Apple and TRS-80) are likely to be developed as software packages begin to catch up to the hardware technology. Computerized games are now one of the hottest selling products available in the marketplace. The potential of this market seems inexhaustible, yet many major manufacturers are moving with caution because they lack a complete understanding of the market. Decisions on how these products will be marketed, to whom they will be marketed (homeowners or small/large businesses), and what related services will be offered have yet to be finalized. Marketing executives will have to carefully monitor the life cycle of these products to minimize the risk of large capital investments in a volatile market.

MARKET MYOPIA

When a product reaches market maturity and saturation (see Fig. 1–1), firms are often forced to introduce new products. This is most easily done by modifying or extending the existing product line. The danger in doing this is that many firms take a myopic view of their markets and tend to gravitate to the lowest-risk line extension or modification to stimulate continued growth in company sales. Although relatively safe, low-risk decisions often have a low return. Decision-making of this type could also lead to a cannibalization of existing brands, resulting in a small or negative net increase in total company sales.

An example of market myopia occurred when the railroad and movie industries failed to recognize the full potential of their markets.[4] If railroads

---

[3]See *Wall Street Journal*, 18 October 1977, 1.
[4]Theodore Levitt, "Marketing Myopia," *Harvard Business Review* 38, no. 4 (July–August 1960): 24–47.

had perceived themselves as being in the transportation business in general, they would have foreseen opportunities to diversify and to provide new services to meet the transportation needs of their customers. Instead of attempting to fight off competition from television, the movie industry might have provided new movies and services to meet the needs of this market. Many failures resulted from the myopic view taken by the railroads and movie industries.

WHAT IS A NEW PRODUCT?

Part of the problem of defining a "new" product is in identifying what is actually new about it.[5] For example, fashion jeans were very popular in the 1980s, yet the concept of blue jeans is not new. What is new is the use of fashion names such as Sasson, Vanderbilt, and Chic to promote the jeans. The first home permanent product introduced by Toni is a similar example. Of course, hair permanents had long been available at hairdressing shops throughout the world. However as a product available for home use, the home permanent was a new concept.

In these examples the newness was in the consumer concept. Other types of products, not necessarily new in concept, have also been defined as new. Nescafé recently introduced a new and improved instant coffee in which the only change in the product was the size of its granules. Yet the initial promotional campaign stressed the word *new* in its copy.

Other old products have simply been marketed in new packages or containers but have been identified as new products by the manufacturer. In 1969 an innovatively packaged food product called Zonkers was introduced in supermarkets. This product was actually caramel corn sold in a psychedelic package of bright colors, comic figures, and riddles. Obviously, caramel corn was not a new concept in 1969.

The discovery of the aerosol can was also an example of a change in the package or container that updated old, established products, such as whipped cream, deodorants, and hair sprays. Flip-top cans, plastic bottles, and screw-on caps also contributed to this perceived image of newness in old products. Drain cleaners are now marketed in pressurized containers to emit bursts of air to clean drains and pipes. Other firms, such as detergent manufacturers, have merely changed the colors of their packages and added the word *new* to the package and to their promotional copy to convey the illusion of a new product.

Panty hose is another example of a product that has undergone significant changes in marketing strategy in the past few years. L'eggs was the first product to take advantage of supermarket merchandising, new packaging, lower prices, and new displays. Now competitors are using similar marketing strategies to create the image of a new product.

In the industrial market, firms often call their products new when only slight changes or modifications have been made. For example, improvements in metallurgical techniques have increased the precision and strength of many

---

[5]For a more comprehensive discussion of this important issue, see Chester Wasson, "What Is New About a New Product?" *Journal of Marketing* 25 (July 1960): 52–56.

raw materials used in industrial products, such as machinery. These improved characteristics have led firms to market products containing the improved metals as new. Computer manufacturers have expanded into the target markets by modifying their hardware slightly and by adapting computer software techniques to meet the needs of new customers not previously serviced, such as firms engaged in specialized wholesale operations.

In the process of expanding their sales volume, many companies add products to their product line that are already marketed by other companies. For example, a drug company that added a cold tablet to its product line and a long-time manufacturer of soap pads that entered the dishwasher detergent market both advertised their products as new. In both cases the product was new to the manufacturer but not to the consumer. With the increased emphasis on diversification in our economy, this situation is quite common today. Firms are constantly looking for new markets to exploit to increase profits and make more effective use of their resources.

Although there are many products that have some new variation, there are relatively few truly new products, considering the sophisticated technological improvements in our present society. Examples of truly new products are the electronic computer, the polio vaccine, television, the transistor, and contact lenses. All of these products were based on new technology and represented something far different from anything else previously available in the marketplace.

CLASSIFICATION OF NEW PRODUCTS

New products may be classified from either the viewpoint of the consumer or the firm. Both points of view must be carefully analyzed since both the ability of a firm to establish and attain its product objectives and consumer perception of these objectives may determine the success or failure of any new product.

From a Consumer's Viewpoint

There is a broad interpretation of what may be labeled a new product from a consumer's viewpoint. Robertson has attempted to schematically identify these new products based on given criteria for interpreting newness.[6] New products are classified according to how much behavioral change or new learning they require of the consumer. This approach looks at newness in terms of its effect on the consumer rather than whether the product is new to a company, is packaged differently, has changed in physical form, or is an improved version of an old or existing product.

The continuum proposed by Robertson (shown in Fig. 1–2) contains three categories, all based on the disrupting influence the use of the product has on established consumption patterns. Most new products tend to fall in the continuous end of the continuum. Examples include annual automobile

---

[6]Thomas Robertson, "The Process of Innovation and the Diffusion of Innovation," *Journal of Marketing* 31 (January 1967): 14–19.

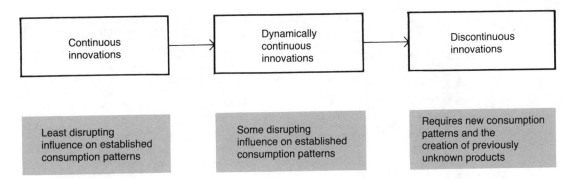

FIGURE 1–2
Continuum for classifying new products

changes, fashion style changes, package changes, and size or color changes in existing products. Products such as the electric carving knife, electric toothbrush, and electric comb tend toward the dynamically continuous portion of the continuum. The truly new products, referred to as *discontinuous innovations* in Figure 1–2, are rare and require a great deal of new learning by the consumer because they generally perform either a previously unfulfilled function or an existing function in a new way.

Identifying new products based on their effect on consumer consumption patterns is consistent with the marketing philosophy that satisfaction of consumer needs must be a prime justification for the firm's existence. Thus when defining what a new product is, one must refer to the effects it will have on consumption or behavioral patterns. This kind of operational definition is needed for consistency in analysis of research data and projections or conclusions made from different categories of new products.

From a Firm's Viewpoint

In addition to recognizing the consumer's perception of newness, the innovative firm may also classify its new products on some other similar dimension. One way to classify the objectives of new products is shown in Figure 1–3. These dimensions, first suggested by Johnson and Jones, provide an important basis for determining the extent of the marketing strategy needed to achieve company objectives.[7]

Classification is most difficult when both a new technology and a new market are involved. In this case, the firm will need a new and carefully planned marketing strategy. Replacements, extensions, product improvements, reformulations, and remerchandising require less complicated marketing strategies, since the firm will have had prior experience with a similar product or with basically the same target market. The minicomputer provides a good example of the possible combinations of new markets and new technology that can affect the classification of the product for the firm. With little

---

[7]Samuel C. Johnson and Conrad Jones; "How to Organize for New Products," *Harvard Business Review* 35 (May–June 1957): 52.

Product Objectives	No Technological Change	Improved Technology	New Technology
No market change		Reformulation: Change in formula or physical product to optimize costs and quality	Replacement: Replace existing product with new one based on improved technology
Strengthened market	Remerchandising: Increase sales to existing customers	Improved product: Improve utility to customers	Product life extension: Add new similar products based on new technology to serve more customers
New market	New use: Add new segments that can use present products	Market extension: Add new segments modifying present products	Diversification: Add new markets with new products developed from new technology

Increasing Technological Newness →

Increasing Market Newness →

FIGURE 1–3

Classifying new products based on product objectives

or no new technological development, the manufacturer of a minicomputer may choose to exploit the marketplace further. The same system may be marketed through original equipment manufacturers (OEMs) to reach new markets, such as banks and insurance firms. OEMs buy computer hardware and modify it and add to systems to reach special consumer markets. Another option would be to increase the sales of the same system to existing customers (i.e., customers can link two systems together as part of a data communication network). These decisions would be classified as remerchandising or new use by the developing firm (see Fig. 1–3).

Technological improvements in the same market or new markets can also be implemented. For example, IBM's reduction in cost of a memory kit or Data General's reduction of power needed for computer operation has allowed for more efficient costs and improved quality of these systems to the same market (reformulation). Improved user capacity for a distributive network system, such as the Honeywell Level 60 series, expanded sales to existing customers as well as to new customers. Higher-speed printers, expanded memory, disk, cassette, or batch adaptability all would allow the manufacturer to expand its capability in existing markets (improved product) or reach totally new markets (market extension).

High technological development in the computer industry has recently involved the personal or microcomputer. New technology affecting price, quality, reliability, and simplicity has resulted in the potential mass marketing of these products for households. These systems could be marketed to existing customers (replacement); small businesses, such as medical and legal practices and real estate brokerage firms (product line extensions); or private individuals (diversification).

This classification system allows the firm to recognize from the category the complexity of the market decision-making that is required. The less experience the firm has with the market and the technology, the more complex the decision process will be.

Figure 1–4 illustrates some of the effects new markets and technologies have on the product line of a firm. The firm in this example has an initial product line of one product—A2. With no technological change in the product but improved marketing strategy, the firm may increase the size of its existing market. The improved marketing strategy may consist of an increased sales force, increased advertising, new channels of distribution, or perhaps a lower price. The remerchandising or change in the marketing strategy in the example in Figure 1–4 results in greater sales or an expanded market.

Product improvement may result in an increase in sales of the product. In our earlier discussion of the Honeywell Level 60 series, these improvements consisted of such changes as higher speed and increased memory, which satisfy the needs of the same market but give the firm Product A1. It is conceivable that product improvements could also increase the sales of the product to the same market. Thus customers may shift from other manufacturers in light of the product improvements in A1.

Significant developments in new technology are likely to lead to the fourth situation in Figure 1–4, where the manufacturer replaces product A1 with A2. In this situation A1 would no longer have any demand because of the improved technology. A good example would be the technological development of pocket calculators, which left little or no demand for slide rules. In the computer industry, firms stop manufacturing a system when it is replaced with an improved system.

Market extension can occur with or without new technology. Product A2 could be introduced to a new market without having undergone any changes. For example, hair dryers are now marketed to men and women, providing a new market for the same product. Industrial products introduced as consumer products or expansion to an international market would

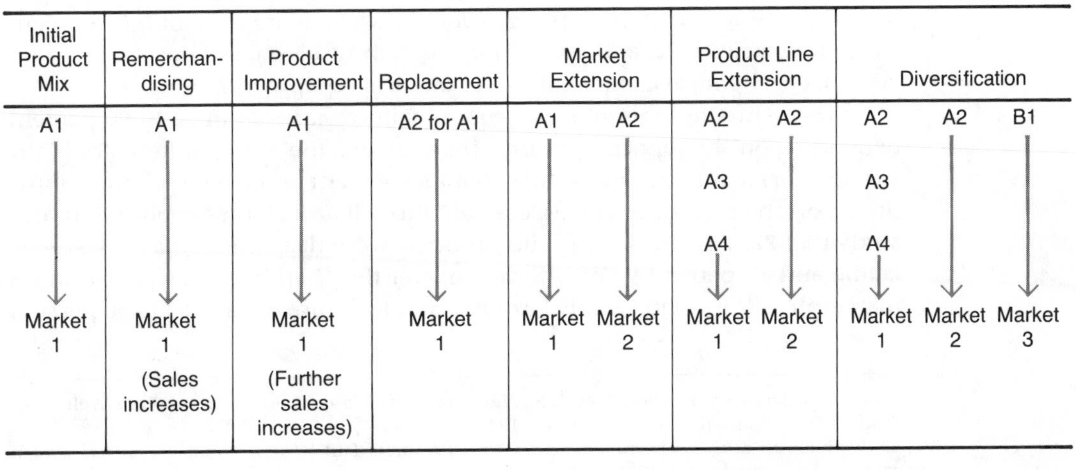

Initial Product Mix	Remerchandising	Product Improvement	Replacement	Market Extension		Product Line Extension		Diversification		
A1	A1	A1	A2 for A1	A1	A2	A2	A2	A2	A2	B1
						A3		A3		
						A4		A4		
Market 1	Market 1	Market 1	Market 1	Market 1	Market 2	Market 1	Market 2	Market 1	Market 2	Market 3
	(Sales increases)	(Further sales increases)								

FIGURE 1–4
Examples of new product classifications

also be examples of market extension. Each firm's classification could be different, depending on the market it presently reaches and the level of its technology.

Product line extension generally provides the company with an opportunity to use the same communication and distribution channels to service the same market. This product line extension could conceivably take two directions. One possibility would be to provide consumers with some variations in style, color, quality, or other features. This may increase sales by adding new customers in the same market. Another possible product line extension would increase the inventory of the firm's products that any one customer would buy. For example, Heinz added a new barbecue sauce to its tomato catsup product so that a consumer loyal to its catsup would also purchase the new product.

Diversification in Figure 1–4 consists of the addition of a new product—B1—that is significantly different from A2. This strategy also adds a new market of customers to those presently being serviced. This diversification strategy can be achieved by acquiring the firm that developed B1 or by licensing a product invented by someone else or through internal development. In all cases the risk to the manufacturer is highest since it may have little or no expertise in this market.

SUCCESS AND FAILURE OF NEW PRODUCTS

Of course not all new products quickly become spectacular profit or sales successes. New products are often removed from the marketplace at one of the earlier stages of their life cycle because of low sales volume (see Fig. 1–1). Yet research and development in 1983 will be about $75 billion, which is about a 9 percent increase over 1982. The government spent about half of this as the new administration continued its efforts to support business.

It is expected that most of this increase will occur in electronic computing equipment with forecasted growth rates expected at about 18 percent. Other industries with anticipated increases in research and development are aerospace, chemicals, machinery, and medical equipment.[8]

The estimated rate of new product failure varies from 10 to 90 percent, depending on the reporting source. For example, the National Industrial Conference Board, Inc., indicated that about 30 percent of the new products introduced on the market are unsuccessful.[9] Buzzell and Nourse reported in their study that 22 percent of new food products were discontinued after test marketing and 17 percent were withdrawn after they had been introduced in the marketplace.[10] In a later study Lazo referred to failure rates of 80 to 90 percent,

---

[8]"Business Recovery Predicted by U.S.: New Product-Related Companies to Fare Well," *New Product Development* 3, no. 2 (February 1983): 4–5.
[9]Betty Cochran and G. Thompson, "Why New Products Fail," *The National Industrial Conference Board Record* 1 (October 1964): 11–18.
[10]Robert D. Buzzell and Robert E. M. Nourse, *Product Innovation in Food Processing, 1954–1964* (Boston: Division of Research, Harvard Business School, 1967), 10.

which included all products introduced by firms of varying size and competence.[11] Percent analysis supports the premise that failure rates today remain as high as they were 25 years ago.[12]

The disparity in these findings can be explained by examining the method used to define a new product, the manner in which a failure was determined, and the sample of the firms used in each study. Variances in these three factors resulted in different conclusions regarding the percentage or the number of new product failures. Regardless of the method or the varied results, marketing executives agree that the failure rate is too high.

REASONS FOR NEW PRODUCT FAILURE

The National Industrial Conference Board study suggested the following reasons and their percentage of occurrence for new product failures:

1. Inadequate market analysis	32%
2. Product defects	23%
3. Higher costs than anticipated	14%
4. Poor timing	10%
5. Competition reaction	8%
6. Inadequate marketing effort (includes weaknesses in sales force, distribution, and advertising)	13%
	100%[13]

A breakdown of these figures indicates that marketing inadequacies cause most failures (inadequate market analysis, poor timing, competition and sales force, distribution, and advertising). The Buzzell and Nourse study also substantiates these findings by concluding that approximately 80 percent of the reasons given for discontinuing the product resulted from marketing misjudgments or inadequacies.[14]

Thus although organizational and operational problems can often impede the success of a new product, products usually fail because of marketing problems. Figure 1–5 lists comparable products whose degree of success or failure is attributable to a marketing problem. The promoters of the successful products seemingly recognized the importance of critical marketing variables and developed an appropriate marketing plan that focused on these issues.

The next two sections discuss and analyze some of the more well known failures and successes.

---

[11]Hector Lazo, "Finding a Key to Success in New Product Failures," *Industrial Marketing* 50 (November 1965): 74–75.
[12]Merle Crawford, "Marketing Research and the New Product Failure Rate," *Journal of Marketing* 41, no. 2 (April 1977): 51–61.
[13]Cochran and Thompson, "Why Products Fail," 14.
[14]Buzzell and Nourse, *Product Innovation*, 18.

Highly Successful Product	Less Successful or Failed Product	Key Marketing Reason for Failure
Lipton's Cup-A-Soup	Knorr Soup	Positioning
Mustang	Edsel	Product design
Taster's Choice Freeze-dried Coffee	Maxim Freeze-Dried Coffee	Advertising
Soaky Bubble Bath	Mr. Bubbles	Packaging
Prestone Antifreeze	Zerex Antifreeze	Distribution
Clorox Bleach	Action Bleach	Pricing

FIGURE 1–5

Effect of marketing factors on comparable products

Source: F. Beaver Ennis, "Improving the Success Ratio of New Products," *Product Marketing* 1976.

PRODUCT FAILURES: SOME EXAMPLES

There are many examples of well known product failures. A few of these are considered classics and are continually discussed and analyzed by business people and students of marketing.

The Ford Edsel The impact of new product failures on society and the manufacturer raises important concerns about more effective marketing decison-making. One of the most famous failures was the Ford Edsel, originally destined to be Ford's entry into the medium-priced field. Ford spent ten years in planning and researching it, invested approximately $50 million in advertising and promotion in 1957, and set up new, independent Edsel dealers. The competition was firmly entrenched with numerous models in this medium-priced growth market, while Ford had only Mercury. During the first three years of its existence, only 109,466 Edsels were sold—far below forecasts. In 1959 production was halted. Even though Ford used some of the Edsel plants and tools in other divisions, the firm still lost more than $100 million on the original investment and an estimated $100 million in operating losses.

Although many factors contributed to the failure of the Edsel, certain marketing factors were primarily responsible. First of all, 1958 turned out to be a recession year that negatively affected sales of all medium-priced cars. Second, two of the Edsel's most important features, power and performance, were under extreme criticism and scrutiny by the National Safety Council and Automobile Manufacturing Association. In fact, Ford was not even allowed to advertise these two features. Third, because the demand for smaller cars was beginning to take hold, the style and size of the Edsel did not have the impact on consumers that had been expected. In addition, because Ford rushed to get the product to the market, many cars were defective. Other factors, such as the dealer organization and a lack of an innovative design, also contributed to the Edsel's demise.

Dupont's Corfam Another classic failure was Corfam, a synthetic material manufactured by Dupont as a substitute for leather in shoes. This product, which took Dupont some 13 years to develop, was taken off the market after only seven years. With a $2 million advertising campaign in the first year, Corfam was promoted as a light material that breathed and flexed easily, did not lose its shape, was highly resistant to abrasion and water, and did not have to be polished. The leather industry, understandably concerned about a synthetic product becoming a substitute for leather, attacked Corfam as a poor substitute. Problems with quality control also occurred, causing Dupont's costs to rise higher than anticipated. However there was extreme optimism as problems were resolved and volume increased.

But in 1969, less expensive vinyls and fabrics entered the market, the leather industry began promoting soft leather, imports continued to infiltrate the United States market and Corfam sales declined by 25 percent. A new version of Corfam was introduced in 1970 but to no avail. The synthetic market had ceased to grow. After seven years of losses of $80 to $100 million, Dupont abandoned Corfam in 1971. The higher anticipated costs had put Corfam in the $15 to $20 market, which comprised only 10 percent of the shoe market. If Corfam had been in the lower-price market comprising 80 percent of the total market, it might have succeeded.

As with the Ford example, we see that Dupont misjudged the foreign market and competitive reaction, as well as consumer preferences. Like the Edsel, years of prior research and planning did not ensure the success of the new product.

Campbell's Red Kettle Soups   Another noted example of an unsuccessful product occurred when Campbell introduced its line of Red Kettle dry soups in 1962 with lavish advertising expenditures. In 1965, after spending $10 million in advertising, Campbell finally dropped the product. Competition from Lipton in particular proved too difficult to overcome.

Brown-Forman's Frost 8/80   In 1971, amidst a great deal of fanfare, Brown-Forman Distillers Corporation introduced Frost 8/80, a dry, white whiskey with a lower alcohol content than its competitors. Research indicated that there was widespread acceptance of the product. About two years later Frost 8/80 was taken off the shelves at a loss of about $2.5 million to Brown-Forman. Despite research findings, the concept of a white whiskey was apparently too drastic a change for consumers to accept. Also, because other distillers were promoting light whiskeys of their own at the time, consumers confused Frost 8/80 with other brands, causing the product to fail.

Analoz   The list of familiar new product failures is, of course, quite extensive. Many other new products fail much earlier in their life cycles and thus are much less familiar to the average consumer. These failures generally go unnoticed, except by the manufacturer who must absorb the financial loss. A good example occurred a number of years ago when a drug company decided that the success of antacid pills and analgesics warranted the development of a combination antacid and pain killer that could be taken without water. The company's laboratories finally came up with a cherry-flavored combination called Analoz. The product was tested by a consumer panel, which was also given competing products to compare with Analoz. The panel overwhelmingly

preferred Analoz. Then, with the support of heavy advertising, the company introduced the product in a number of test markets where it was advertised as a combination analgesic-antacid that worked without water. Impact analysis indicated that the ads were reaching a desirable number of consumers. However sales were so low that the company withdrew the product within a matter of months. Later research indicated that the fatal flaw was emphasizing that the product worked without water. People who used pain killers felt that water was necessary to cure pain and had no confidence in taking the pill without water. This example illustrates a firm's failure to consider consumer attitudes toward taking nonprescription drugs for pain relief. Analoz thus failed as a result of what was mentioned earlier as one of the main causes of new product failures—inadequate market analysis.

For the following reasons, it may be even more difficult to successfully introduce new products in the future:

1. Shortage of new ideas: The discontinuous-type innovations (see Fig. 1–2) seem to occur rarely. Most ideas are modifications or repositioning decisions.
2. Social and government pressures: The criteria for bringing a product through development to the marketplace have become more rigid. Some of these factors are discussed in Chapter 2.
3. Shorter life spans of products: Some estimates indicate that future consumer packaged goods will have about one-half the life span of existing products.
4. Cost of new product development: Since the risks are so high, costs to ensure that the idea has been thoroughly analyzed before commercialization become prohibitive. Inherent is the high cost of money, which places a burden on the return on investment (ROI) to maintain cash flow and minimum profitability requirements.

Management must face these financial constraints and pressures and seek ways to reduce the risk of new product development.

NEW PRODUCT SUCCESSES: SOME EXAMPLES

In spite of the risks, new products can succeed if their development is carefully researched and if marketing strategies are well planned. The following are examples of such successes.

Polaroid Corporation   In 1947–48 Polaroid startled the photographic industry with its Model 9's "instant photography." This developing process, although it has been refined and improved, is still used in Polaroid cameras and has proven to be one of the major innovations in the photographic industry in the last 25 years. Polaroid has continued to develop successful new products. For example in April 1972, founder and president Edwin H. Land demonstrated the SX-70 pocket camera, which could produce color pictures at the rate of one every 1.2 seconds. More recently the firm has added extensive modifications and improvements to its product line with the development of a single-lens reflex concept and improved color processing.

Tandy Corporation   In 1977 Tandy introduced the TRS-80 personal computer without cost data, sales plans, or profile of potential customers. However the company soon realized that it had produced an innovation that satisfied an important need in business, professional, instructional, and home markets. Since introduction of the TRS-80, Tandy has sold more than 200,000 units and has since added numerous accessories and software to their system. As the firm gained more experience, Tandy Corporation realized that its stores and personnel were better qualified to handle, display, and sell the computers. Because the original selling price for the basic system was $599, it was difficult for competitors to follow Tandy into the market. With sales reaching over 180 units per day, generating $180 million in annual sales, it is clear that the future for this market is excellent. Although initial marketing efforts were haphazard, recent efforts indicate a clear and sound marketing plan.[15]

Other successes   Gillette also has a history of successful new product introductions, including such recent examples as Silkience conditioner, the Trac II razor, Cricket lighters, Dry Idea deodorant, and Earth Born shampoo.

On a smaller scale a Massachusetts firm, Illuminations Inc., recently embarked on a new trend in the United States—the rainbow decal. This decal has become a new fad, as people have placed them on the rear windows of automobiles, on sliding glass doors, and in windows. The rainbow, the universal symbol of peace, became the rage of teenagers. This small firm faced extensive competition but, through careful market planning and the introduction of spin-off products, has maintained a significant share of the major market segments. Successful anticipation of the competition and improved marketing techniques turned this small company into a multimillion dollar firm.

In many instances the small innovators of the past became the giants of today. Xerox, IBM, and Digital Equipment are just a few examples of firms that were started as new ventures.

These examples of successes prove the importance of developing a sound marketing plan to avoid or reduce the probability of failure. Through the use of sound marketing strategies, a firm can successfully launch new products. Further discussion of market planning and strategy will occur in later chapters of this book.

THE PRODUCT DEVELOPMENT PROCESS

In addition to sound marketing strategy, a company must also plan an effective development process for its new product.

Given some of the major issues historically confronted in developing products and the continued proliferation of new consumer and industrial products in the 1980s, a firm must develop some mechanism— either formal or informal—to evolve new products for commercialization.

Typically, the product development process consists of the following areas.

---

[15]"Brute Force Strategy Swells Sales of Radio Shack's Home Computer," *Marketing News*, 12 December 1980, 21.

- Idea generation: The process of generating new ideas may consist of brainstorming, reverse brainstorming, attribute listing, groups, or problem inventory analysis.
- Screening: Techniques for evaluating new ideas may consist of check lists or open discussion where ideas are either eliminated or considered further.
- Business analysis: Focus groups and concept testing techniques can provide further insight to the idea prior to development of the prototype. This analysis should also provide further evaluation of ideas to eliminate any of those not considered favorably at this point.
- Development: Development of the prototype must be evaluated in light of production problems, safety requirements, costs, and modifications necessary prior to entering any test market.
- Testing: Test markets can provide valuable data on the nature of the market and needed marketing strategy changes or product modifications necessary to ensure a successful launch.
- Commercialization: The new product is launched into the market at full-scale production with a significant commitment of the firm's resources and reputation.

During the product development process, top management must watch for inadequate market analysis, product defects, higher development costs, poor timing, or poor marketing strategy. These problems can be controlled through a carefully organized planning process that meets the firm's objectives.

Figure 1–6 shows the product development process, the launch, and the subsequent product life cycle. This illustration shows that the firm does not begin to earn profits on the product until the growth stage. Up until this point, money has been spent on developing and launching the product. During the introduction stage of the product life cycle, negative profits can usually be expected until sales have reached the break-even point.

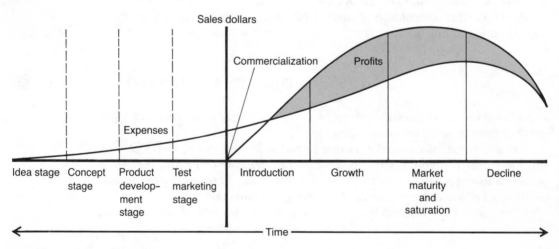

FIGURE 1–6
Sales and profits during product development and product life cycle

SUMMARY

In the near future marketing executives will be faced with their greatest challenges and pressures. Increased competition and social, cultural, and economic changes will necessitate new industry.

New product innovations range from modification of existing products to truly new products based on new technology that are different from anything else available in the marketplace. The degree of innovation in new products can be classified according to how much new learning is required of the consumer in order to use the product. This process of classification is consistent with the marketing philosophy that satisfaction of consumer needs must be one of the primary purposes for the firm's existence.

There are many dramatic successes and failures of new products. Analysis of product failures reveals the importance of adequate marketing analysis and decision making to assure new product success. The success of Ford, General Foods, Polaroid, Tandy and Gillette may be attributed to sound marketing planning and strategy. Thus, building a better mousetrap does not ensure instant success.

The remainder of this book provides insight into the process of new product development and management so that the most effective and successful market introduction may be achieved.

SELECTED READINGS

CARSON, J. W., and RICHARDS, T. *Industrial New Product Development.* New York: John Wiley & Sons, 1979.
Provides a systematic approach called *SCIMITAR* to the development of new products. The approach discussed throughout the book provides the basis for understanding the role of innovation, particularly in industrial settings where research and development traditionally play an important role in producing new products.

DAY, GEORGE. "The Product Life Cycle: Analysis and Applications Issues." *Journal of Marketing* 45 (Fall 1981): 60–132.
A special section on the product life cycle, including eight articles on its application and conceptual and empirical foundations. The first four articles deal with specific applications of the product life cycle in the houseware market, appliance industry, high technology industrial markets, and international trade pattern. The last four articles relate to the soundness of the theoretical framework for product life cycle analysis.

HARTLEY, ROBERT F. *Marketing Mistakes.* 2d ed. Columbus, Ohio: Grid Publishing, Inc., 1981.
Describes some of the classic marketing mistakes made by a variety of firms. The time span of these cases ranges over several decades and includes an indepth discussion and analysis of such firms as Gillette, World Football League, W. T. Grant, Dupont, Ford, and Coors.

HOPKINS, DAVID S. "New Product Winners and Losers." Report No. 773. New York: The Conference Board, 1980.
Based on information provided by a cross section of 148 executives in the United States. The report discusses the relevance of new products for company growth, the reasons for success and failure, and procedures and strategies to enhance the rate of success.

PESSEMIER, EDGAR A. *Product Management: Strategy and Organization.* 2d ed. New York: John Wiley & Sons, 1981.

A comprehensive text that provides a scholarly view of the process of managing innovation. It emphasizes the use of quantitative techniques and models to support the management decision process for a new product.

SACHO, WILLIAM S., and BENSON, GEORGE. *Product Planning and Management.* Tulsa, Oklahoma: Pennwell Books, 1981.

Provides thorough coverage of the critical process of new product development, along with treatment of product life cycle analysis, strategic planning, product elimination, organization, and branding and packaging. In addition, the book focuses on environmental issues related to new product development.

U.S. DEPARTMENT OF COMMERCE. *1982 U.S. Industrial Outlook.* U.S. Government Printing Office, Washington, D.C., 1982.

Discusses the role of new products in the growth of United States companies. The findings indicate that the development of new products was the most significant growth factor in more than 75 percent of the firms studied. Product substitution and modification was the second most important growth factor. The report provides detailed information on the results of the study.

WEBSTER, FREDERICK E., JR. "Top Management's Concerns About Marketing: Issues for the 1980's." *Journal of Marketing* 45 (Summer 1981): 9–16.

Reports the findings of a study of top managers' views of the function of marketing in their particular businesses. These executives indicated that marketing was the most important management function in their businesses and would become more important in the future. The discussion focuses on the product management system, the need for more marketing orientation, the lack of understanding of financial dimensions by marketing managers, and the lack of management innovation and creative thinking.

Social and Environmental
Influences on Product
Planning and Development

The growing public awareness of social and environmental factors is now of primary importance to product planners when they make decisions about their product line. Failure to consider the social and environmental impact of a new product invites government intervention, which will not only negatively affect that product but could also be detrimental to the reputation of existing products. Figure 2–1 shows the relationship between the four environmental elements that may affect new product development.

A number of specific forces stimulate concern for social and environmental factors. Some of the most noteworthy of these are discussed below:

1. The growing affluence of our society has caused consumers to become more interested in social needs. As a nation grows in wealth, some of the attention originally given to economic needs can be focused on social needs. The Arab nations, whose wealth is derived from the increased world-demand for petroleum, provide a good example of this. Initially their only purpose was to improve their profit position; they were not concerned with the social impact of these decisions. Now these countries are beginning to recognize the need for careful evaluation of their position—lest they destroy their environment—and are spending millions of dollars to improve and protect their natural resources.

2. The growing interest among consumers in preserving the environment and protecting scarce resources has fostered new government

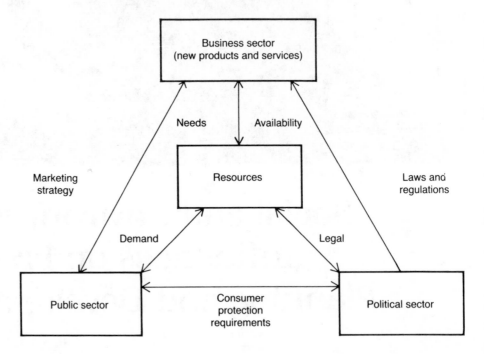

FIGURE 2–1
Social and environmental influences on new product development

regulations, as well as greater public awareness of hazards to the environment.

3. Consumer organizations, led by individuals such as Ralph Nader, have heightened the general public's awareness of social and environmental problems.

4. The mass media's increased coverage of environmental and social problems has helped to educate the general public.

5. Events in the early and late 1970s, such as the Watergate scandal and the energy crisis, focused consumers' attention on the welfare of society and the environment in which we live.

6. Agencies such as the Food and Drug Administration, Federal Trade Commission, Environmental Protection Agency, and the Office of Consumer Affairs have been empowered to enforce regulations on business and industry. If a firm fails to adhere to these regulations, it may suffer financial losses and possible deterioration of its competitive position.

7. Increased inflation has caused consumers to become more conscious of the quality and price of products and services. Consumers are now more sophisticated and more knowledgeable regarding their market needs.

Increased regulation for consumer protection will have long-term effects on both the consumer and the development of new products. It is imperative that both the benefits and costs of any government social marketing program be carefully weighed prior to the implementation of new regulations. Sometimes such cost/benefit assessment can provide more appropriate remedies that will have a less restrictive impact on industry yet protect the consumer.

Figure 2–2 illustrates a framework for evaluating consumer protection regulation. In this framework the objective is to identify the least restrictive remedies that will be the least disruptive and the least costly for consumers, industry, and government.

One example of how this framework operates can be found in the attempt to reduce the incidence of cigarette smoking. The following options available to the government vary in the amount of restriction and disruption they will cause.

1. Pass new legislation making the sale of cigarettes or cigarette smoking illegal or difficult. Prohibiting smoking in public places and increasing taxes are possible restrictive remedies. The costs of directly legislating smoking would be very high, and the impact of such laws on the rights of smokers and the tobacco industry would be extreme.

2. Encourage industry to develop new products with less tar and nicotine to reduce the possible harmful effects. Alternatively it may be conceivable to develop a nonharmful cigarette. The costs to industry would be high, depending on the financial support provided by the federal government. The costs to the consumer would be relatively low.

3. Raise prices of cigarettes to consumers through taxes or insurance rates. Although not very restrictive, except possibly to industry, this process has already been proven to be ineffective.

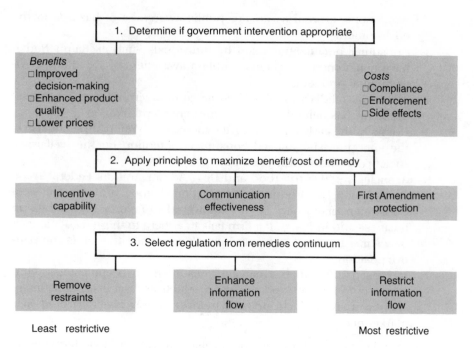

FIGURE 2–2

A framework for evaluating consumer information regulation

Source: Michael B. Mazis, Richard Staelin, Howard Beales, and Steven Salop, "A framework for Evaluating Consumer Information Regulation," *Journal of Marketing* 45 (Winter 1981) 12.

4. Disseminate information about the harmful effects of smoking. Based on past experience, effectiveness of such an option is questionable. Who pays for the information and what is actually stated in the information become difficult questions.[1]

The effects of each of these four alternatives can be carefully evaluated using a framework for assessment, instead of haphazardly making judgments (often in the form of new legislation) based on cost/benefits to only part of the people and industries involved.

Many executives of major firms in the United States have issued strong warnings about the impact on innovation of government policy and regulations. In particular they are concerned that (1) basic research performed by industry has dropped from 38 percent of the national total in 1956 to about 16 percent of the total in 1982, (2) labs are no longer as committed to new ideas as they once were, (3) the amount of research and development conducted by foreign labs has increased, (4) red tape required to get new products to commercial market has increased, (5) the antitrust laws have been unfavorably interpreted and reinforced, and (6) risk capital is in short supply due to new research and development accounting procedures.

Bell Labs has pointed out that excessive federal regulatory policy has made it nearly impossible to develop scientific innovations. In 1938 an appli-

---

[1]Karen F. A. Fox and Philip Kotler, "The Marketing of Social Causes," *Journal of Marketing* 44 (Fall 1980): 24–33.

cation for adrenaline in oil was presented to the Food and Drug Administration (FDA) in 27 pages. In 1958 it took 439 pages for an application for a new treatment of pigworms. In 1972 a muscle relaxant application to the FDA required 456 two-inch-thick volumes and weighed one ton.

In 1977 the Justice Department moved against perceived monopolies by forcing firms such as Xerox and Industrial Electronic Engineers, Inc. to make public certain designated patents. Xerox was forced to open its portfolio of 1,700 copier patents to competitors.

President Reagan is expected to reduce the impact of government on businesses. Already some of these influences have been observed. OSHA has indicated that many of its burdensome requirements on industry will be eliminated or eased. The arguments in favor of this philosophy center on the ability of United States industry to think in terms of progress, rather than having to first look up all the reasons why a particular product should not or cannot be developed.[2]

The arguments in favor of or against the increased role of government are endless. One thing however is clear: the new product development process, as well as the marketing of mature products, will require sophisticated marketing personnel and strategy. Understanding the implications of the environment are necessary and should be used to aid the development process rather than hinder it.

HISTORY OF CONSUMER RIGHTS

In 1962 President Kennedy issued his Consumers' Bill of Rights message, proposing that the nation be more concerned with four basic rights of consumers:

1. Right to safety
2. Right to be informed
3. Right to choose
4. Right to be heard

This message had far-reaching effects on new product development, as well as on the management of existing products. Since Kennedy's message, the federal government has been strongly committed to ensuring these rights. New legislation was passed to protect these rights and to establish consumer agencies. As more and more legislation is passed, however, we see important implications for the new product development process. Neglect or ignorance of these forces may contribute to recalls or subsequent new product failures. But a firm can profit by marketing its new products to satisfy these external demands more effectively than existing competing products. Thus through environmental analysis a firm can enhance its ability to introduce a new product successfully. Some of the major issues and concepts necessary to understanding social and environmental influences are discussed in the following paragraphs.

---

[2]"OSHA's Latest Recall of Carter Era Rules," *Business Week*, 27 April 1981, 48.

PRODUCT DEVELOPMENT AND THE MARKETING CONCEPT

The traditional market concept requires that a firm consider customer needs before designing a new product.[3] New social pressures have altered the meaning of the marketing concept so that protection may be imposed on the consumer to an undesirable extent. Protection provided by seat belts, auto bumpers, removal of lead in gasoline, removal of phosphates from detergents, and packaging restrictions are examples of changes imposed by the government because of public pressures. But the average consumer may not feel that such protection is desirable. Thus public pressures, in some instances, induce nonessential product changes that result in higher production and consumer costs.[4] Protection of this sort presents a difficult decision for many consumers, especially those who have been impoverished by inflation and unemployment. Product trace and recall programs enforced by consumer protection groups, for example, result in increased costs to the firm, which, in turn, are passed to the consumer via increased prices. Other examples exist where economic, social, and environmental factors are in conflict. Some changes required by recently passed laws on emissions standards for automobiles have an adverse effect on fuel economy, yet reduction of gasoline consumption and good air quality are both important goals of consumer protection groups.

Problems with the marketing concept do not mean that firms should revise their thinking about consumer needs in the planning and development of new products. On the contrary consumer preferences must continue to be the primary concern. What firms must *not* do is rely on preconceived concepts (either their own or those of public policy groups) of what is best for the consumer. Only when a firm emphasizes consumer preferences will it be able to introduce new products successfully.

In addition to modifying products, companies are also asked to provide more information so that customers are better informed about available alternatives. However, the consumer has been uninterested in or confused by labeling and unit pricing meant to provide more information at the point of purchase.[5] To avoid additional unnecessary costs to the consumer, more thought should be given to decisions to provide increased consumer protection or information.

Raw materials and energy shortages in recent years have led the public to demand products that are necessary because of shortages, rather than those that are merely wanted. If shortages become more severe, the government, with the support of consumer protection organizations may pressure firms to curtail production of products and packagings that are deemed nonessential. This rationing, if forecasted early by firms using these materials, will force the development of substitutes or some other solution. The company that can anticipate such events will gain a competitive advantage by modifying

---

[3]Philip Kotler, *Principles of Marketing* (Englewood Cliffs, N.J.: Prentice-Hall, Inc., 1980), 22.

[4]Douglas L. MacLachlan, "Editor's Comments," *Journal of Contemporary Business* 4, no. 1 (Winter 1975): *ii*, and Jerry McAfee, "How Society can Help Business," in *Marketing and Society: Cases and Commentaries,* eds. Roy D. Adler, Larry M. Robinson, and Jan E. Carlson (Englewood Cliffs, N.J.: Prentice-Hall, Inc., 1981), 10–12.

[5]Kert B. Monroe and Peter J. LaPlaca, "What are the Benefits of Unit Pricing?" *Journal of Marketing* 36 (July 1972): 22.

product and packaging strategy. Companies that fail in this forecasting are likely to suffer serious declines in market share.

A few firms have already established environmental study groups whose responsibility is to analyze changes in the environment and their effects on new product development, as well as their influence on existing marketing strategy. Unfortunately many firms do not concern themselves with these external variables until a crisis occurs that forces them to make costly and disruptive decisions.[6]

ADVANTAGES OF ENVIRONMENTAL AND SOCIAL EVALUATION

Considering social and environmental factors in evaluating new product alternatives will result in a product line that will gain wider consumer acceptance. It may also result in less objection by nonusers of the product because there are no ill effects on the natural and social environment.[7] One study found that new products that seem to alleviate environmental problems are more likely to be accepted.[8]

Concern with social and environmental factors has recently become a requirement of government agencies that provide funds in the form of contracts, grants, or loans to firms engaged in new product development. For example the Environmental Protection Agency will withhold funds from violators of water or air pollution standards—someimes even after a contract, grant, or loan has been made. In these cases, the company in violation is required to rectify the problem before funds are released. Equipment for pollution reduction can be very costly. In most cases, careful scrutiny of the pollution problem from the start would have saved the firm the cost of adding pollution-reducing equipment later. Early recognition of undesirable environmental effects of a new product manufacturing process can prevent having to make major design changes at a later point. Firms must not evaluate a new product idea purely on economic terms without considering environmental and social factors.

DISADVANTAGES OF ENVIRONMENTAL AND SOCIAL EVALUATION

Evaluation of environmental and social factors in new product decisions result in the following disadvantages to the firm:

1. The costs incurred in product development will substantially increase—especially as the scope of the evaluation process increases.
2. Timing of new product introductions will be affected because this evaluation must be included as part of the development process.

---

[6]George Fisk, "Impact of Social Sanctions on Product Strategy," *Journal of Contemporary Business*, 4, no. 1 (Winter 1975): 2.

[7]Dale L. Varble, "Social and Environmental Considerations in New Product Development," *Journal of Marketing* 36 (October 1972): 14.

[8]Harold H. Kassarjian, "Incorporating Ecology into Marketing Strategy: The Case of Air Pollution," 35 *Journal of Marketing* (July 1971): 61.

3. Ideas for new products will need to be more carefully screened due to increased costs, thus possibly eliminating products that are otherwise potentially successful.

These disadvantages may seem quite severe on the surface, but failure to conduct an environmental and social evaluation could result in far greater costs if the new product does not meet government and consumer requirements.

PRODUCT SAFETY

The Consumer Product Safety Act was signed into law by President Nixon on October 28, 1972.[9] This act went into effect on December 26, 1972, creating an independent, five-member Consumer Product Safety Commission that has the power to prescribe safety standards for nearly all consumer products, except where regulation already exists. This legislation contains some important implications for the new products development process. It may also have significant effects on existing product lines if any of the offerings are subject to these regulations.

The concern for safety in consumer products has existed for many years, but early standards were vague. As a result of 20 million serious injuries attributed to unsafe products per year—with annual costs of more than $5.5 billion—product safety legislation was initiated.[10] It is argued that many injuries could have been prevented if manufacturers had been more conscious of product safety. Regardless of where the fault lies, firms must realize that the government and consumer protection groups intend to press the product safety issue in order to alleviate the large number of injuries. This realization must be manifested at an early stage in the product development process in order to prevent costly modifications or complete withdrawal of a new product. Modification will require lead time to redesign the new product and inevitably could place the firm in an unfavorable market position.[11]

More than 10,000 products are now subject to the legal powers of the Consumer Product Safety Commission, which has control over "any article, or component part there produced or distributed (i) for sale to a consumer for use in or around a permanent or temporary household or residence, a school, in recreation or otherwise, or (ii) for the personal use, consumption or enjoyment of a consumer in or around a permanent or temporary household or residence, a school, in recreation or otherwise . . ."[12] This definition is broad and could conceivably affect hundreds of thousands of retailers, distributors, and manufacturers who will be policed to ensure compliance with product safety regulations.

---

[9]Consumer Product Safety Act of 1972, Publication L, no. 92 (28 October 1972) 41 and U.S.L.W. 57 (31 October 1972).

[10]U.S., Congress, Senate, Commission on Commerce, *Hearings on National Commission on Product Safety*, 91st Cong., 2d sess., 1970, 37.

[11]Rajan Chandran and Robert Linneman, "Planning to Minimize Product Liability," *Sloan Management Review*, 20 (Fall 1978): 33–45.

[12]Public Law 92-573, 92d Congress, S. 3419, 27 October 1972, 2

PRODUCT SAFETY STANDARDS AND PROCEDURES

The Consumer Product Safety Standards are the means by which the objectives of the Consumer Product Safety Act are attained. These standards "consist of (1) requirements as to performance, composition, contents, design, construction, finish, or packaging, or (2) requirements that a consumer product be marked with or accompanied by clear and adequate warnings . . . or (3) any combination of (1) and (2)."[13] According to the specifications of the Consumer Product Safety Act, the development of standards begins with a majority agreement by the five commissioners that a product poses "an unreasonable risk of injury," thus necessitating some criteria for safety. The Commission then publishes a notice in the *Federal Register* and informs the press, inviting any person or group to submit an existing standard or an offer to develop a mandatory standard. The Commission extends this offer to ensure representative public participation in the development of these standards. The act allows any trade association, consumer organization, professional society, testing laboratory, university or college department, wholesaler, retailer, government agency, ad hoc association, company, or person to submit an offer to develop a proposed consumer product safety standard.

Once the Commission evaluates any responses, it can then proceed in one of three ways: it can (1) publish an existing standard as a proposed mandatory standard, (2) accept one or more offers to develop a standard, or (3) independently proceed to develop a standard. These standards fulfill the main objective of the Consumer Product Safety Act: reduce unreasonable risks of injury associated with the use of consumer products.

SUBSTANTIAL HAZARDS AND THE COMMISSION

During the first year of operation, the Consumer Product Safety Commission received about 130 defect notifications involving 14 million product units. These complaints included such products as baby cribs, soft-drink bottles, bicycles, swimming pools, minibikes, snowmobiles, power lawn mowers, sliding doors, and artificial turf.[14]

Under Section 15 of the Consumer Product Safety Act, the Commission has a great deal of responsibility and power to identify what it considers to be substantial hazards and to ban products it considers unsafe. The Commission feels that, by using its power with discretion, it can eliminate what it considers to be unreasonable product risks, although it will not be able to eliminate all product risks. In spite of this, some people feel that the authority of the Commission is too extensive because it can determine the fate of many new or existing products.

The Commission has made several significant decisions since the law's inception. Its decision to ban TRIS-treated children's sleepwear and require the statutory repurchase of these products caused much legal confusion. At

---

[13]15 U.S.C. 2052 (a) (1) 1972
[14]U.S. Consumer Product Safety Commission, *Fact Sheet*, no. 32, (Washington, D.C.: Government Printing Office, 1974), 1.

issue was whether the Commission's statements were enforceable as interpretations of the Federal Hazardous Substances Act (FHSA) or whether it was taking action under the authority of the FHSA. The clarification came as a statement that the Commission was acting under the authority of the FHSA.

Many firms attempted to dispose of their TRIS-treated clothing products by exporting them. This was also prohibited.

To offset the tremendous inventory losses suffered by some firms, a bill was introduced in Congress to reimburse losses resulting from the TRIS ban, provided that the firm showed "good faith" by destroying its TRIS inventory.[15]

Regardless of how extensive the powers of the Commission, success depends significantly on cooperation from business. Business has the responsibility of reporting to the Commission any product defect that could be a substantial risk to consumers. The term *business* includes, specifically, manufacturers, importers, distributors, and retailers who market any product coming under the jurisdiction of the Commission. Failure to report a defect or furnish required information could result in civil and criminal penalties.

Notification to the Commission must be made within 24 hours of discovery of the violation and must be certified by the chief executive officer of the notifying company or by some other designated person. This initial notification must include the following information:

1. Identification of the product
2. Name and address of the manufacturer, if known
3. Nature and extent of the defect
4. Name and address of the person informing the Commission

When a company notifies the Commission of the defective product, it must also explain what corrective action will be taken, including whether buyers have been notified and whether the company will repair the product or provide a refund or replacement.

In Section 15 the Commission is responsible for determining whether a product defect may pose a substantial risk of injury. The Commission contacts the party making the notification to gather details about the product and any associated injuries. If the risk of injury is found to be substantial, a letter is sent to the manufacturer, distributor, or retailer indicating the Commission's knowledge of the situation. This letter requests the firm to provide an estimate of the number of products involved and of any corrective action that may have been taken. If the Commission and the firm concur that no substantial hazard exists, no further action is taken.

Section 15 of the Consumer Product Safety Act enables the Commission to ensure that the public is quickly notified of product hazards and that industry will repair, replace, or repurchase the defective products. Section 15 places the responsibility on manufacturers, importers, distributors, and retailers to monitor the safety of their products.

Although the Commission has established itself as an important element in product safety, its members feel that the agency needs the following to accomplish its responsibilities:

---

[15]See *Federal Register* 61593 and 61621 (6 December 1977) and 25711 (14 June 1978).

1. Power to file civil and criminal cases against violators independently, rather than having to work through the Justice Department, which has in many instances declined to pursue referred cases
2. Authority to seize products judged to pose substantial hazards
3. The right to eliminate public participation in setting standards so that delay and confusion can be avoided.

Firms involved in the development of new products must pay close attention to the requirements of the Consumer Product Safety Act. The act is meant to offer firms in the new development process incentives to introduce safe products, so federal intervention will be unnecessary.[16]

NEW PRODUCTS AND WARRANTY POLICY

A warranty is an affirmation by the seller of the quality or performance of the goods. Express warranties make such an affirmation in writing, ranging from simple statements about product attributes or product performance (e.g., satisfaction guaranteed) to the complex documents limiting the legal responsibilities of manufacturers of such products as appliances and automobiles. The Uniform Commercial Code also implies that each product is fit for the purpose for which it is sold.

In 1975 the Magnuson-Moss Warranty Act was passed, stipulating that any warranty of a consumer product must conform to the rules of the Federal Trade Commission (FTC). All aspects of the warranty must be (1) disclosed fully and conspicuously in simple and understandable language, (2) labeled as full or limited, and (3) made available to read and compare before purchase. Most recent cases regarding this law focus on the inadequate disclosure of the warranty. In a recent case involving Montgomery Ward & Co., the FTC ruled that the company must make the text of the warranty available prior to the sale of products over $15.[17]

According to an FTC study published in 1980, the warranty protection benefits provided to American consumers have improved since the inception of the Magnuson-Moss Warranty Act. The FTC study supported two other studies, conducted for them by Arthur Young and Company, that show that manufacturers have increased warranty coverage since passage of the act. Of the warranties investigated (comparing 1974 with 1978), no company dropped their warranty, and the scope and remedies were improved. The study also concluded that 1978 warranties were more readable than 1974 warranties but that some are still difficult for consumers to understand.[18]

---

[16]For further discussion of product safety and innovation, see Walter Guzzardi, Jr., "The Mindless Pursuit of Safety," *Fortune*, 9 April 1979, 54–64 and Lawrence M. Kushner, "Product Safety Standards and Product Innovation Too," *Harvard Business Review* 99 (July–August 1980): 36–38.

[17]See Commerce Clearing House's (CCH) *Trade Regulation Reporter*, no. 21,648 (January 1980) and no. 21,282 (April 1977).

[18]FTC Studies on Warranty Coverage Charges, CCH *Trade Regulation Newsletter*, no. 424 (February 1980), 1.

TABLE 2–1
New warranty disclosure
regulations (Effective
January 1, 1977)

- Describe and identify the products, parts, or components covered by—and, when necessary for clarification, not covered by—the warranty.

- State what the manufacturer or store (warrantor) will do in the event of a defect in the product, including a description of what items or services the warrantor will pay for—and, when necessary for clarification, those items that the warrantor will not pay for.

- Identify the person to whom the written warranty is extended. For example, is the warranty extended to second owners of the product?

- State the duration of the warranty, including whether the term commences at the purchase date or when the company receives the owner registration card.

- Provide a step-by-step explanation of what the consumer should do to get repair or replacement. This should include such information as (1) the name of the warrantor, (2) the address of the warrantor, (3) the name, title, and address of the department responsible for warranty obligations, and (4) a telephone number that consumers can use, without charge, to get warranty information.

- Provide information about any settlement mechanism to resolve warranty complaints.

- Provide consumers an opportunity to read and compare warranties at the time of sale. This can be done by either displaying the warranty near the product or maintaining a binder for different competing products.

The new disclosure requirements tend to lengthen warranties, making it difficult for a firm to simplify them. The new warranty disclosure regulations that became effective in 1977 are summarized in Table 2–1.

The last statement in the table is one of the most revealing for firms introducing new products. The implication is that consumers will make comparisons of competitive products and select the one that provides the best warranty. Firms marketing products subject to these new disclosure laws will be able to differentiate their product offerings. The warranty, if not at least equivalent to the competition's, may result in an unfavorable consumer reaction. It is conceivable that a firm attempting to introduce a new product in a highly competitive market may be able to enhance its early adoption by providing a more comprehensive warranty.

Typically, written warranties have not been offered for frequently purchased products, such as consumer packaged goods, because complaints are unlikely. In case of complaint, the producer generally will replace, refund, and/or apologize to the consumer. Figure 2–3 depicts an example of an expressed warranty for a Toshiba color television set.

FIGURE 2–3
Example of expressed warranty

LIMITED WARRANTY
TOSHIBA COLOR TELEVISION

Toshiba America, Inc. ("TAI") and Toshiba Hawaii, Inc. ("THI") make the following limited warranties. These limited warranties extend to the original consumer purchaser or any person receiving this set as a gift from the original consumer purchaser and to no other purchaser or transferee.

Limited One (1) Year Warranty
TAI and THI warrant this product and its parts against defects in materials or workmanship for a period of one year after the date of original retail purchase. During this period, TAI and THI will repair a defective product or part, without charge to you. You must deliver the entire set to an Authorized TAI/THI Service Station. You pay for all transportation and insurance charges for the set to and from the Service Station.

Limited Two (2) Year Warranty of Picture Tube
TAI and THI further warrant the picture tube in this product against defects in materials or workmanship during the second year after the date of original retail purchase. During this period, TAI and THI will replace a defective tube, but you must pay labor charges. You must also deliver the entire set to an Authorized TAI/THI Service Station. You pay for all transportation and insurance charges for the product to and from the Service Station.

Owner's Manual and Warranty Registration
You should read the owner's manual thoroughly before operating this product. You should also insure that your name and address are on file as owners of a TAI/THI product by completing and mailing the attached registration card within ten days after you, or the person who has given you this product as a gift, purchased this product. This is one way to enable TAI/THI to establish the date of purchase of the product, as well as to provide you with better customer service and improved products. Failure to return the card will not affect your rights under this warranty so long as you retain other proof of purchase such as a bill of sale.

Your Responsibility
The above warranties are subject to the following conditions:
(1) You must retain your bill of sale or provide other proof of purchase. Completing and mailing in the attached registration card within ten days after the original retail purchase is one way of providing such other proof of purchase.
(2) You must notify an Authorized TAI/THI Service Station within thirty (30) days after you discover a defective product or part.
(3) All warranty servicing of this product must be made by an Authorized TAI/THI Service Station.
(4) These warranties are effective only if the product is purchased and operated in the U.S.A.
(5) Labor service charges for set installation, set up, adjustment of customer controls and installation or repair of antenna systems are not covered by this warranty. Reception problems caused by inadequate antenna systems are your responsibility.
(6) Warranties extend only to defects in materials or workmanship as limited above and do not extend to any product or parts which have been lost or discarded by you or to damage to products or parts caused by misuse, accident, improper installation, improper maintenance or use in violation of instructions furnished by us; or to units which have been altered or modified without authorization of TAI/THI or to damage to products or parts thereof which have had the serial number removed, altered, defaced or rendered illegible.

Step-By-Step Procedures—How to Obtain Warranty Service
To obtain warranty servicing, you should:
(1) Contact an Authorized TAI/THI Service Station for warranty service within thirty (30) days after you find a defective product or part. You can find an Authorized TAI/THI Service Station listed in the enclosed directory or by contacting one of the TAI/THI Service Centers listed below, or any TAI/THI Authorized Dealer.
(2) Arrange for the delivery of the product to the Authorized TAI/THI Service Station. Products shipped to the Service Station must be insured and safely and securely packed, preferably in the original shipping carton, and a letter explaining the defect and also a copy of the bill of sale or other proof of purchase must be enclosed. All transportation and insurance charges must be prepaid by you.
(3) If you have any question about service, please contact one of the following TAI/THI Service Centers:

Toshiba America, Inc.	Toshiba America, Inc.	Toshiba America, Inc.	Toshiba America, Inc.	Toshiba Hawaii, Inc.
Service Center	Service Center	Service Center	Service Center	327 Kamakee Street
41-06 Delong Street	19515 South Vermont Ave.	3225 East Carpenter Freeway	2900 Mac Arthur Blvd.	Honolulu, Hawaii 96814
Flushing, N.Y. 11355	Torrance, Cal. 90502	Irving, Texas 75062	Northbrook, Ill. 60062	Phone Number:
Phone Number:	Phone Number:	Phone Number:	Phone Number:	(808) 521-5377
(212) 939-7400	(213) 538-9960	(214) 438-5814	(312) 564-5110 •	

ALL WARRANTIES IMPLIED BY STATE LAW, INCLUDING THE IMPLIED WARRANTIES OF MERCHANTABILITY AND FITNESS FOR A PARTICULAR PURPOSE, ARE EXPRESSLY LIMITED TO THE DURATION OF THE LIMITED WARRANTIES SET FORTH ABOVE. Some states do not allow limitations on how long an implied warranty lasts, so the above limitation may not apply to you. WITH THE EXCEPTION OF ANY WARRANTIES IMPLIED BY STATE LAW AS HEREBY LIMITED, THE FOREGOING EXPRESS WARRANTY IS EXCLUSIVE AND IN LIEU OF ALL OTHER WARRANTIES, GUARANTEES, AGREEMENTS AND SIMILAR OBLIGATIONS OF MANUFACTURER OR SELLER WITH RESPECT TO THE REPAIR OR REPLACEMENT OF ANY PRODUCT OR PARTS.
IN NO EVENT SHALL TAI OR THI BE LIABLE FOR CONSEQUENTIAL OR INCIDENTAL DAMAGES. Some states do not allow the exclusion or limitation of incidental or consequential damages so the above limitation may not apply to you.

No person, agent, distributor, dealer, service station or company is authorized to change, modify or extend the terms of these warranties in any manner whatsoever. The time within which an action must be commenced to enforce any obligation of TAI and THI arising under this warranty or under any statute, or law of the United States or any state thereof, is hereby limited to one year from the date you discover or should have discovered, the defect. This limitation does not apply to implied warranties arising under state law. Some states do not permit limitation of the time within which you may bring an action beyond the limits provided by state law so the above provision may not apply to you. This warranty gives you specific legal rights and you may also have other rights which vary from state to state.

TOSHIBA AMERICA, INC. TOSHIBA HAWAII, INC.
Keep this card for your record. Printed in U.S.A. Part No. 23959127   **TOSHIBA**
CTV-79-S

WARRANTY REGISTRATION
(PLEASE FILL IN AND MAIL THIS CARD WITHIN 10 DAYS)

MODEL NO._____   SERIAL NO._____

DATE PURCHASED_____

YOUR NAME in full_____

ADDRESS_____

    CITY_____ STATE_____ ZIP CODE_____

DEALER'S NAME _____

    CITY_____ STATE_____ ZIP CODE_____
CTV-79-S

WARRANTIES AND THE INDUSTRIAL PRODUCT

Many industrial products do not have express warranties, but they are generally purchased as a competitive bid or under some type of written contract. These written contracts will state clearly the requirements and/or responsibility of the seller regarding any defects in the product. For example a firm buying coffee beans for processing into various blends of coffee requires that the raw materials meet certain quality standards. Thus the warranty is actually implied as part of the written contract.

In many cases, particularly in the equipment industries, a guarantee or warranty is explicitly stated as part of the standard terms, rather than being implied. Figure 2–4 depicts a portion of a standard contract used by a large equipment manufacturer. As shown in capital letters at the bottom of the first section, the firm makes no warranties other than what is explicitly stated in the paragraph. In effect this warranty provides little protection for the buyer other than for defective workmanship or materials.

One possible strategy for differentiating a new product from its competition would be for an industrial firm to provide an express warranty that offers the buyer greater protection. It is possible that the growing concern for protection of the buyer and the increasing intensity of competition may necessitate a change in the use of express warranties by industrial firms.

OTHER LEGAL REQUIREMENTS FOR NEW PRODUCTS

In addition to the major laws and governing agencies mentioned previously, there are many other regulations that firms must consider before introducing a new product. Laws relating to labeling, packaging, advertising, and pricing, for example, may affect any new product and should be carefully surveyed to ensure that the firm has complied with all requirements.

PATENTS

A patent is a grant given by the government to provide an investor protection for 17 years from others who may seek to make, use, or sell a particular invention. By law "any person who invents or discovers any new useful process, machine, manufacture or composition of matter, or any new and useful components thereof may obtain a patent" subject to the conditions and requirements of the law.[19]

To be patented, an invention must be new, as defined in the statute. That is, the invention must not have been used for more than one year prior to the date of application, no other person must have known of or used the invention, and the invention must not have been described in a printed publication in the United States or a foreign country prior to application for the patent.

---

[19]*Technology, Assessment and Forecast*, U.S. Dept. of Commerce, Patent and Trademark Office, 7th Report, March 1977, p. 7.

Standard Terms and Conditions of Sale

1. *Guarantee.* The Company warrants all equipment manufactured by it or bearing its name plate to be free from defects in workmanship and material, under normal use and service, as follows: (a) Equipment not installed by the Company which is returned transportation prepaid to the Company's originating factory within 12 months after date of shipment and is found by the Company's inspection to be defective in workmanship or material will be repaired or replaced, at the Company's option, free of charge and return-shipped lowest cost transportation prepaid. (b) Equipment installed by the Company, or under the direct supervision of the Company, which within 12 months after date of installation is found by the Company's inspection to be defective in workmanship or material will be repaired or replaced, at the Company's option, free of charge. If inspection by the Company does not disclose any defect in workmanship or material the Company's regular published rates will be charged as they apply. Items such as Thermocouples, Electrodes, Glassware, Chart Paper and similar items subject to wear or burnout through usage, shall not be deemed to be defective by reason of wear or burnout through usage. Equipment not manufactured by the Company or bearing its name plate is guaranteed in accordance with the published guarantee of the manufacturer. WITH EXCEPTION OF THE 12 MONTH WARRANTY, SET FORTH ABOVE, THE COMPANY MAKES NO EXPRESS WARRANTIES, NO WARRANTY OF MERCHANTABILITY AND NO WARRANTIES WHICH EXTEND BEYOND THE DESCRIPTION ON THE FACE HEREOF. In no event will the Company be liable for indirect, special or consequential damages of any nature whatsoever.

2. *Patents.* The Company agrees that it will at its own expense defend any suit that may be instituted against the Purchaser for alleged infringement of United States patents relating to products of Company manufacture furnished the Purchaser hereunder, provided such alleged infringement shall consist only in the use of such product by itself and not as a part of any combination of other devices and/or parts, and provided the Purchaser gives the Company immediate notice in writing of any such alleged infringement and of the institution of any such suit and permits the Company, through its counsel, to answer the charge of infringement and to defend such suit, and provided the Purchaser gives all needed information, assistance and authority to enable the Company to do so, and thereupon in case of a final award of damages in any such suit the Company will pay such award, but shall not be responsible for any settlement made without its written consent.

3. *Delivery.* Delivery of equipment not agreed on the face hereof to be installed by or under supervision of the Company shall be F.O.B. at the Company factory, warehouse or office selected by the Company. Delivery of equipment agreed on the face hereof to be installed by or under supervision of the Company shall be C.I.F. at the site of installation.

   The Company shall not be liable for any delay in the production, delivery, supervision or installation of any of the equipment covered hereby if such delay shall be due to one or more of the following causes: fire, strike, lockout, dispute with workmen, flood, accident, delay in transportation, shortage of fuel, inability to obtain material, war, embargo, demand or requirement of the United States or any government or war activity, or any other cause whatsoever beyond the reasonable control of the Company. In event of any such delay, the date or dates for performance hereunder by the Company shall be extended for a period equal to the time lost by reason of the delay.

4. *Damage or loss.* In the case of equipment not to be installed by or under supervision of the Company, the Company shall not be liable for damage to or loss of equipment after delivery of such equipment to the point of shipment. In the case of equipment to be installed by or under supervision of the Company, the Company shall not be liable for damage or loss after delivery by the carrier to the site of installation; if thereafter pending installation or completion of installation or full performance by the Company, any such equipment is damaged or destroyed by any cause whatsoever, other than by the fault of the Company, the Purchaser agrees promptly to pay or reimburse to the Company, in addition to or apart from any and all other sums due or to become due hereunder, an amount equal to the damage or loss so occasioned.

5. *Claims for shortages.* Each shipment shall be examined by the Purchaser immediately upon his receipt thereof, and any claim for shortage or any other cause must be reported to the Company promptly after such receipt.

6. *Taxes.* With regard to sales of equipment not installed by the Company, the amount of all present and future taxes and governmental charges upon the production, shipment, sale, installation or use of the equipment covered hereby shall be added to the price and paid by the Purchaser. With regard to contracts for the installation of equipment by the Company, the amount of all present or future taxes and governmental charges upon labor or the production, shipment, sale, installation or use of the equipment covered hereby are not included in the price and, unless otherwise stated in the proposal, shall be added to the proposal price and paid by the Purchaser.

FIGURE 2–4
Sample of terms of standard contract

Application for a Patent

Applying for and obtaining a patent requires knowledge of patent laws and practices and thus necessitates the services of a patent attorney or patent agent.[20] The inventor may prepare the application and file it with the Patent Office without the services of an attorney or agent, but this would likely result in a patent that does not adequately protect the invention. Patent attorneys or agents are recognized by the Patent Office and are admitted to a special register that allows them to render patent services to inventors. To qualify for admission to this register, the attorney or agent must pass a special examination.

The formal application, once filed and accepted by the Patent Office, is followed by a patent search to determine whether the invention is indeed new. During the period between filing the application and issuance of the patent, the firm may mark the product *patent pending*. This has no legal effect other than to indicate the existence of a formal application for a patent. Once the patent is granted, the firm is required to mark all products with the word *Patent*, followed by the patent number. If this is not done, the firm may not be able to recover damages from infringement.

Figure 2–5 illustrates the patent application oath for an individual inventor. This document is short and reasonably simple. The critical aspect of the patent is the written document that comprises the product specification. This document generally requires the services of an attorney.

Patent Infringement

Infringement of a patent allows the person or firm holding the patent to sue for redress in a federal court. The patentee may also seek an injunction from the court to prevent the continuation of the infringement. The court may award damages if it finds that infringement has occurred.

The major concern with most court cases is the cost of any suit for damages. Many smaller firms fear patent infringement and feel that should a larger firm attempt to copy their new product, there is little they could do. The court costs and the cost of losing the suit could force them out of business. This factor probably prevents many infringments from ever reaching the courts. On the other hand, larger firms are concerned with the continued pressure placed on them by the government (through antitrust laws) and, in many instances, would rather not involve themselves in a court case that may affect the competitive structure of the market in such a way that they may be considered a monopoly.

Patent Costs

The costs involved in the application and issuance of a patent generally range from $200 to $300. However the cost of a patent attorney will, on the average, increase the overall cost of the patent to approximately $2000 to $3000. More complicated patents would, of course, result in much greater costs. Thus the cost for a well written patent, which gives maximum protection, is not prohibitive and would, in most cases, be a valuable investment.

---

[20]For general information, see *Patents*, U.S. Dept. of Commerce, revised June 1974.

To the Commissioner of Patents:

Your petitioner, _____, a citizen of the United States and a resident of _____, state of _____, whose post office address is _____, prays that letters patent may be granted to him for the improvement in _____, set forth in the following specification and he hereby appoints _____, of _____ (Registration No. _____), his attorney (or agent) to prosecute this application and to transact all business in the Patent Office connected therewith. (If no power of attorney is to be included in the application, omit the appointment of the attorney.)

[The specification, which includes the description of the invention and the claims, is written here.]

_____, the above-named petitioner, being sworn (or affirmed), deposes and says that he is a citizen of the United States and resident of _____, State of _____, that he verily believes himself to be the original, first and sole inventor of the improvement in _____ described and claimed in the foregoing specification; that he does not know and does not believe that the same was ever known or used before his invention thereof, or patented or described in any printed publication in any country before his invention thereof, or more than one year prior to this application, or in public use or on sale in the United States more than one year prior to this application; that said invention has not been patented in any country foreign to the United States on an application filed by him or his legal representatives or assigns more than twelve months prior to this application; and that no application for patent on said invention has been filed by him or his representatives or assigns in any country foreign to the United States, except as follows: _____.

State of _____
County of _____
ss:

_____
(Inventor's full signature)

Sworn to and subscribed before me this

_____ day of _____, 19_____

_____
[seal]   (Signature of Notary Officer)

_____
(Official Character)

FIGURE 2–5
Sample of patent oath
Source: U.S. Government Patent Office.

SUMMARY

To develop an appreciation for all the regulations that could affect a new product would require a text of its own. This chapter has focused on a few of the more recent and perhaps more important regulations and their affect on new products.

It is apparent from the recent experience of selected industries that the development of new products in the future will require extensive testing procedures and documentation. FDA requirements are so extensive in some industries that a drug company needed 456 volumes of data in order to request commercialization of a new muscle relaxant.

The Consumer Product Safety Act signed into law in 1972 created a five-member commission with the power and responsibility to establish safety standards as well as review complaints regarding product defects. Despite the extensive powers of the commission, the success of the act depends on cooperation from business to report any possible defects.

In 1975 the Magnuson-Moss Warranty Act was passed stipulating that any warranty of a consumer product must conform to the rules of the FTC. New warranty disclosure regulations became effective in 1977 that provided clear statements of procedures to be followed by manufacturers involved in providing warranties.

Patent regulations and protections are also important to any firm developing a new product. The initial aspect of the patent application is the written documentation that comprises the product specifications.

SELECTED READINGS

ABERNATHY, WILLIAM J., and CHAKRAVARTHY, BALAJI S. "Government Intervention and Innovation in Industry: A Policy Framework." *Sloan Management Review* 20 (Spring 1979): 3–17.

Discusses two other dimensions to government intervention: technology creation action and market modification action. A framework relating combinations of these actions to their joint consequences for technological innovation is developed. The general applicability of these dimensions is then tested using other known examples of technological significance.

GUZZARDI, WALTER, JR. "The Mindless Pursuit of Safety." *Fortune* 99 (9 April 1979): 54–64.

Discusses the cost/benefit associated with governmental regulations and consumer protection. The author believes that the true cost is misrepresented in court and that the allocation of risk is shifted too much away from the consumer and back upon the manufacturer. Taking a pro-business viewpoint, the author questions the cost and effectiveness of product recalls.

MASIZ, MICHAEL B.; STAELIN, RICHARD; BEALES, HOWARD, and SALOP, STEVEN. "A Framework for Evaluating Consumer Information Regulation." *Journal of Marketing* 45 (Winter 1981): 9–16.

Develops a framework for evaluating alternative consumer information regulations. The approach taken is to provide a coherent structure by integrating the theories of three diverse disciplines: economics, consumer behavior, and law. Classifying regulations from least to most restrictive in marketplace forces is then used to select the most appropriate regulatory approach.

PEPPERELL, H. C., and TURNER, R. W. "Barriers to Entry: Antitrusts's Search for A New Look." *California Management Review* 23 (Spring 1981): 29–40.
Discusses the doctrine of barriers to entry, the centerpiece of antitrust intervention in the market. Although some believe that anything that makes entry into a market difficult is unfair, and that government has the responsibility to eliminate this unfairness, the authors argue that it is a serious mistake for antitrust to tamper with endogenous economic phenomena.

U.S. CONSUMER PRODUCT SAFETY COMMISSION. "Product Safety Fact Sheet." Washington, D.C., 1981.
Briefly describes several of the important government agencies that have jurisdiction over product safety. Information in the pamphlet is valuable for those industries developing new products that will come under the jurisdiction of these agencies. Responsibilities of the agencies are also briefly described.

U.S. DEPARTMENT OF COMMERCE. *Patents.* Washington, D.C.: Government Printing Office, 1974.
A comprehensive discussion of the procedure for applying for a patent. Basic requirements for patent application as well as the role of patents in protecting the holder are discussed. This is a valuable document to familiarize new venture management with the necessary procedures.

WALKER, DEAN. "Hitting Out at Government." *The Executive* 21 (September 1979): 85–86.
Author argues that government has intervened too much in business, causing the cost of small business compliance to become too excessive and encouraging monopolies among large corporations.

WEIDENBAUM, MURRAY L. *Business, Government and the Public.* Englewood Cliffs, N. J.: Prentice-Hall, Inc., 1981.
Discusses the role of government in business, focusing particularly on necessary changes taking place in corporations in order to adapt to increased regulation. The changes proposed provide significant insight to enhance effective management decision-making.

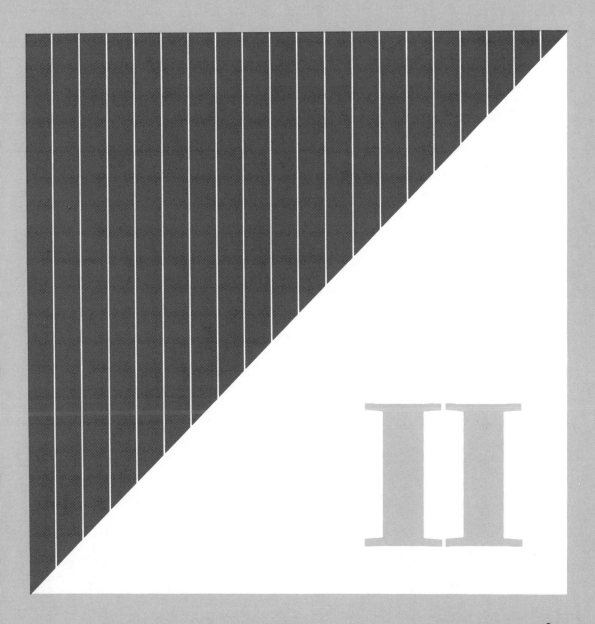

II

Strategic
Market Planning

3

Strategic Planning

Perhaps the most significant change in the corporate planning process in the past few years has been the increasing awareness and use of strategic planning. Today firms involved in multiple products and markets require an overall planning method that emphasizes the long-term goals of the corporation, rather than the potential of a specific product. This is particularly important because of the constant change in the environment that affects the corporate decision-making process.

Strategic planning can be defined as the managerial process of developing and maintaining a strategic fit between the organization and a changing environment. Through the establishment of a clear company mission, objectives and goals, growth strategies, and product mix strategies, a firm can take advantage of its strengths and aggressively pursue opportunities that will provide the highest return.

This text divides the planning process into corporate or long-term planning (referred to as *strategic planning*) and short-term market or product planning. Generally the strategic plan is the responsibility of the chief executive and other senior executives in the corporation (exceptions exist in smaller companies, these will be discussed in Chapter 6). Various assessment techniques and programs —such as PIMS, Boston Consulting Group, product life cycle analysis, profit analysis, product portfolio analysis, and perceptual mapping—are available to top management to aid them in developing the corporate strategic plan. These programs and techniques not only serve as input to the long-term corporate or strategic plan but also provide important direction to lower levels of the organization in preparing the market/product plans. In contrast to the strategic plan, the market/product plan (see Chapter 5) outlines the marketing mix activities, tasks, and decisions for each product or market within the framework of the long-term corporate or strategic plan.

Levi Strauss & Co. considers strategic planning to be the basis for all decision-making in the corporation. A few years ago the company undertook a major change in its planning, taking advantage of its strongest assets. Diversification had brought Levi Strauss from a one-product company to one that had divisions in women's wear, sportswear, and youthwear, as well as international operations. With the new program, major emphasis was placed on women's wear, which represented a $40 billion annual domestic market and was twice as large as the men's wear market. All products were changed to the Levi name, and a mission was established "that consumers would be ours from cradle to grave, regardless of sex, regardless of whatever they may be interested in."[1] Separate marketing divisions, with the intention of aggressively pursuing new market opportunities, were established within each operating division. This strategic planning process has resulted in larger profits, with sales increasing from $1 billion in 1975 to nearly $3 billion in 1980.

This one example demonstrates that strategic planning is very important in today's changing environment. The strategic plan thus has an important impact on market/product planning and hence on marketing mix and new

---

[1]"Levi Strauss' Success Due to Coordination of Marketing, Strategic Planning," *Marketing News,* 10 July 1981, 1.

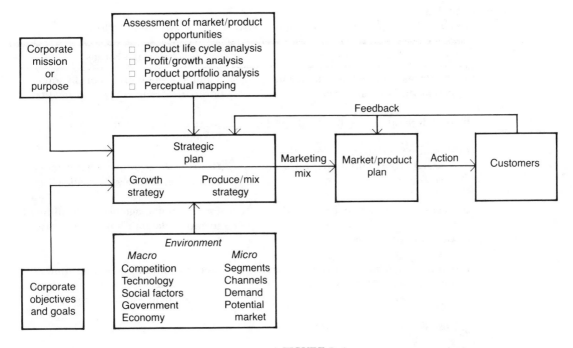

FIGURE 3–1
The strategic planning and market planning process

product development. Figure 3–1 illustrates the corporate-level planning process, assessment of product line opportunities, and the marketing plan.

NATURE OF THE STRATEGIC PLANNING PROCESS

The nature of the strategic planning process will vary with each company, depending on its product line(s) and market(s). Two basic levels of strategies may be identified: the corporate strategy and the strategy for each product/market. In the large corporation it is likely that the emphasis of the strategic plan will be much broader, with less attention paid to the specific product/market strategy. Strategy at the product/market level would be the responsibility of management at the division level. Note that for the small company, the strategic plan is much more specific, relating detailed strategy for its products. Small firms tend to concern themselves with short-term issues, where existing products are critical. Regardless of the size of the corporation, its performance depends on its ability to match the organization with its environments. Ideally the firm should access the environment and set appropriate objectives—then determine what framework will meet these objectives and provide a feedback system to ensure effective implementation of the strategy.

The most difficult part of the strategic planning process is the assessment and evaluation of the environment, since it changes so rapidly. Many industries that try to plan five years or more in advance face the problem of attempting to forecast what the environment will be like in the future. This difficult task is typical to high-technology organizations, such as computer

TABLE 3–1
Forces exerting pressure for a more effective long-range planning effort

External Forces	Internal Forces
a. Shortages of resources, especially capital and raw materials	a. Limited growth prospects in some industries, requiring diversification
b. High inflation rates that erode profits	b. More complex organizational structures
c. Pressure from stockholders for increased profits and higher return on investment	c. Growth in size, requiring greater coordination of diverse activities
d. Technological advances	d. Increased management commitment to and understanding of long-range planning
e. Government regulation	
f. Political instability, especially in international activities and environmental areas	e. More sophisticated organizational skills through better techniques and younger, better-trained managers
g. Sophistication of information systems	

Source: Karal E. Said and Robert E. Seiler, "An Empirical Study of Long-Range Planning Systems: Strengths–Weaknesses–Outlook," *Managerial Planning* 28, no. 1 (July–August 1979): 27.

companies. Even though some industries are in a more stable environment, the 1980s will probably find more and more firms operating in a volatile environment.

Table 3–1 summarizes some of the major internal and external forces that make more effective corporate strategic planning an important responsibility of top management. These forces are evident across many industries and markets and represent important evidence for the need to develop an effective planning process.

Corporate Mission

The strategic planning process begins with the corporate mission and purpose. Basically every organization needs to define what business it wants to be in. Just as important is the business that the company does *not* want to be in. Examples of mission statements that provide corporate direction are shown in Table 3–2. BIC's mission to sell low-cost disposable products has been evident from their marketing of ball-point pens, disposable lighters, and disposable razors. Polaroid's mission seems to be in a state of flux. Edwin H. Land, the company founder, believes that Polaroid should be a research company that relies on specific research and development to refine his initial inventories and to boost the instant photo market. However the new chief executive officer (CEO), William J. McCure, Jr., feels that Polaroid must broaden its base and reduce its reliance on amateur instant photography. Some diversification into commercial and industrial products has already occurred, but

Company	Basic Corporate Mission
Levi Strauss & Co.	Provide clothing to individuals, from cradle to grave, regardless of sex or interests
Polaroid	Diversify to build a broader and more balanced product line
Bendix	Develop its high technology industrial manufacturing experience through a dedication to research and acquisitions
Sony	Diversify to balance revenues in consumer electronics equipment, run consumer electronics hardware and other businesses
Societé BIC	Manufacture and sell low cost disposable products

TABLE 3–2
Example of corporate mission statements

its future may depend on Land and the new management coming to a mutual agreement on the company's mission.[2]

Unfortunately many firms lose sight of the company mission, often because of growth in the organization and its products and markets. The mission can also be lost because of changes in the attitude of management or in the environment. Polaroid is a good example of how management's changing interest has affected the corporate mission.

In many cases the company mission is too general to be of any guidance to management in strategic planning, however. Statements such as "to be socially responsible in a profitable manner" or "to maximize long-term profits" are much too broad and simply do not provide direction to the organization. The company mission should provide answers to such questions as "What is our business?" "Who is the customer?" "What value should we provide to the customer?" "What customer needs do we—or should we—satisfy?" Because future direction should be implicit in the company mission, it is not something that should be revised every few years as a result of changes in the environment. However it is also necessary for management to monitor carefully the course being taken as a result of the company mission and to be prepared to change that course if it is no longer effective.

The company mission should take into account the following factors:[3]

- History of the organization—Given that the organization is not a new venture, it is important that the mission statement consider the past history of the company, particularly with regard to the business it has been in and the needs it has satisfied. It would not make much

---

[2]Polaroid: Turning away from Land's One-Product Strategy, *Business Week*, 2 March 1981, 108–12.
[3]Philip Kotler, *Marketing Management: Analysis, Planning and Control*, 4th ed. (Englewood Cliffs, N.J.: Prentice-Hall, Inc., 1980), 66–67.

sense for BIC to consider electric razors when its success has been in disposable products, for example.

- Preferences of management—It is important that management believe in the mission it delineates. The fact that new executives of Polaroid may not believe in the past mission set down by its founder may create disorder and confusion in the organization. Resolution of this conflict is necessary to provide a successful and effective strategic plan.

- Environmental considerations—Social and cultural trends and changes must also be considered. A company mission to produce pollutants or to waste scarce resources would not be sensible in today's environment.

- Internal resources—Concern should be given to financial as well as labor resources within the firm in preparing the statement of company mission. Tandy Corporation (manufacturer of the TRS-80) should not consider becoming the leading large computer manufacturer, since they do not have sufficient skilled labor or financial resources.

- Capability of the firm—It is senseless for any firm to attempt to do something that it is not capable of doing. BIC should probably not attempt to manufacture small electronic appliances, since this would not be making the most effective use of their capability or competence.

In Chapter 1 we discussed the topic of market myopia and explained how many firms in certain industries (e.g., railroads and movies) failed to recognize their main business. In many of these instances companies failed because they had not effectively established their company mission.

Company Objectives and Goals

Once the mission is established, the firm must focus on a set of long-term objectives and goals that will support it. These, in turn, should translate into more specific goals and objectives for each division at lower levels of management.

The company's objectives and goals should be based on (1) specific decisions on sales growth and (2) the desired percentage of growth. Generally every goal or objective has a bottom line or profit consideration. However if these goals or objectives are achieved higher profits often result. Therefore, rather than defining the goal as profits, other factors are used that contribute to the profit position of the firm.

Once these corporate objectives and goals are defined, they can be translated by management at other levels in the organization. Let's suppose, for example, that BIC set a goal of increasing corporate sales by 10 percent. This goal could be translated in many ways by division or product managers. Does this mean that BIC pens will seek a 10 percent increase in sales? Or will disposable razors be expected to achieve a sales increase of 20 percent, while sales remain the same or drop in other markets? Alternatively it is possible to translate the 10 percent sales growth as international sales in new foreign markets. In any case the management at other levels of the organization must be brought into the strategic planning process at this point to determine where the expected or forecasted changes will take place. How these objectives are to be carried out in each market or by each product will be delineated in the market-product plan (see Chapter 5).

In addition to setting the objectives and goals, management must establish priorities. For example it may be more important for BIC to increase sales by 10 percent than to enter new or foreign markets. Management must also project realistic levels of attainment. Trying to increase sales by 10 percent in a mature or decreasing market, for example, may not be feasible.

Company Growth Strategy

Once the company objectives and goals are established, mangement must begin to develop specific growth strategies. Chapter 1 (Fig. 1–3), explained how various options for new product development could be developed by combining new technology and new markets. Figure 3–2, a shortened version of this figure, indicates the role of these specific growth strategies in the strategic planning process. Each of the blocks in the matrix represents alternative growth strategies for achieving the company's objectives and goals. Earlier we presented feasible sales-growth objective of 10 percent for BIC. How can BIC meet this objective? By using the four alternatives listed in Figure 3–2:

1. Market penetration—a 10 percent sales growth can be achieved through an increase in sales to existing customers by either affecting the frequency of purchase, rate of consumption, or product use. As an alternative, nonusers may be converted to users, or it may be possible to increase marketing efforts to enhance brand-switching by the competitors' customers.

2. Market development—The firm can find new markets for the existing products to achieve the 10 percent sales growth. International markets, new regions, and new market segments would represent possible areas of opportunity.

3. Product development—Through imitation or research and development, new products could be developed to achieve the stated objectives. A formal explanation of this process is discussed in detail in Chapters 8–11.

4. Diversification—Many alternatives exist for achieving the 10 percent sales growth through diversification. New products may be developed or licensed in order to reach completely new markets. Mergers, acquisitions, or joint ventures also represent opportunities for diversification.

	Existing Technology	New Technology
Existing Markets	Market penetration	Product development
New Markets	Market development	Diversification

FIGURE 3–2
Market/technological development matrix

Management must carefully assess the environment in determining the most effective growth strategy to achieve the corporate objectives and goals.

Product Mix Strategy

Management must consider the firm's product mix when developing the strategic planning process. Once growth strategies have been examined, the appropriate product mix can be assessed. This assessment may involve product life cycle analysis, cash flow analysis, positioning analysis, product profit analysis, and strategic management of the product mix. Chapter 4 discusses these analysis techniques in detail.

EXAMPLES OF STRATEGIC PLANNING

The following paragraphs describe two somewhat diverse examples that illustrate the strategic planning process.

Fiat Fiat recently announced a new five-year strategic plan to stimulate sales and bring the firm out of a serious negative profit position. The corporate mission in part emphasized that the company return to what it once did best—making light, fuel-efficient cars. The goals and objectives involved massive capital investments, a rejuvenated product line, and big savings from greater productivity. One such objective was to increase productivity by 3 percent. (It has reached 5 percent). The growth and product mix strategies required an investment of $6.2 billion by 1985 to develop and produce the new generation of cars; reduce debt and financial costs, especially in Latin America; increase the number of international joint ventures to produce components for all lines of Fiat's business; and foster the spread of robots and automation to increase profitability.[4]

IBM IBM has developed a four-pronged growth plan for the 1980s to reestablish its prominence in the total information processing business. Although still the industry giant, IBM has not taken advantage of the growing minicomputer market. IBM's size and internal competition amongst its divisions seemingly caused the company to miss some major opportunities in this market. Management has four goals intended to re-establish IBM's position: (1) reduce production costs by investing $4 billion in a new plant and equipment, (2) reduce distribution costs by augmenting the direct sales force in the small computer market with independent distributors, mail orders, catalogs, and company-owned retail stores, (3) reorganize the management structure so that it will allow more effective movement to attack the market place segment-by-segment, (4) target high-growth markets, such as home computers, for future penetration.[5]

In the previous examples, both Fiat and IBM have defined specific objectives and goals for long-term growth. Changes in the product mix to achieve

---

[4]"Fiat: Going Back to the Basics to Make It Through the 1980s," *Business Week*, 12 January 1981, 46–48.
[5]"No. 2's Awesome Strategy, *Business Week*, 8 June 1981, 84–90.

the stated goals and objectives are also identified. It is this type of long-term strategic planning that provides an important framework for large corporations to maintain their strong market positions.

IMPLEMENTING THE STRATEGIC PLANNING PROCESS

Once all phases of the strategic plan are completed, including the company mission, goals and objectives, growth strategy, and product mix strategy, top management will document its decisions for future reference. All of these strategic decisions must then be put into action plans (i.e., market/product plans) that will direct the marketing decision-making process for the next year.

The marketing plan specifically delineates all marketing mix variables, including product, price, promotion, and distribution. The nature of the marketing plan is discussed in Chapter 5, following an in-depth discussion in Chapter 4 of techniques for the strategic management of the product mix.

SUMMARY

Strategic planning is a long-term planning process with several integral elements that affect new product development, planning, and strategy. Prior to developing the strategic plan, management at high levels in the organization will first assess and evaluate the environment to determine its long-term impact on product development and strategy. This is not always easy to accomplish, given the rapid changes that have recently taken place in the environment.

The strategic plan contains the company mission, company objectives and goals, growth strategy, and product mix strategy. These decisions are made with the support of strategic management techniques, such as product life cycle analysis, product portfolio analysis, and perceptual mapping.

The strategic plan provides an important framework for long-term marketing decision-making. Based on the strategic plan, lower levels of management in the organization can begin to formulate market/product plans that identify specific short-term marketing strategies on product pricing, distribution, and promotion.

The strategic plan should be flexible and subject to review by top management based on the achievement of certain goals and objectives. However even with this flexibility, the strategic plan should be recognized as an important implement in guiding management decision-making at all levels in the organization.

SELECTED READINGS

COX, KEITH K., and McGINNIS, VERN J., eds. *Strategic Market Decisions: A Reader.* Englewood Cliffs, N.J.: Prentice-Hall, Inc., 1982.
Articles provide insight on strategic planning, management, and markets. The strategic planning articles focus on the use of the planning process. The strategic management articles emphasize an approach that can be used for future strategy decisions. The strategic market articles provide viewpoints that match opportunities to the strength of the organization.

CRAVERS, DAVID. *Strategic Marketing*. Homewood, Ill.: Richard D. Irwin, Inc., 1982.
Emphasizes strategic marketing planning from the point of view of the enterprise rather than as a functional area of the business. The topics covered are of interest to anyone seeking a greater exposure to strategic analysis, planning, and the use of marketing strategies in executing business strategies. While considerable attention is given to strategic analysis, portfolio and screening analysis, profit impact of marketing strategy (PIMS), and market segmentation and positioning strategies, this book provides a strategic framework rather than an in-depth coverage of each analysis area.

FRENCH, GUY P. "The Payoff From Planning." *Managerial Planning* 28 (September/ October 1979): 6–12.
An excellent assessment of corporate planning from the chief executive's point of view. The article takes the reader through the basic elements of sound planning, the payoffs from planning, common planning failures, and the post-audit phase. In addition, the author discusses the problems in government planning and what can be done to improve it.

KOTLER, PHILIP. "Strategic Planning and the Marketing Process." *Business* 30 (May/ June 1980): 2–9.
Provides an in-depth discussion of the marketing process and its evolvement from corporate strategy. The strategic planning process is outlined as consisting of defining the corporate mission, objectives and goals, growth strategy, and portfolio plans. From the strategic plan, the marketing process can be developed through a process of evaluating market opportunities, determining the product/market structure, and identifying the best target market.

NAYLOR, THOMAS H., and NEVA, KRISTIN. "Design of A Strategic Planning Process." *Managerial Planning* 28 (January/February 1980): 3–7.
Describes the necessary steps to translate external and internal environmental information into the design of a strategic planning system. The role of the executives is considered important in the design of this system and is outlined in the article. Eight steps are described to achieve the desired strategic planning system.

STEINER, GEORGE A. *Strategic Planning*. Oxford, Ohio: Planning Executives Institute, 1979.
Provides a clear and simple explanation of the information that every manager should use to achieve growth and corporate goals. Discussion of the goals, the best methods to achieve them, the recognition of possible problems, and the need to adapt if conditions change are covered in detail as part of the strategic planning process.

Strategic Management
of Product Mix

T he strategic plan is an overall "roadmap" for the corporation. It describes what business the company is in, where the company is going, and what the company hopes to achieve. In addition, all aspects of the strategic plan reflect environmental changes that will have a major impact on the firm's new product development program.

This chapter illustrates numerous management techniques and programs that can be used in developing the strategic plan. These techniques are especially valuable in assessing the product line and identifying opportunities in the marketplace for new product development, product modification, and product deletions. By thoroughly assessing the product mix, the company can identify opportunities for long-term rewards, as well as provide lower levels of management (e.g., product managers) with guidelines for preparing marketing/ product plans.

PRODUCT COSTS: THE EFFECTS OF EXPERIENCE WITH A PRODUCT IN THE MARKETPLACE

In the 1960s the Boston Consulting Group led a research trend to determine the effects of cumulative production on costs. It observed and analyzed industries involved in such products and services as automobiles, semiconductors, petrochemicals, long-distance telephone calls, synthetic fibers, and airline transportation. The results of the analysis indicated that each time the cumulative volume of a product doubled, total value added costs (manufacturing, administration, sales, marketing, distribution, and so forth) fell by a constant and predictable percentage. In addition suppliers reduced prices as their costs fell, due to the effects of experience. This relationship was called the *experience curve*.[1]

Figure 4–1 illustrates an 85-percent experience curve, which means that each time experience (cumulative volume) doubles, costs per unit drop to 85 percent of the original. Alternatively this means that costs decrease 15 percent each time the cumulative production or volume doubles.

Table 4–1 shows the actual effects of an 85-percent experience curve. At one million units, the cost per unit for this product was $10.50. Each time the volume doubled, the cost per unit was reduced approximately 15 percent; this means that at a volume of 64 million, the cost per unit was reduced to $3.75. This effect is plotted in Figure 4–2. The use of a logarithmic scale in both the horizontal and vertical axes (cumulative volume and cost per unit), presents the experience curve as a straight line.

The occurrence of the experience curve can be explained by a number of factors:[2]

1. Labor efficiency—As production workers repeat and learn a particular task, their efficiency increases so that it takes them less time to

---

[1]The Boston Consulting Group, *Perspectives on Experience* (Boston: Boston Consulting Group, Inc., 1972).
[2]Derek F. Abell and John S. Hammond, *Strategic Market Planning: Problems and Analytical Approaches* (Englewood Cliffs, N.J.: Prentice-Hall, Inc., 1979), 1–12.

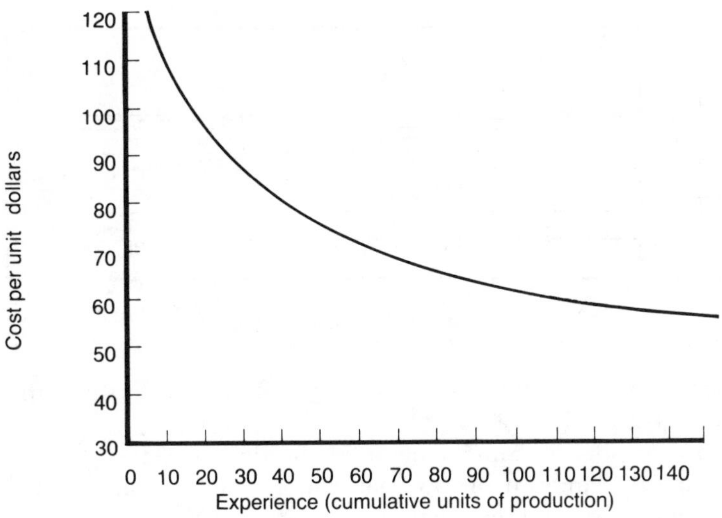

FIGURE 4–1
Example of 85-percent experience curve

produce the same number of units. Labor efficiency also increases in other areas of the firm, such as marketing, sales, and administration.

2. Specialization of work activities—As employees gain knowledge of the work tasks, they become more specialized and hence take less time to perform a particular activity. Efficiency increases as responsibility can be more effectively assigned to the most experienced people.

3. Production innovations—Improvements and innovations in production processes can also result in cost reductions, especially for labor-intensive products, such as semiconductors and automobiles.

4. Substitution of resources—As the firm gains experience in the production process, it may be able to substitute less expensive labor and raw materials that will still meet the product's design specifications. Automation can also affect the cost structure.

5. Standardization of production—In the automobile industry standardization—using a single part in many different products—was used in production of the Model T, which used a standard engine, body type, and color. The learning process of the workers assembling the car was enhanced by the repetition of tasks. Even today the automobile industry produces only a few engines, transmissions,

Volume (millions of units)	Costs Per Unit
1.0	$10.50
2.0	8.50
4.0	7.10
8.0	6.25
16.0	5.25
32.0	4.70
64.0	3.75

TABLE 4–1
Sample cost reductions with increased cumulative volume

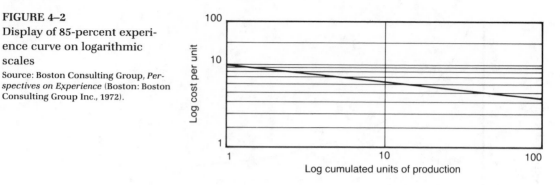

FIGURE 4–2
Display of 85-percent experience curve on logarithmic scales

Source: Boston Consulting Group, *Perspectives on Experience* (Boston: Boston Consulting Group Inc., 1972).

and chassis, which enhances replication and standardization and reduces costs.

6. Redesign of product—Improvements in the design of the product may reduce the amount of raw materials needed or simplify the production process and reduce costs.

It is important to recognize that management cannot reduce costs without a great deal of effort. Experience does not automatically lead to cost reductions; it merely represents an advantage that the firm might have over the competition.

THE PRODUCT LIFE CYCLE: SALES/CONTRIBUTION ANALYSIS

The concept of the product life cycle introduced in Chapter 1 will be considered here as a means for aiding management in the strategic planning process. The basic premise involved in a product life cycle assessment is that the firm should ensure its long-term existence by producing a balanced product line, with a blend of old and new products. A blend of products at different stages in the life cycle also represents a continuous flow of cash, so that more mature products will support the growth and development of new products.

Figure 4–3 represents the product mix of Kimchris Electronics, Inc., a hypothetical manufacturer of household small appliances and electronic products. This firm has a product mix that includes items at each stage in the life cycle—including products still at the early stage of evaluation and testing for possible commercialization. The dollar contribution performance for each product is also represented for assessment purposes.

Each product can be located in a given stage of its life cycle by matching its performance, competitive history, and current position with known characteristics of a particular stage of the life cycle. The following performance factors can help determine the stage of the life cycle: sales-growth changes, sales and profits for similar products in industry, number of years the product has been on the market, design problems, and history of similar products.[3] Competitive history analysis will include ease of market entry, profit history, the number of competitors currently in the market, and the number of com-

---

[3]Subhash C. Jain, *Marketing Planning and Strategy* (Cincinnati, Ohio: South-Western Publishing Co., 1981) 410–11.

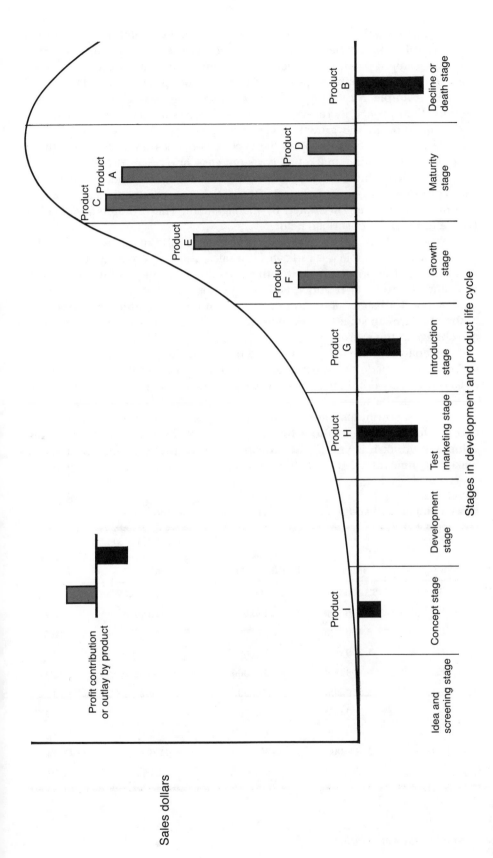

FIGURE 4-3
Life cycle status of sample product line for Kimchris Electronics, Inc.

petitors that have dropped out of the market. An analysis of the firm's current position will include whether sales are rising, leveling off, or decreasing; probable replacement products or technological improvements; consumer needs; the number of new distributors; and the amount of sales effort needed.

For example, a relatively new product with increasing sales, new competitors, high profits, increasing distribution, and price reductions (cost savings) would be in the growth stage of its product life cycle. The product reaches the maturity stage of its life cycle when sales level off, competitors drop out of the market, marketing costs increase, and prices drop.

In the hypothetical example of Kimchris Electronics, Inc. (see Fig. 4–3), seven products are being marketed with another new product in a test market, and one idea is being evaluated by management. The products are labeled from A to I and are described below.

Products C, A, and D are at various levels of the maturity stage of the product life cycle. Table 4–2 indicates the sales, profit, and percentage change in sales form the previous year for these products, as well as for products B, E, F, and G. Products H and I, which have not yet reached the commercial market, are not listed, since no sales contribution exists. Products F and E— both in the growth stage—show increasing sales, as evidenced by the percentage change in the sales column.

By analyzing the product mix and data in Table 4–2, a number of observations can be made that are relevant to strategic planning. Product C (a coffee maker) presently dominates the total sales for the company; representing over 40 percent of the total sales. Product E (a portable stereo cassette/radio recorder) also contributes significantly to the total sales (about 27 percent). Thus two products represent more than two-thirds of the company's total sales volume. Product E is still in the growth stage and probably will continue to make a significant sales contribution. Product C however is in the maturity

TABLE 4–2
Recent annual sales and profits for Kimchris Electronics, Inc.

Product	Sales	Gross Profit (or Loss)	Profit (or Loss) as a percent of Sales	Change in Sales
A	$3,300,000	$ 960,000	29%	+ 4%
B	570,000	(150,000)	(26%)	− 25%
C	8,400,000	1,350,000	16%	+10%
D	1,200,000	255,000	21%	− 4%
E	5,400,000	675,000	13%	+20%
F	1,050,000	90,000	9%	+16%
G	330,000	(36,000)	(11%)	+ 8%
Totals (actual)	$20,250,000	$3,144,000	15.5%	+10.5%
Objectives	22,500,000	$3,500,000	15.6%	+12.0%
Gap	($2,250,000)	($356,000)	(.1%)	(1.5%)

stage, with sales beginning to level off, and has limited future growth opportunities.

Product A (a food processor) is beginning to reach the later part of the maturity stage. Its profit contribution during the past year however was the highest (29 percent) of all of the products in the product line. Product D (an electric carving knife) also has a high profit contribution (21 percent) but is reaching the declining stage of its product life cycle.

Product B (a blender) incurred a loss during the past year and its sales declined by 25 percent. Management should carefully examine this product to determine its future feasibility.

Product F (a miniature, portable stereo radio with headphones) showed significant sales growth and turned a profit for the first time. Product G (a new home burglar alarm) is still in the critical introductory stage of its life cycle and lost money during the last year. However sales did increase by 8 percent over the previous year, and future opportunities are anticipated for this product.

Because the firm did not meet its sales and profit objectives during the past year, management must evaluate and analyze its marketing strategy and make appropriate changes to achieve the long-term goals.

The firm must immediately decide on the future of product B, determine whether product G can be successful, and assess the impact of declining contributions from products C and A. New products must be developed and sales of other products improved to make up for possible sales declines or deletions of other products. Because the product mix as a whole did not achieve the desired goals, management must determine whether (1) the objectives were too high, (2) uncontrollable circumstances affected market conditions, or (3) the marketing mix strategy decisions for one or more products were inappropriate.

Table 4–3 presents the product mix and a three-year sales forecast for each product in the Kimchris product line. The sales forecasts were provided by each product manager, based on current sales, profits and market conditions. Again it is apparent that the objectives will not be reached, necessitating immediate changes in the marketing strategy.

In particular products D, H, and I are of some concern. H and I are new products in the introductory stage of their product life cycle. Although losses are not unexpected at this stage, steps should be taken to ensure rapid growth and a favorable profit picture. Possible strategies to enhance sales and profits include distribution changes, cost reductions, increased-awareness advertising, sales promotion, and price reductions.

Product D may be dropped or eliminated, depending on its effect on the sales of other products in the line and new products being developed to replace it. Management may find that it needs to enhance its product development process to replace product D and to achieve the profit and sales performance objectives.

This product life cycle and performance analysis can aid management in the planning and strategy of its product line. Another alternative to this technique involves cash flow analysis, which is discussed in the following paragraphs.

TABLE 4–3

Projected sales and profits for Kimchris Electronics, Inc. (three-year forecast)

Product	Sales Projections	Gross Profit (or Loss) Projections
A	$1,700,000	$ 350,000
B	0	0
C	9,300,000	2,100,000
D	450,000	(170,000)
E	7,800,000	1,350,000
F	4,900,000	580,000
G	1,200,000	70,000
H	700,000	(30,000)
I	800,000	(65,000)
Totals projected	$26,350,000	$4,185,000
Objectives	$30,000,000	$5,500,000
Gap	($3,750,000)	($1,315,000)

PRODUCT PORTFOLIO ANALYSIS

As an alternative to sales and profit contribution performance, management may want to evaluate the product mix based on the amount of cash generated. Profits do not always give a clear indication of the firm's position and capabilities in a competitive market, since they could be tied up in receivables, capital equipment, or inventories and not reflect the firm's liquidity or its ability to generate cash quickly in light of sudden changes in the environment.

The Boston Consulting Group conceived a matrix to categorize products according to their cash flow.[4] This matrix is presented in Figure 4–4 and describes four cash quadrants based on a product's market growth rate and market place position. Each quadrant is discussed below.

"Star" These products have high market growth and a large share of the market. Significant amounts of cash are required to maintain their growth rate. Special promotions, high sales costs, lack of economy and high competition (new market entries) force the firm to spend a great deal of cash to enhance or maintain its market position. Products F (the miniature, portable stereo with headphones) and E (the portable stereo cassette tape recorder) could be classified as stars for Kimchris Electronics, Inc., depending on their market share position.

"Problem Child" These products may have been stars that lost their market share position to competition or "cash cows" that had their position eroded

---

[4]For a complete description of this model, see Bruce D. Henderson, *The Experience Curve Reviewed: The Growth Share Matrix or the Product Portfolio* (Boston: Boston Consulting Group, 1973).

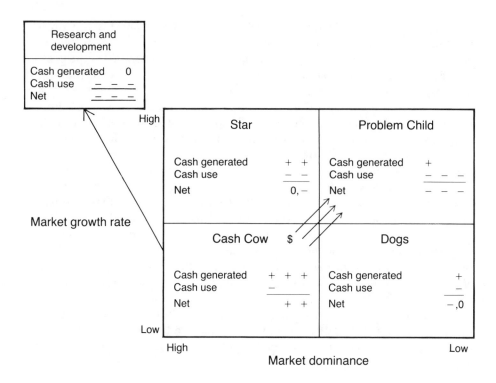

FIGURE 4–4
Product portfolio
Source: Henderson, Bruce D., *The Experience Curve Reviewed: The Growth Share Matrix of the Product Portfolio* (Boston: Boston Consulting Group, 1973).

by superior competitive products. Depending on the market share, either Product F or E from Kimchris, Inc. might be an example of a problem child.

"Cash Cow"   As a product moves out of the growth stage of the product life cycle, it no longer needs large amonts of cash to support its growth and defend its market position. At this point the product has reached maturity (high experience) and generates cash that can be used to support new product development (stars) and enhance the position of a problem child. For Kimchris Electronics, Inc. products C (the coffee maker), A (the food processor), or D (the electric carving knive) may fit this category, depending again on their actual market share position.

"Dogs"   These products are faced with no growth potential and low market share. Since future opportunities for these products are unlikely, management may give serious consideration to eliminating them, depending on the impact on sales of other products in the line. Cost-cutting or a fine-tuning of the target market may salvage some dogs and provide a return to the firm, but careful analysis and control of these products is required at this stage. Chapter 19 fully discusses when to eliminate products of this type. An example of a dog in the Kimchris product line would be product B (the blender).

Each of the Kimchris Electronics products can be categorized in one of the four quadrants to determine the appropriate long-term strategy needed to meet the company objectives and goals. Figure 4–5 illustrates the present location of each of the products and also indicates the outcome of the three-year forecast. This procedure gives management a better picture of the present and future product mix situation so they can effectively balance the mix to optimize cash flow.

As seen in Figure 4–5 the firm will have (based on forecasts) two important products (E and C) that will primarily support the rest of the product line. Both products will have low growth, requiring less cash, and a dominant market share position. Product B must be analyzed to determine whether it should be eliminated or divested. Product D may be of some concern because it represents a low-growth product and holds only a small market share. Product A, which uses a cash cow, will lose some of its market share position unless a plan can be developed to prevent this change. Product G will lose some of its market share position, even though the growth opportunities are very favorable. (This kind of loss in market share occurs when sales for the product increase but not as rapidly as for the industry as a whole.) Product F is projected to move closer to a star position and seems likely to become a cash cow eventually.

The new products H and I are also shown in Figure 4–5. Based on preliminary forecasts, product H is expected to achieve a better market share position because of the advanced technology of Kimchris Electronics and because

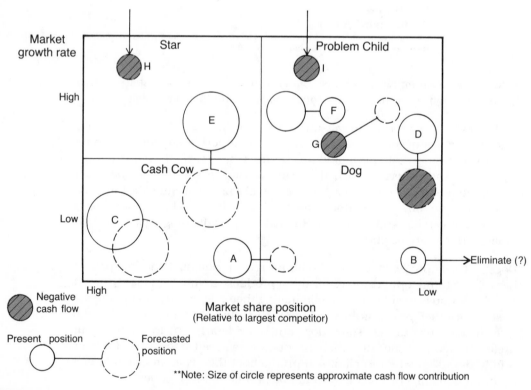

FIGURE 4–5
Product portfolio: Present and projected for Kimchris Electronics, Inc.

of limited competition. Product I, on the other hand, is an imitation product and is unlikely to give the firm a dominant market share position. Management was aware of this from the start, however, and may be very satisfied with the projected position.

Management can analyze the product portfolio to determine what marketing strategy it might undertake to optimize the projected positions of its product mix. Examples of specific strategies that can be formulated for each product are given in the following paragraphs.

MARKETING STRATEGIES

Product portfolio analysis, like the product life cycle and contribution analysis, is designed to demonstrate the importance of balancing the product mix and to suggest possible strategies that management can use to ensure that the long-term corporate objectives and goals are achieved. Using the matrix, management can try to build the market share, hold the market share position, milk the product, or divest the product. These strategies may depend again on the overall long-term objectives and goals, as well as the firm's present environmental conditions (both internal and external).

Building Market Share

This particular strategy can be pursued either as an offensive or defensive strategy. As an offensive strategy, the firm may aggressively change existing products or decide to introduce a new spin-off product to capture a larger share of a particular market. Gillette has pursued an aggressive market share build-up strategy in selected product categories. Dry Idea and Silkience allowed Gillette to recapture their lost market share in two of these particular product categories.[5] McDonald's marketing strategy to alter its advertising and promotion budget from a 50–50 split to a 60–40 split in favor of sales promotion was an attempt to build market share in a very competitive and soft restaurant market.[6]

Philip Morris has recently undertaken an aggressive strategy to build market share in the tobacco industry. This strategy would appear to be quite unusual considering the increase in public apprehension over cigarette smoking and the barring of advertising from radio and television. However the company has completed a $350 million deal with British-based Rothmans International, which combines the world's second- and fourth-largest cigarette companies. This deal would make it the largest company in the world. Philip Morris has continuously built up its market share by introducing low-tar Virginia Slims and Merit cigarettes in the United States. Because of the high return on assets, Philip Morris has chosen a build-up strategy in the cigarette industry.[7]

For the Kimchris Products presented earlier, possible market share build-up strategies might be considered for Products A, G, and F. (Product I was an

---

[5]"Gillette Is Looking Sharp Again," *The Boston Globe*, 14 April 1981, 35.
[6]"McDonald's: The Original Recipe Helps It Defy a Downturn," *Business Week*, 4 May 1981, 161.
[7]"Philip Morris Undiversifies," *Fortune*, 29 June 1981.

imitation and would probably work better with a hold market share strategy.) Market share increases can be obtained by reducing the price, increasing advertising, concentrating marketing segments, providing sales incentives, increasing distribution, and improving delivery and customer service. Whether this strategy is appropriate depends on the past market growth rate. When market growth rate is low, increasing market share is generally difficult and costly. However in a high-growth market, experience, product improvements, and the like can have a pronounced effect.[8]

This build-up strategy can be used defensively when some minimum market share is needed to provide long-run visibility for a product that would otherwise have to be withdrawn from the marketplace.

Building up the market share can require a large amount of cash. As a result short-term profits may suffer and further drain the firm's overall cash position.

Holding Market Share

When a product is in the mature stage of its product life cycle and holds a strong market share position relative to the competition, management will continue to emphasize strategy to maintain this position. Building the market share for a product in this situation would be costly and unnecessary, since it is already a cash cow that supports many other products, as well as the development of new products. To maintain the market share the strategy should entail investment in technological leadership, price leadership, strong distributor network, and quality image-building through advertising and promotion. For example, by heavily advertising and emphasizing product quality, experience, and the distribution network, Polaroid has attempted to prevent Kodak from gaining market share.[9] Caterpillar Tractor Co. has been successfully using a holding share strategy by supporting and assisting dealers in side businesses, keeping tight control over parts' inventories, repurchasing parts or equipment that dealers can't sell, and introducing new products only after building up a two-month supply of spare parts. Thus in a mature market, Caterpillar has maintained its market share position and increased sales and profits.[10]

For Kimchris Electronics holding market share is important for products C and E. These products are the cash cows of the company, and such strategies as dealer incentives and support and some price reductions (based on experience) should be emphasized.

Milking

When market share is declining in a low-growth market, the firm may consider a "milking" strategy that involves any activities to lower costs to enhance profit margins. Cost cutting, particularly in advertising, sales promotion, selling, and product quality, might maximize short-term earnings and cash flow. A com-

---

[8]Paul Bloom and Philip Kotler, "Strategies for High Market Share Companies," *Harvard Business Review* 53, no. 6 (November–December 1975): 63–72.

[9]"Polaroid: Turning Away From Land's One Product Strategy," *Business Week*, 2 March 1981, 108–12.

[10]"Caterpillar: Sticking To Basics To Stay Competitive," *Business Week*, 4 May 1981, 74–80.

pany can use a milking strategy in an attempt to save its problem children or dogs. Hoffmann–La Roche's patents, which recently expired on Valium and Librium, have led to a milking strategy where competitors and producers of generic drugs gain significant market share. For the Kimchris example, a milking strategy might be especially suitable for product D. Product A, which will lose market share based on the forecast, might require the building share strategy discussed earlier.

Hoffmann–La Roche was thus left with declining market shares of these two products in a saturated market. The result was a milking strategy which consisted of major cost cutting to maintain a reasonable profit margin given the competitive pressure to lower prices.

Divesting

Management should analyze products categorized as dogs (products that have reached the last stage of their product life cycle) to determine whether withdrawal is appropriate. Alternatively there may be instances where lost market share in a growth market is so significant and the costs of rebuilding the market share so prohibitive that a firm also considers divesting. RCA, Xerox, and General Electric all chose to withdraw from the computer market when the resources needed to build market share were prohibitive. Standard Brands has undertaken a similar strategy to trim products from their line that are losing market share and have limited or no growth opportunities, and where competition is firmly extended.[11]

Kimchris Electronics, for example, should consider divesting B and D. Product B (the blender) appears to be the best candidate for divesting, since substitutes have entered the market and its market share is very low. Product D may be either withdrawn, milked, or built up. Concentration on a specific market segment, with a balanced marketing effort, can make the product profitable.

Determining the most effective market strategy requires much analysis and consideration by management. The ability of the firm to adjust and to resolve market conflicts in a changing environment may determine its long-term success in meeting corporate goals and objectives.

USE OF POOLED EXPERIENCE: PIMS

Product life cycle analysis and product portfolio techniques are somewhat limited in that they consider only a few variables as a basis for outlining market strategy to meet long-term corporate goals or objectives. The PIMS program (Profit Impact of Marketing Strategy) overcomes this limitation by pooling empirical evidence from a large number of businesses in a large number of situations.[12] PIMS uses a computer model to analyze factors to determine their relationship to performance (particularly return on investment [ROI]). The idea

---

[11]"Standard Brands: When A New Product Strategy Wasn't Enough," *Business Week*, 18 February 1980, 142–6.
[12]Sidney Schoeffler et al., "The Impact of Strategic Planning on Profit Performance," *Harvard Business Review* (March–April 1974).

was conceived as part of a project initiated in 1960 by General Electric. After several years it became evident that certain identified variables could be used to explain variations in performance of various divisions within the company. Later the model was further developed at the Harvard Business School and the Marketing Science Institute. In 1975 the Strategic Planning Institute was formed to manage the PIMS program. This Institute was designed to be a nonprofit organization operated by member firms who would benefit from the program. At present there are over 200 members with more than 1,000 business units or divisions (all considered independent businesses).[13] Each member receives the results of the analysis of key performance variables in many industry situations. They can review these findings to isolate key factors that may influence profitability and cash flow and to generate key questions about current and future strategy. The following are questions sometimes asked:

1. What profit rate is *normal* for a given business, considering its internal and external environment?
2. If the business (unit or division) continues on its current track, what will its future operating results be?
3. What strategic changes in the business are likely to improve the operating results?
4. What will be the short-term and long-term effects of a specific defined strategy on profitability and cash flow?

Management uses the PIMS data to establish the overall objectives for a business and to identify strategies that may be employed to reach these goals and objectives. The data also assists in forecasting profits; allocating capital, manpower, and other scarce resources; measuring managerial performance; and appraising new business opportunities.

Table 4–4 provides a sample list from a Strategic Planning Institute brochure of the type of information given to each business unit in the PIMS data base.

PERCEPTUAL POSITIONING: EVALUATING THE PRODUCT MIX

In developing strategic plans, market plans, and strategies, a perceptual map can be used to represent an existing product in relationship to the competitive market. A perceptual map is a graphic representation of how consumers in a particular market perceive a competing set of products relative to each other. It serves two important functions for management in preparing strategic and market plans. First it allows management to relate existing products in the mix with competitive products along important market dimensions. Second, the perceptual map provides insight to marketplace opportunities for new products or suggests possible changes in the strategy to reposition an existing product.

Figure 4–6 shows an example of a perceptual map for some hypothetical candy bars. Two important dimensions are used: sweetness and chewyness.

---

[13]"The PIMS Program" in *The Strategic Planning Institute Brochure*, (Cambridge, Mass.: 1979) 17.

TABLE 4–4

Sample of information provided to each business unit in
the PIMS data base

Characteristics of the Business Environment

- Long-run growth rate of the market
- Short-run growth rate of the market
- Rate of inflation of selling price levels
- Number and size of customers
- Purchase frequency and magnitude

Competitive Position of the Business

- Share of the served market
- Share relative to largest competitors
- Product quality relative to competitors
- Prices relative to competitors
- Pay scales relative to competitors
- Pattern of market segmentation
- Rate of new product introductions

Structure of the Production Process

- Capital intensity (degree of automation, etc.)
- Degree of vertical integration
- Capacity utilization
- Productivity of capital equipment
- Productivity of people
- Inventory levels

Discretionary Budget Allocations

- R&D budgets
- Advertising and promotion budgets
- Sales force expenditures

Source: The Strategic Planning Institute, Brochure (Cambridge, Mass.: The Strategic Planning Institute, 1979) 17.

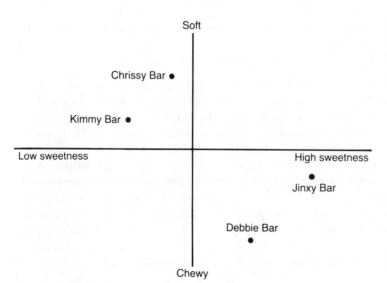

FIGURE 4–6

Example of perceptual map for
four candy bars

It can be seen from this map that consumers perceived the Jinxy Bar as being more sweet than any of the competing brands. The Debbie Bar was perceived as more chewy than any of the other three brands. The Chrissy Bar and Kimmy Bar are relatively close on these two dimensions, indicating the possibility for significant consumer substitution that may be of some concern to management of the firms producing these brands. This particular map oversimplifies the evaluation process, since most consumers use more than two criteria when choosing a product. However, because the map uses two significant product-choice attributes, it does indicate market opportunities, since no one brand appears to be relatively low in sweetness but chewy, or high in sweetness but soft.

Mapping Procedure

In order to develop a perceptual map, management must first identify the attributes or dimensions of the product that are important to consumers. For example it is conceivable that consumers purchase candy bars based on such attributes as sweetness, chewyness, amount of chocolate, shape, size, nutrition, and so on. Each of the important attributes can then be placed along a scale like those shown in Figure 4–7, and consumers can be queried to determine their perception of each of the competing brands. Each point in the scale can be assigned a value, and the aggregate of the scores can be used to determine the location on the two-dimensional map used in Figure 4–6.

The formula used to determine consumer preference for an existing brand is shown below:[14]

$$P_{jk} = \sum_{i=1}^{n} (A_{ijk}) W_{ij}$$

where:

P_{jk} = the jth customer's preference for product k

A_{ijk} = the level of attribute i perceived by customer j to be possessed by product k

W_{ij} = the salience (importance) of attribute i to customer j for choice in the product market

n = the number of salient attributes for choice in the product market.

Management should carefully evaluate two important factors in preparing perceptual maps: the relevant criteria used by a consumer to choose a particular product and the importance of those criteria in choosing a particular brand. Both of these factors are an inherent part of the formula described above.

These factors can be determined in different ways, depending on the level of quantitative sophistication it is practical to use. The simplest way to identify the relevant criteria is from the experience of management or from prior research. This may be appropriate in situations where the attributes are obvious or limited by the nature of the product. The importance of the criteria

---

[14]Allan D. Shocker and V. Srinivasan, "Consumer-Based Methodology for Identification of New Products," *Management Science* 20, no. 6 (February 1974): 927.

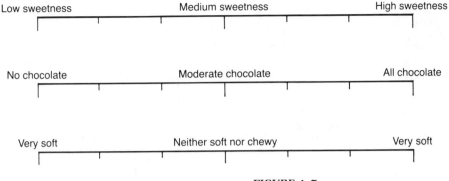

Low sweetness   Medium sweetness   High sweetness

No chocolate   Moderate chocolate   All chocolate

Very soft   Neither soft nor chewy   Very soft

FIGURE 4–7
Example of product positioning for selected attributes

could also be assigned by management based on prior experience or knowledge of the marketplace.

The extreme alternative is to determine the important criteria and their importance by using a multivariate statistical model, such as factor analysis, discriminant analysis, or regression/correlation.[15] These techniques in general will "load" or identify those criteria most important in the buying process, as well as provide an importance value for each criterion. Those criteria found to be unimportant in the consumer's selection process or those found to be highly correlated with another criterion (and therefore unnecessary) can be dropped from consideration in locating brands on the maps.

In addition to preference markings for competing brands, consumers may be asked to provide a preference rating for an "ideal brand." Thus each person would be asked to identify along a scale the ideal amount of chewyness, sweetness, and so forth. These ideal points for the entire sample of customers can then be clustered for comparison with the competing four brands.

Figure 4–8 illustrates the ideal points and four clusters for the sample customers along two of the important criteria: sweetness and softness/chewyness. These ideal clusters may lead to the conclusions that clusters 2 and 4 are not being satisfied in the marketplace and that they may represent opportunities for developing new products or repositioning existing products.

The information provided from the perceptual maps can be valuable in developing strategic and market plans and assessing the product mix. More specifically perceptual mapping helps management to do the following:[16]

- Understand how competing products or services are perceived in terms of strengths and weaknesses by various consumer groups
- Understand the similarities and dissimilarities between competing products and services
- Reposition a current product in the perceptual space of consumer segments

---

[15]For more information on these techniques, see Paul Green and Donald Tull, *Research for Marketing Decisions*, 4th ed. (Englewood Cliffs, N.J.: Prentice-Hall, Inc., 1980), 303–41, 381–459.
[16]William D. Neal, "Strategic Product Positioning: A Step-By-Step Guide," *Business* (May–June 1980), 34–41.

FIGURE 4–8
Clustering of ideal points on salient attributes

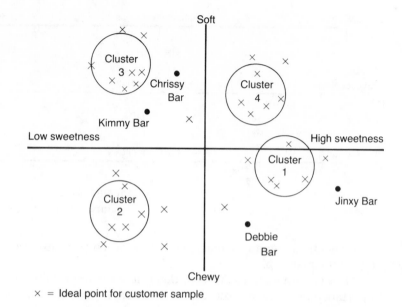

× = Ideal point for customer sample

- Position a new product or service in an established marketplace
- Track the progress of a promotional or marketing campaign on the perceptions of targeted consumer segments

COMPARISON OF STRATEGIC MANAGEMENT TECHNIQUES

It would be impossible to judge what approach should be used by a particular firm to assess its product mix and develop strategic and market plans. Each technique is valuable in its own way and should be considered by management only if they feel comfortable in the analysis and believe it is helpful in making planning and strategy decisions. Using all of the techniques to some degree can be most helpful to management in assessing the product mix for short-term market planning and long-term strategic planning.

SUMMARY

This chapter has presented numerous valuable management techniques that can be used in developing the strategic plan. These techniques are particularly relevant in assessing the product mix and identifying opportunities in the marketplace for new product development, product modification, and product deletions.

Sales/contribution analysis provides management with a perspective of the product mix at various stages of the life cycle, including the development phase. This analysis ensures that appropriate marketing decisions are made regarding the need to develop new products.

Product portfolio analysis is an alternative to sales/contribution analysis and consists of an evaluation of the product mix based on the amount of cost generated from a product. The matrix approach is used to classify by market growth and market share potential.

PIMS and perceptual mapping are other techniques presented in the chapter. The PIMS approach uses a large number of variables and information pooled from many diverse businesses. Perceptual mapping provides a graphic representation of how consumers in a particular market perceive a competing set of products relative to each other.

These techniques use different approaches but ultimately have the same purpose: to provide management with the basis for evaluating existing product problems and identifying new product opportunities.

SELECTED READINGS

ABELL, DEREK F., and HAMMOND, JOHN S. *Strategic Market Planning: Problems and Analytical Approaches.* Englewood Cliffs, N.J.: Prentice-Hall, Inc., 1979.
Presents both text material and cases relating to contemporary management problems that require the integration of marketing planning into the strategic planning process. A number of analytical techniques and formal strategic planning methods are compared and contrasted. The final section discusses how to prepare a strategic plan.

DAY, GEORGE S. "Diagnosing the Product Portfolio." *Journal of Marketing* 41 (April 1977): 29–38.
One of the best explanations of the product portfolio concept. The article clearly presents the nature of the matrix approach and provides the critical assumptions, measurement, and application issues that affect the strategic planning process.

ENIS, BEN M. "GE, PIMS, BCG, and the PLC." *Business* (May–June 1980), 10–18.
An excellent review and comparison of four approaches to strategic planning. The author presents the argument for increasing the use of these approaches not as a substitute for but as a supplement to managerial judgment.

"General Electric: The Financial Wizards Switch Back to Technology." *Business Week* (16 March 1981), 110–18.
Discusses the strategy chosen by General Electric for the 1980s. It provides a good example of the use of a cash cow to support other products and new product development. The drain in the cash cows actually caused the firm to make an extensive effort through acquisitions and financial control to ensure that more technology would be brought to the market place.

HASPESLAGH, PHILIPPE. "Portfolio Planning: Uses and Limits." *Harvard Business Review* (January–February 1982): 58–73.
Shows that product portfolio planning helps managers with problems in diversified industrial companies. The most important contribution that it can add is to the management process based on the nature of the business, its competitive position, and its strategic mission.

HAUSER, J. R., and KOPPELMAN, F. S. "Alternative Perceptual Mapping Techniques: Relative Accuracy and Usefulness." *Journal of Marketing Research* (November 1979): 495–506.
Presents both theoretical and empirical arguments and evidence for alternative perceptual mapping techniques. Consumer images of alternative shopping-location analysis indicates that factor analysis was superior to discriminant analysis for developing a measure of consumer perceptions.

KIECHEL, WALTER. "The Decline of the Experience Curve," "Three (or Four, or More) Ways to Win," "Oh Where, Oh Where Has My Little Dog Gone? Or My Cash Cow? Or My Star?," "Playing the Global Game." Four part series. *Fortune* (5 October 1981, 139–145, 19 October 1981, 183–88, 2 November 1981, 148–154, 16 November 1981, 111–126).

A four-part series on new management strategies. Each article discusses an important topic related to the formulation of corporate strategy, including the experience curve, generic strategies, the product portfolio matrix, and the threat of foreign competition and management strategies to meet these threats.

NEAL, WILLIAM D. "Strategic Product Positioning: A Step-By-Step Guide." *Business* (May–June 1980), 34–42.

Presents an example of perceptual mapping used to position a new product or service.

The PIMS Program. *The Strategic Planning Institute Brochure."* Cambridge, Mass.: The Strategic Planning Institute, 1980.

Provides a complete description of the PIMS program including the input data of about 100 items obtained from each participating company. The data cover 1,700 products and services over a long period of time.

The Market/
Product Plan

Each product or group of products (categorized by market or segment) needs a well defined plan. Market/product planning is an annual or short-term planning cycle that provides action on the marketing mix variables needed to support long-term corporate goals and objectives defined in the strategic plan. Particularly in large corporations, it differs from the strategic plan in that it is usually formulated by lower-level managers, such as product or marketing managers. Regardless of the level or time dimension, the marketing plan requires an assessment of environmental variables and defines objectives and goals for the product in question. Whereas strategic planning is concerned with the welfare of all products, rather than the strengths and weaknesses of the company, the market/product plan is concerned only with a single product or market and describes the appropriate product strategy—branding, packaging, distribution, pricing, and promotion. Thus the market/product plan serves as more of a day-to-day guide for management.

THE MARKETING SYSTEM

To understand the nature of the market/product plan, let's first examine the marketing system. The marketing system identifies the major components that interact in the firm's environment to enable the firm to successfully provide products and/or services to the marketplace. Figure 5–1 provides a simplified view of the marketing system.[1] We can categorize the relevant variables in the marketing system as being external and internal to the firm. These variables are also interactive in determining the nature of the market/product plan.

As with strategic or corporate plans, management must evaluate the environment as it specifically relates to the market/product in question. The environment that can be categorized as external or internal provides an important framework for the marketing decision-making process.

External Environment

The external environment as a whole may be described as extremely volatile and generally uncontrollable by management. In fact it is becoming increasingly difficult to develop effective marketing plans, even for the short-term, because environmental changes occur so often and with so little warning. A sudden change in economic policy (decontrol of prices), new technology (a low-cost solar-powered car), political changes (assassination of Anwar Sadat), shortages of raw materials (oil boycott), and new legislation (pollution controls) all represent examples of changes that have or could have a significant impact on market/product planning. These examples show how uncontrollable such forces can be.

---

[1]For in-depth discussion of the marketing system see William Stanton, *Fundamentals of Marketing*, 6th ed. (New York: McGraw-Hill Inc.) 19–33, or Philip Kotler, *Principles of Marketing* 2d ed. (Englewood Cliffs, N.J.: Prentice-Hall, Inc., 1983), 31–52.

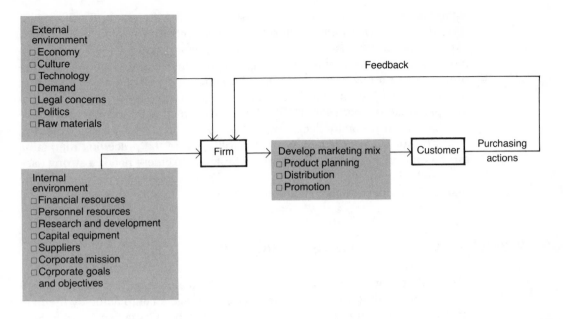

External
environment
□ Economy
□ Culture
□ Technology
□ Demand
□ Legal concerns
□ Politics
□ Raw materials

Internal
environment
□ Financial resources
□ Personnel resources
□ Research and development
□ Capital equipment
□ Suppliers
□ Corporate mission
□ Corporate goals
  and objectives

Feedback

Firm

Develop marketing mix
□ Product planning
□ Distribution
□ Promotion

Customer

Purchasing
actions

FIGURE 5–1
The marketing system

Internal Environment

The internal environment represents a group of variables that are generally
controllable by the firm, although some—such as personnel resources and
suppliers—may be uncontrollable under certain circumstances (for example,
a shortage of certain types of technical managers). Certainly these variables
are subject to the influence of the external environment. The corporate mission, long-term objectives and goals inherent in the strategic plan, represent
part of the internal environment because they also influence the marketing
mix and the development of the market/product plan.

THE MARKETING MIX

The marketing mix describes the interaction of four factors that represent the
core of the firm's marketing system: product, pricing, distribution, and promotion. Each of these factors varies in its importance, depending on the industry, corporate mission, objectives and goals, or size of the firm.

For example Hershey chose at one time not to spend any funds on mass
media advertising, whereas competitors (such as Nestlé) spent millions of
dollars on mass media advertising. Some firms emphasize research and development and like to be the first in their industry to introduce new products.
Others in the same industry (perhaps because of their size) may choose to
imitate or take a wait-and-see attitude before introducing a new product.

Within each of the four factors that make up the marketing mix there are
countless other variables. For example the product area encompasses packaging, branding, product design, and product development. In pricing, man-

agement must be concerned with costs, discounts, freight, and other price related factors. Distribution represents those activities related to providing place utility to the customer—that is, providing a product where it is convenient to buy and when it is needed. The type, number, and location of distributors are just a few of the countless decisions that must be made as part of the firm's distribution plan. The final factor—promotion—includes advertising, personal selling, sales promotion, and publicity. Within each of these variables there are many decisions that must also be made by management.

The market/product plan must clarify each of the preceding marketing mix factors. Although flexibility is important, management needs a strong base to provide direction for day-to-day marketing decision-making. The marketing plan is the framework for guiding these actions.

MANAGEMENT'S EXPECTATIONS OF MARKET/PRODUCT PLAN

The market/product plan should be viewed by management as an instrument to guide marketing decision-making. When managers fail to abide by the plan or decide that "planning is a waste of time," they have usually misunderstood what market planning can and cannot do. Table 5–1 provides a list of things that can and cannot be expected from the market/product plan. The mere organization of the thinking process of management at different levels in the firm is a step in the right direction. However to develop successful plans, it is necessary to formally document and describe as many marketing details as possible that will enter into the decision process during the following year. Even though firms have generally become more dependent and committed to market/product plans than ever before, it is still important to recognize that there are common problems that make designing of an effective market/product plan difficult.[2] The following paragraphs discuss some of these obstacles.

Common Problems

A recent study of 267 manufacturing firms and service companies revealed a number of difficulties in preparing effective market plans.[3] Of course, the difficulties vary by industry, the size of the firm, and other factors, but the problems discussed here were most common across the entire sample.

Forecasting

The ability to set realistic forecasts for a particular market or product has become a difficult task in today's environment of economic turbulence and inflation. With rapid changes also taking place in government regulation and the competitive marketplace, management must be prepared to move quickly to maintain their market position and achieve the forecasts established. Various control mechanisms to allow for the modification of marketing strategy

---

[2]Hopkins, *The Marketing Plan* (New York: The Conference Board, Inc., 1981), 1.
[3]Ibid, 2.

TABLE 5–1
What market planning can and cannot do

Can Do	Cannot Do
▪ It will enhance the firm's ability to integrate all marketing activities to maximize efforts toward achieving the corporate goals and objectives.	▪ It will not provide a crystal ball that will enable management to predict the future with extreme precision.
▪ It will minimize the effects of surprise from sudden changes in the environment.	▪ It will not prevent management from making mistakes.
▪ It establishes a benchmark for all levels of the organization.	▪ It will not provide guidelines for every major decision. Judgment by management at the appropriate time still will be critical.
▪ It can enhance management's ability to manage since guidelines and expectations are clearly designated and agreed upon by many members of the marketing organization.	▪ It will not go through the year without some modification as the environment changes.

are also necessary to ensure that forecasts are achieved. Various forecasting techniques are discussed in Chapter 14.

Obtaining Needed Information

In order to develop an ideal market plan, information regarding market trends, consumer needs, technology, market share changes, competitor reactions, and the like are necessary. Generally management is not able to obtain all of the information it needs. Even with sophisticated market research techniques and information processing using advanced computer systems, management often falls far short of perfect information. The information needed and the ability to obtain this information, of course, varies from firm to firm and industry to industry. For example in a stable, predictable market, such as institutional vegetable shortening, management may find that they need only information on competitors' prices and consumer attitudes. In a potentially volatile market, such as oil, gas, computers, coffee, lumber, and meats, much more information is needed and is harder to obtain, making the planning process more complicated.

Time Constraints

Marketing managers often complain about the lack of time to adequately and effectively prepare the market/product plan. The added responsibility of the planning process places another burden on marketing departments. The argument for market/product planning seems to address the problem of limited time, since the plan's ultimate objective is to save time by providing a quick line for management to follow during this short-term period. However insufficient time spent on preparing the market/product plan can result in a superficial plan that may lead to market strategy decisions and actions that do not take into consideration all possible alternatives and their outcomes.

Coordination of the Planning Process

Chapter 6 examines this problem more thoroughly. Basically product managers, marketing managers, and other middle managers are hampered in the planning process because they lack the authority to obtain necessary cooperation and commitment from various parts of the organization. Corporate politics, a lack of commitment by line managers, or even poor lines of communication can damage the planning process.

It is also important that the market/product plan(s) be coordinated with the strategic plan (see Fig. 5–2). A lack of commitment to longer-term goals or poorly defined objectives and goals will affect the integration of the market/product plan. For example marketing managers often complain that the management hierarchy sets unrealistic goals and objectives for short-term earnings, which restricts the direction of the market/product plans.

Implementation of Market Plan

Even though members of the marketing team may participate in the development of the plan, some may not be committed to it. Thus the plan becomes a mere formality and is not effectively implemented. This is particularly evident in those situations where senior management is also not totally committed to the market/product planning process. A market or product plan is a waste of valuable time and effort if it is not implemented. Thus the plan should be comprehensive and detailed so that there is no question as to how the goals and objectives are to be achieved.

CRITERIA FOR AN EFFECTIVE MARKET/PRODUCT PLAN

Although every market/product plan will be different, there are certain criteria that are generally required for an effective plan:

- It should be short—A voluminous plan will be tossed in a file drawer and never used. Some firms set requirements for length, or they request a summary of a more detailed market planning document.
- It should be specific—Although length is often a concern, the document should not be so short that specific details of the marketing mix are not thoroughly conveyed. It is useless to state that "we need to increase distribution," without stating how. The market/product plan must provide a specific outline of strategy and actions to achieve the increased distribution.
- It should be written—From the discussion in this chapter so far it should be clear why this requirement is necessary for effective planning.
- It should be formal yet flexible—Although smaller firms may choose to keep their market/product plans informal, most firms will formalize their plans. Usually firms with no formal market/product plan rely on more formality in budgeting for sales and marketing expenditures.

Where formal plans do exist it is important to recognize that they are not "etched in gold" and that they should allow for flexibility given sudden changes in the environment.

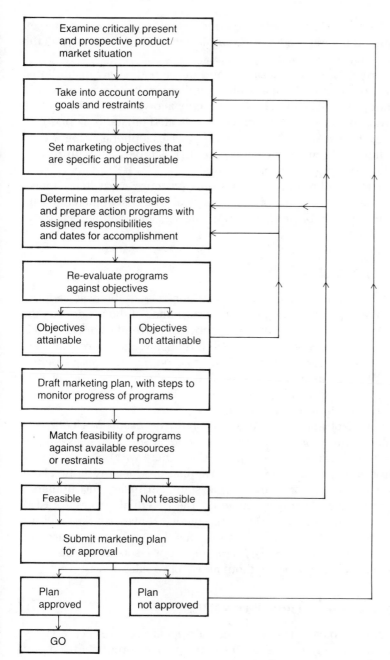

FIGURE 5–2

Sample flow chart for a marketing plan

Source: David S. Hopkins, *The Marketing Plan* (New York: The Conference Board, Inc., 1981) 17.

PREPARING THE MARKET/PRODUCT PLAN

The market/product plan is a document that attempts to answer the following questions: (1) Where are we now? (2) Where do we want to go? (3) How do we get there? Figure 5–2 illustrates the stages involved in the planning process that will provide a response to these questions.[4]

---

[4]Hopkins, *Marketing Plan*, 16.

An example will be used to enhance understanding of how the market/product plan should be prepared. This example will be carried through all the steps described below in the planning process.

Example: Grill-Clean Corp. designed and developed a new institutional grill cleaner that could outperform any other similar product because it could be used on a hot grill. This eliminated downtime, allowing a clear grill at all times of the day. Initial testing of the product indicated outstanding performance at a low cost. Demand for the product was exceptionally high, even though production had not reached full potential. Samples left with restaurants had resulted in a large number of orders.

The product could be produced in a milk pasteurizer vat that allowed for the proper order and amounts of each chemical. The cost of materials and labor was very low. Space for production and storage was readily available at a low cost.

DEFINING THE BUSINESS SITUATION

The business situation analysis involves a careful review of the past performance/history of the product(s), the present performance of the product(s), the firm and industry as a whole, and the prospective opportunities or changes in the marketplace. In the Grill-Clean example the business situation was very significant because the product was new and had no past or present performance to be evaluated. In this case, this phase of the market/product plan should review trends in the marketplace, product benefits, and strengths and weaknesses of the firm in its environment.

Grill-Clean had obvious advantages: it could be used on a cold grill, eliminating grill downtime; it was odorless and noncaustic; it was easy to use; and it was relatively inexpensive per application. There was no other firm at the time of introduction in the market with a similar product, presenting large sales opportunities in the company's home state. No major environmental variable appeared to be likely to impinge on the potential success of Grill-Clean, since the product met all FDA requirements. However it was still important to recognize that the customers in general were long-time users of traditional grill cleaning methods (i.e., scraping, stones, screens, pickle juice) and would not be likely to change without an effective sales effort.

DEFINING MARKET SEGMENTS AND OPPORTUNITIES

The market segments to which the product will appeal also must be described and analyzed in terms of size, needs, location, present products used, and other relevant factors. Opportunities to reach one or more market segments can then be documented.

Grill-Clean identified a number of potential market segments: restaurants (fast food and standard menu), hospitals, schools, and federal government and corporation cafeterias. Based on its own strengths and weaknesses, Grill-Clean chose and documented the restaurant market as the area with the most potential.

Specifically Grill-Clean's management chose the restaurant market in Massachusetts as their target market. Analysis revealed that there were over

10,000 operating grills in this state. Since there was no direct competition, this represented a good opportunity for a full-scale marketing effort.

ESTABLISHING GOALS AND OBJECTIVES

Before marketing strategy decisions can be determined for the marketing mix variables, management should establish *realistic* marketing goals and objectives. These goals and objectives should indicate where the firm is going and provide specific formal goals for each product, group of products, or market. Objectives relating to such factors as market share, profit, sales (by territory and region), market penetration, number of distributors, awareness level, new product launchings, pricing policy, sales promotion, and advertising support can be established.

Grill-Clean's objectives were to achieve 10 percent market penetration at the end of the first year; sample (leave a sample of the product) with 60 percent of the potential users; and have distribution coverage in 75 percent of the market. All of these goals focused on only the restaurant market.

In the Grill-Clean case each objective was easy to quantify and measure for control purposes. However not all objectives or goals must be or can be fully quantified. For example a firm may establish objectives to complete a market research study, improve customer service, set up a sales training program, or improve packaging. Clearly any objective or goal that management feels is important to satisfy corporate goals and objectives must be considered relevant to the market/product plan.

Although the nature of the objectives will differ for industrial products (with more emphasis placed on sales to major accounts and customer service plans) and consumer products (with more emphasis placed on advertising, sales promotion, and product planning) it is a good idea to set a limit of six to eight prime objectives. Limiting the number of objectives ensures that key areas receive the necessary concentrated attention.

DEFINING MARKETING STRATEGY AND ACTION PROGRAMS

Once the marketing goals and objectives are established, management can begin to develop the instructions (marketing strategy) regarding the marketing mix that will lead to achieving these objectives. These strategy decisions and action programs indicate how the firm is going to get where it is going. Thus the marketing strategies and action programs run parallel to the specific statement of objectives. Again, as in the case of the market objectives, the strategies and action programs should be specific and detailed. Examples of poor and good strategies and action programs are as follows:

- Poor—We will increase market share for product A by increasing the advertising budget.
- Good—We will increase product A's market share by 5 to 6 percent by (1) adding two additional sales representatives, (2) doubling the number of free samples given out and (3) enhancing packaging attractiveness with new design and colors.

The marketing strategy and action programs developed for Grill-Clean emphasized sales, distribution, and sales promotion to achieve the desired goals and objectives. A schedule was set up showing dates certain accomplishments (for example, contacting and negotiating with ten manufacturers' representatives in the western part of the state) would be completed. A missionary salesman was hired to build awareness and contact and sample 60 percent of potential users in a designated market area. New packaging and production efficiencies were also outlined that would hopefully improve the pofit structure and image of the product. Quantity discounts were established to encourage larger orders. To enhance awareness and to demonstrate the effectiveness of the product to potential users, attendance at major restaurant trade shows was also included in the strategy and action programs.

DESIGNATING RESPONSIBILITY FOR STRATEGY IMPLEMENTATION

The implementation of each strategy and action must be the responsibility of someone in the organization. For example sales strategies may be the responsibility of the sales manager and advertising/promotion strategies may be the responsibility of the advertising manager. It is unlikely that the implementation of all these strategies would be the responsibility of one person. It is possible that a product manager may have the responsibility of coordinating all of these programs but not necessarily implementing them.

Since Grill-Clean is a very small company, the responsibility for implementing the marketing plan would be in the hands of the owner or CEO.

MONITORING PROGRESS OF PROGRAMS

It is important for management to establish specific procedures to monitor progress and execution of marketing strategy. Tracking data such as sales (by product, customer, territory, and so forth), distribution (number of distributors, retailers) and promotion (awareness-level research or recall surveys) is just one kind of monitoring that can be done to review the effectiveness of the marketing strategy and its probable attainment of the marketing goals and objectives. Any "weak signals" from the monitoring process provides management with the opportunity to redirect portions of the marketing program to enable the firm to achieve its original objectives.

In some cases firms may require the responsible party to document and submit a formal explanation whenever some action has been delayed or an objective has not been met. Corrective actions that were instituted would also be noted in this report to senior management. This requirement provides top management with feedback on the company's ability to reach corporate goals and objectives established in the corporate strategic plan.

The management of Grill-Clean is able to track sales data, distribution intensity, and the sampling program to determine how likely they are to achieve the desired goals and objectives. For example the sampling program had only been able to reach 3 percent of the potential end-users. In reviewing this problem management found that many of the manufacturers' representatives were forgetting to leave a sample with the customer. To correct this, incentives were introduced whereby a manufacturers' representative would get bonus credit for every customer given a free sample who followed with an order. The bonus credits would then be accumulated and prizes or cash bonuses awarded. Distribution coverage was also found to be below the expected level. However it was felt by management that this particular problem was due to the limited time of the CEO and missionary salesmen during the first few weeks of operation. It was felt that this factor needed to be more critically evaluated in the second month after product introduction when these individuals would have more time to negotiate with potential distributors of the product.

CONTINGENCY PLANNING

It was mentioned earlier that a good marketing plan should be specific but flexible, particularly given the sudden changes that have occurred in the environment. Generally market plans do not contain alternative plans of action should changes in the environment take place. The mere combination of possible changes or the ability to forecast these changes for some firms represents a most difficult task. However in some cases contingency plans are spelled out in detail so that management can move quickly in meeting an unexpected change in the environment.[5]

A key factor in contingency planning is the extent to which management felt it could control a change in the environment. When the firm cannot control these changes, it is often committed to developing contingency plans. The following was pointed out in the Conference Board study by one vice president of an oil company:

> The uncertainties in the supply of crude, compounded by the vagaries of government regulations, have complicated our ability to plan our future course, and increased the number of external factors over which we have little or no control . . . causing us to move from a simple long-range plan reference to several plans with appropriate tactics covering the various contingencies."[6]

If contingency plans are instituted they will also be monitored and inducted in the same way as the original market/product plan.

MARKET/PRODUCT PLANNING: A RECIRCULATING PROCESS

The market/product planning process should not be an annual exercise that has no use in guiding the firm through its market decision-making process. Even though these plans are annualized, management must be wary of all possible factors that will contribute to future planning decisions. Feedback at every stage is critical to integrating, as well as enhancing, the ability of management to make effective marketing decisions.

---

[5]The Conference Board, Inc. study reported that two in four of those reporting had detailed contingency plans. Hopkins, *Marketing Plan*, 28.
[6]Hopkins, *Marketing Plan*, 16.

SUMMARY

The market/product plan is an annual short-term planning process that provides a framework for action in the marketing mix variables needed to support long-term corporate goals and objectives. Like the strategic plan, the market/product plan must evaluate the environment, which may be categorized as external or internal. The external environment includes such factors as the economy, technology, politics, raw-material shortages, legislation, and so on. The firm cannot control these factors. However the firm can control the internal environment, which includes such factors as the corporate mission and goals, suppliers, capital equipment, personnel, research and development, and financial resources.

The market/product plan includes a definition of the business situation, definition of market segments and opportunities, establishment of goals and objectives, and marketing strategy and action programs. These areas are generally the responsibility of one person who will coordinate activities and ensure their implementation. Each activity will be monitored to track critical variables necessary in achieving the desired goals and objectives of the firm. Any weak signals from this monitoring process provides management with the opportunity to modify and redirect certain actions identified in the plan.

Common problems do seem to exist in the planning process such as forecasting sales, coordination of various departments, and implementation of the plan. Awareness and sensitivity toward those problems by management can enhance the effectiveness of the planning process.

If sudden changes do occur in the environment, management should be prepared to institute contingency plans. These contingency plans are sometimes spelled out in the market plan so that management can initiate the appropriate action quickly when an unexpected change occurs in the environment. Thus, through a careful monitoring process, the marketing plan becomes a significant guide to management that supports long-term corporate goals and objectives.

SELECTED READINGS

BRIM, JOHM M. *Corporate Marketing Planning.* New York: John Wiley & Sons, Inc., 1967.
Provides a step-by-step guide relating day-to-day actions and activities necessary to support the market planning process. Many sample worksheets and control techniques are illustrated.

HOPKINS, DAVID S. *The Market Plan.* New York: The Conference Board, Inc., 1981.
Report based on findings of the latest in a series of research investigations conducted over a number of years to examine the methods and effectiveness of marketing planning. It describes current practices and approaches and discusses examples thoroughly.

LAWTON, LEIGH, and PARASURAMAN, A. "The Impact of the Marketing Concept on New Product Planning." *Journal of Marketing* 44 (Winter 1980): 19–25.
Describes a study that provides empirical evidence on the impact of the marketing concept on product planning. The results indicate that the marketing concept does not inhibit new product planning.

PETERS, MICHAEL P. "The Role of Planning In The Marketing of New Products." *Planning Review* 8 (November 1980): 24–27.

Discusses the significance of planning as a means of ensuring the success of a new product. Planning for new products is divided into two categories: market development planning (exploiting existing or new markets and technological development planning) and improving the utilization of the firm's existing scientific and production skills or acquiring new skills.

Special Report in *Sales & Marketing Management* "Marketing Planning: How to Get There From Here." (7 December 1981).

Special report consisting of a series of articles that focus on the process of market planning. Organization for planning, the role of computers, step-by-step procedures, and the role of the sales force are discussed. Several examples are provided.

STASCH, STANLEY F., and LANKTREE, PATRICIA. "Can Your Marketing Planning Procedures Be Improved?" *Journal of Marketing* 44 (Summer 1980): 7–15.

Summarizes a study of six consumer package goods companies and compares and contrasts planning practices and establishes "bench marks" and "guidelines" for improving marketing planning procedures.

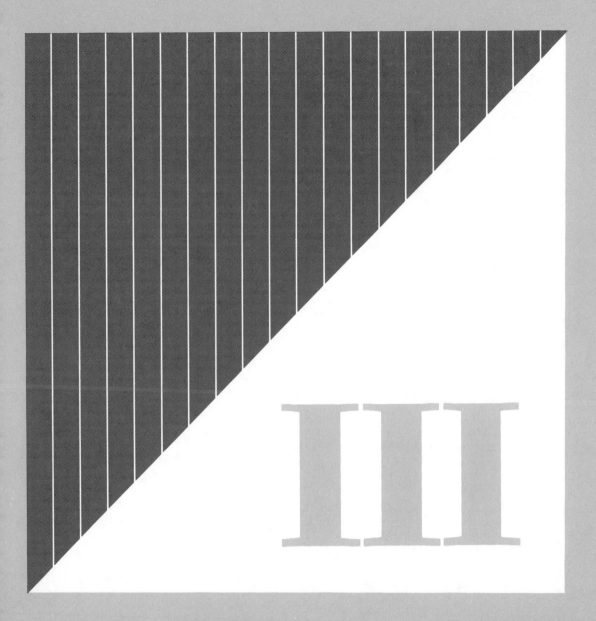

III

Organizing and Planning
New Products

Organizing for Marketing
New Products

Decisions concerning the design, development, and management of new products are crucial and complex as evidenced by the failure rate discussed in Chapter 1. Because of the nature of these decisions, it is of major importance that lines of communication and cooperation within the organization are formally and firmly established. The elements of decision-making must be made into an integrated whole to allow management to function efficiently and effectively in a highly competitive environment. To ensure effective implementation, these complex, highly interrelated tasks for marketing new products must be carefully assigned to competent personnel in the organization.

The responsibilities for the various tasks inherent in the development and marketing of new products or services must be explicitly defined. Certain organizational issues, such as where responsibilities will be absorbed into the present organizational structure and whether the lines in the organization should be established to centralize all new product decision-making, will become apparent as the tasks are assigned. Regardless of the organizational structure, top management must maintain a positive attitude toward new product activity.

ROLE OF TOP MANAGEMENT

Often new product planning and development is assumed to be the undefined responsibility of either the chief executive or a few people in top management. Whether or not the responsibility is defined, these executives must have the final authority over new product decisions, since new product planning and development is vital to a company's future.

In the new product planning and development process, it is generally necessary to establish relationships with persons with new product ideas, both outside the firm and within different departments of the company. Involvement, particularly across formal organizational lines of authority and with groups outside the organization, necessitates top management's participation in the decision-making process because only they have the authority to integrate recommendations from these sources with corporate goals. Indicative of the interdepartmental authority problem is the fact that each department has its own unique interests in any new product development activity. Marketing personnel, for example, are often very sales-oriented and therefore concerned only with new products that will increase sales. Thus it is possible that marketing will give no consideration to the available resources, profits, or corporate objectives. Engineers, on the other hand, tend to be more technically oriented and may be concerned only with the technical aspects of new products, possibly neglecting the importance of customer satisfaction. The need to include top management in the final decision-making is quite clear: these executives must coordinate departmental interests and improve the welfare of the total organization. However, because their time is limited, top-level management would not normally become involved in the details of decision-making. A staff is usually assigned this responsibility and would report directly to some member of top management. The responsibility for this function may go to only one person in a small company or to a committee in a large company. In larger firms product managers are often given the responsibility for a specific product or group of products.

The responsibilities of each member of the organization involved in new product decision-making should be formalized so that effective lines of communication can be established with top management. These individuals should be thoroughly trained to make effective new product decisions so that individual or departmental goals are consistent with company goals.

This chapter discusses the various organizational structures used by firms in the marketing of a new product. The advantages and disadvantages of different structures, the conflicts and constraints that diminish the effectiveness of new product planning are also addressed.

LARGE VERSUS SMALL COMPANY ORGANIZATION

It appears that the smaller the firm, the greater the chance of its being responsible for new product planning.[1] Even when this responsibility rests with top management, it is imperative that the planning, development, and management of the new product be carefully defined to ensure efficient and profitable introduction. One or more top members of management will be charged with generating, screening, and determining the feasibility of new ideas, based on company goals and the difficulty of producing and testing the product. Then these members of management will decide whether to go with a particular idea. In the small firm, the president will generally supervise all these activities and, in many cases, carry them out personally. However small firm managers or executives should be aware of how larger firms organize the development and planning process. Understanding how this process works may provide an insight useful in formalizing their smaller firm's product development process.

ALTERNATIVE ORGANIZATIONS FOR NEW PRODUCT DEVELOPMENT

Primary responsibility for new product planning can be assigned in various ways. The individual charged with these responsibilities can relate with the rest of the organization in different ways to increase compatability within the entire organization. The alternative organizational structures for new product development generally fall into one of the following four types:

1. New product department
2. New product committee
3. New product manager
4. Venture team

The type of organization selected is largely a function of the size of the firm and its resources and objectives within the context of its product and market.

---

[1] Karl H. Tiegjen, *Organizing the Product Planning Function* (New York: American Management Association, 1963), 21.

New Product Department

Multidivisional firms may choose the new product department organization. Other names for these are new product development department, product planning department, or market development department.[2] This organizational structure separates the new product development, planning, and management tasks from the existing divisions in the organization to centralize the new product decision-making process and eliminate redundancy of tasks across divisions. The coordination and control of the new product development process in autonomous divisions creates unique problems in corporate management, including duplication of effort and inefficient use of development funds and other company resources. These problems can be ameliorated by having a central authority direct and assist the divisions to better strive toward corporate objectives. Management of this central authority is generally given to a responsible person who coordinates and controls the required tasks. This new product manager or director holds a staff function at a high level and would likely report to a vice-president or executive vice-president of the firm.[3] An example of how the new product department might relate to other departments within the organization is shown in Figure 6–1. The manager or director of the new product department in this example reports to the executive vice-president who, in turn, reports to the president of the firm. This simplified organization chart shows the new product department positioned as a staff rather than line function.

However many people feel that new product development should be a line function. When it is a line function, the problems of authority become more complex because managers of other departments, such as finance, engineering, production, and marketing, technically rank higher than the new product manager. When new product development is a staff function, this problem is not as great, as long as the new product department is given full support by top management. If support is not provided the new product department has no authority over the line managers and is unable to complete the necessary tasks.

Regardless of how this function is structured, there must be clear lines of communication and authority so that new product planning and development may be performed with the least internal conflict. The continuous support of top management is necessary for this department to implement successful new product programs.

The functions of the new product department are as follows:

1. Recommending new product objectives
2. Planning exploration of new product ideas
3. Screening new product ideas
4. Assisting in development of new product specifications
5. Recommending and implementing test marketing
6. Coordinating interdepartmental effort during the evolution process.[4]

---

[2]Ibid. 41, and Russell W. Peterson, "New Venture Management in a Large Company," *Harvard Business Review* (May–June 1967), 69.

[3]*Management of New Products* (New York: Booz, Allen & Hamilton, 1968), 20.

[4]*Management of New Products*, 21.

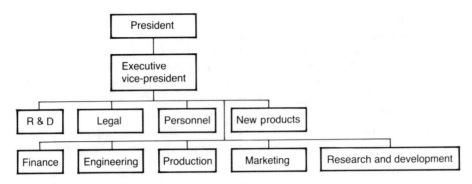

FIGURE 6–1
Relationship of new product department to other departments

New Product Committee

In using this organization structure, new product development decisions would be made by members of various departments or divisions within the existing organization. Thus these individuals are involved in certain aspects of the product development, as well as their regular responsibilities as managers or staff of other departments.

The new product committee may serve special functions, such as brainstorming to generate new product ideas, screening new ideas, evaluating new product proposals, coordinating and controlling the test marketing process, or assisting in the management of the new product introduction. Because this committee exists somewhat informally in the organization and because its membership may be temporary, the responsibilities and roles of the various members are not clearly defined. Even though this type of organization structure has weaknesses, it is probably the most widely used structure for new product planning and development because it is informal and can be used when needed. However this new product committee structure, important particularly to smaller or more technical firms, is not necessarily the most suitable for every firm's development process.

The advantages of the new product committee structure are as follows:

1. Ideas and expertise of key executives may be pooled.
2. Since decisions are made by top management, they are likely to be accepted by the firm.
3. The committee may be organized and utilized when needed.
4. The members may be recruited for special purposes—that is, brainstorming for new product ideas or for the entire new product development process.
5. There is no staff and line conflict because committee members are upper-level management.

Although the advantages are clear and important, this structure has some important weaknesses that limit its functional efficiency:

1. Committee activity takes valuable executive time. Members may limit their involvement so they may return to their regular responsibilities.

2. This type of committee structure lacks clear lines of authority and responsibility that may result in "buck passing."
3. New product planning and development in many instances should be a full-time job and not occur only when there is a need.
4. Members of the committee tend to concern themselves only with their departmental objectives rather than with the firm's goals, resulting in a narrow view of the committee's purpose.
5. Because of the structure's part-time existence, the members are not fully knowledgeable of the new product decision-making process.

It is important to recognize that a larger firm may utilize different executives for different committee functions in the product planning and development process. One committee may exclusively serve the purpose of idea screening, whereas another may be used for business analysis, and so on. Each committee is thus a specialist in a different aspect of the development process. The major problem with this approach is that coordination becomes difficult, unless one person assumes responsibility and authority over committee decisions. The product manager can fulfill this need and may be the coordinator of the new product evolution process.

Product Manager

The product manager concept was innovated by Procter & Gamble nearly 50 years ago. The interpretation of the product manager's responsibility has grown over this time, as each firm institutes its own definition of responsibility and authority.[5] Regardless of the interpretations of the product manager's role, the product manager today typically retains the responsibility for high-level planning and administration in the decision-making for a product or group of products. They are specialists or experts in a particular product market, yet generalists in the sense that they must be concerned with all variables in the marketing mix that are relevant to their products or product lines.

At Procter & Gamble, the product managers are referred to as *brand managers.* Each brand manager is responsible for one brand but is a part of a brand group or product line. Figure 6–2 illustrates the Procter & Gamble conception of the function of the brand manager. This structure optimizes the effectiveness of each brand manager by providing each with the resources needed to perform day-to-day marketing decisions. These resources are shown in Figure 6–3.

It is apparent from Figure 6–3 that the brand manager must work closely with other managers to function effectively. This collaboration of effort is extremely sensitive and for some firms may not provide the desired results. This is true especially if brand managers lack the authority to effectively carry out their responsibilities. With its extremely diverse and extensive product line, Procter & Gamble has made the brand manager concept work and has achieved great success in its product development decisions.

Figure 6–4 illustrates how a firm that uses the product manager concept might structure its organization. As can be seen, product managers report to

---

[5]Product Managers: Just What Do They Do? Special report in *Printer's Ink* (28 October 1966): 16.

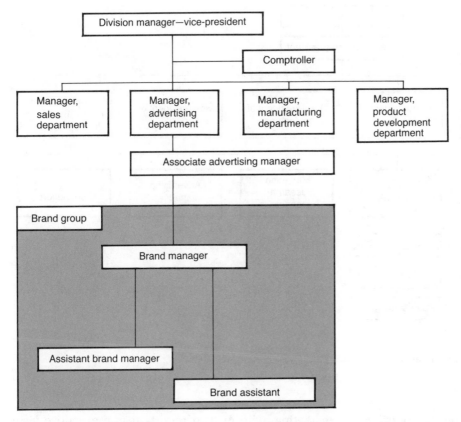

FIGURE 6–2

The brand manager organization at Procter & Gamble

Source: The Procter & Gamble Company, Cincinnati, Ohio.

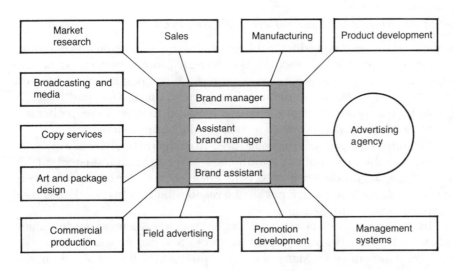

FIGURE 6–3

Interfacing of product manager at Procter & Gamble

Source: The Procter & Gamble Company, Cincinnati, Ohio.

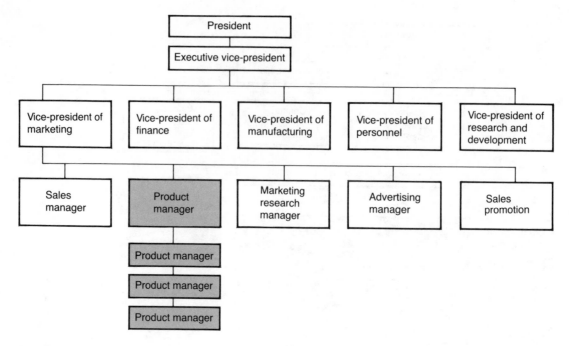

FIGURE 6–4
Traditional organization and the product manager

the vice-president of marketing, who in turn reports to the executive vice-president. The product manager thus assumes a line function, with authority over all decisions regarding his or her product or products.

Role of the Product Manager in Developing New Products

A significant part of the product manager's duties involves marketing decisions for an existing product or products. However the product manager must also assume a role in the development of new products, particularly if the new product idea is an extension of the existing product line.[6] They may conceive of new, unique products, since they are required to have a comprehensive knowledge of the buying process for their product or products.

The role of the product manager varies in each firm but typically includes all aspects of the new product evolutionary process. In an industrial, highly technical market, product managers (who are likely to be the creators or inventors) generally assume most, if not all, the responsibility, especially if they have no existing product to manage.

In consumer markets, such as that of Procter & Gamble, product managers have large staffs and extensive resources that can be used in a coordinated development effort. Many of the new products in the consumer market

---

[6]Philip W. Stein, "The Role of the Product Manager in New Product Development," *Marketing News* (1 October 1972): 4.

are less likely to be unique or discontinuous than those in the highly technical product market. Thus product managers for consumer markets generally plan product line extensions or modifications of existing products. They are not likely to be adept in all aspects of development, as are their counterparts in the highly technical market, but they will assign responsibility for specific tasks to persons whose duties normally encompass these tasks.

In the new product development process, product managers must be in close contact with top management, particularly in the early stages. Approval to send any new product to the development route must be given by top management. Product managers must provide top management with all the relevant data and their recommendations so that the decision to go ahead can be made expediently. As new products move through the development process, product managers must supervise the decisions and coordinate every aspect, especially when other departments in the firm are included. Regardless of their participation in the various stages of new product development, product managers become somewhat expert in such areas as market research and financial analysis.

Figure 6–5 outlines the typical decisions that product managers must make or assist in making. It is evident from this illustration that product managers have numerous duties and responsibilities. In addition to making decisions and recommendations, they generally will have control over significant funds that must be budgeted for the commercialization strategy of a new product. Coordination and supervision of all the major marketing, production, engineering, and financial decisions regarding the new product also fall within their auspices.[7]

Product managers receive the expert advice and recommendations of other departments—particularly sales, production, engineering, accounting, and advertising—in the decision-making process. All decisions must be cleared with top management before product managers are able to continue the development process.

Figure 6–5 lists typical responsibilities for the product manager of a large firm that markets consumer products. Industrial product managers may also have very similar duties in most areas, with the possible exception of the marketing plan.

In addition to duties in new product planning and development, product managers assume the responsibility of marketing decision-making throughout the product's life cycle. Marketing mix strategy decisions to meet competitive pressures and changes in consumer needs are important responsibilities, and in many cases they will determine the success or failure of the new product. The decision to withdraw the new product from the market when it has reached the decline stage is also a job requirement of most product managers. This decision is discussed in Chapter 19.

Duties of the Industrial and Consumer Product Manager

The roles of the consumer and industrial firm's product manager may vary, depending on the nature of the product, company objectives, and the industry

---

[7]For a complete discussion, see Gordon H. Evans, *The Product Manager's Job* (New York: American Management Association, 1964), 23–48.

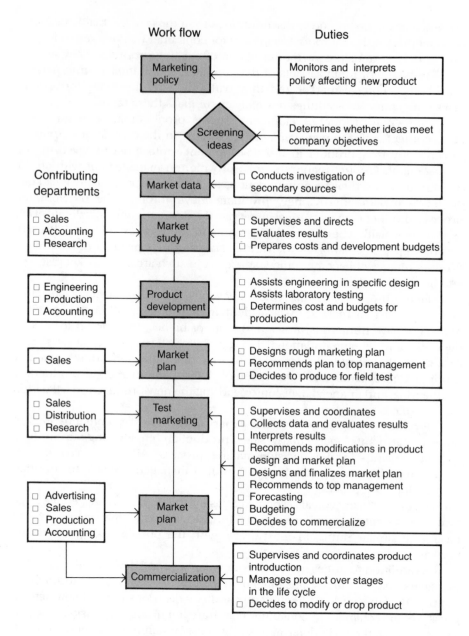

Work flow

- Marketing policy
- Screening ideas
- Market data
- Market study
- Product development
- Market plan
- Test marketing
- Market plan
- Commercialization

Duties

- Monitors and interprets policy affecting new product
- Determines whether ideas meet company objectives
- ☐ Conducts investigation of secondary sources
- ☐ Supervises and directs
- ☐ Evaluates results
- ☐ Prepares costs and development budgets
- ☐ Assists engineering in specific design
- ☐ Assists laboratory testing
- ☐ Determines cost and budgets for production
- ☐ Designs rough marketing plan
- ☐ Recommends plan to top management
- ☐ Decides to produce for field test
- ☐ Supervises and coordinates
- ☐ Collects data and evaluates results
- ☐ Interprets results
- ☐ Recommends modifications in product design and market plan
- ☐ Designs and finalizes market plan
- ☐ Recommends to top management
- ☐ Forecasting
- ☐ Budgeting
- ☐ Decides to commercialize
- ☐ Supervises and coordinates product introduction
- ☐ Manages product over stages in the life cycle
- ☐ Decides to modify or drop product

Contributing departments

- ☐ Sales
- ☐ Accounting
- ☐ Research
- ☐ Engineering
- ☐ Production
- ☐ Accounting
- ☐ Sales
- ☐ Sales
- ☐ Distribution
- ☐ Research
- ☐ Advertising
- ☐ Sales
- ☐ Production
- ☐ Accounting

FIGURE 6–5
Duties of product manager and contributions from other departments

in which the firm operates.[8] Industrial product managers usually have an extensive technical background but lack marketing expertise. Thus industrial product managers may assume only a limited role in decisions such as advertising, sales promotion, merchandising, packaging, branding, and labeling. In the consumer goods firm, product managers are more likely to have sufficient

[8]Evans, *Product Manager's Job*, 26.

marketing background and experience to make marketing mix decisions without having to employ an outside consultant or enlist other internal personnel.

Industrial product managers must maintain a close working relationship with Research and Development (R&D), where changes or new ideas are most likely to occur quickly and require immediate action to determine their feasibility. The product development process in an industrial firm is often initiated by government requests for bids or a request to design a new or improved product. Industrial product manager responsibilities that touch upon marketing or financial accounting will generally be assigned to others in the firm. One significant trend affecting the future role of industrial product managers is that more and more such managers are obtaining graduate business degrees that allow them to expand their responsibilities, particularly in coordinating and implementing the marketing plan.

Role Conflict and the Product Manager

One of the biggest problems confronting product managers is that their authority is not always consistent with their responsibilities. Since the relationships that product managers must develop in order to carry out their responsibilities are so complex, it is sometimes difficult to avoid role conflict.[9] Role conflict will tend to limit effectiveness because product managers will most likely not have the authority to overrule any department head or other manager whose cooperation is needed. For example product managers are responsible for pricing and forecasting, yet they have no control or authority over production costs or the sales force. If product managers must know the exact manufacturing costs for their products, they may have some difficulty in getting a firm quote from production managers. Production managers must account to their superiors for all decisions and, in order to protect themselves, may be reluctant to state exact numbers. Conflicts with various units in the organization may be minimized if product managers can establish good working relationships and trust with their colleagues. Product managers must not only have broad expertise in marketing, accounting, and production, but they must be able to communicate with others in the organization and enlist their trust.[10] The frequent interaction among departments in the new product development process requires patience and diplomacy.

Often product managers must travel in the field with sales personnel to observe markets, customers, competitive activity, and selling tools first-hand. These observations can result in interpersonal conflict, since sales personnel may feel that a product manager is spying on them when, in fact, the product manager has no authority over the actions of the sales force. In order to avoid this conflict it should be clear to these members of the organization that the product manager's role is not to evaluate the sales techniques or the sales personnel but merely to get a better "feel" for the market and customers.

Any interpersonal conflict between sales personnel and product managers must be avoided, since a good working relationship between both parties

---

[9]B. Charles Ames, "Payoff from Product Management," *Harvard Business Review* 41, no. 6 (November–December 1963): 141–152.

[10]For a discussion on how product managers influence action, see Gary R. Gemmill and David L. Wileman, "The Product Manager as an Influence Agent," *Journal of Marketing* (October 1969): 35.

is essential for successful implementation of the marketing plan. Product managers need the input and support of sales personnel to stay attuned to market conditions. Sales personnel also need the support of product managers to implement materials, modifications, and planning suggestions to enhance their ability to successfully sell the company's products.

Product managers usually spend a great deal of time working with an advertising agency. The relationship of product managers with agency personnel is critical in providing promotional materials needed to support the sales force. One of the major problems encountered by product managers is achieving a balanced use of the advertising agency in implementing the market plan. Figure 6–6 indicates how product managers can underuse or overuse the advertising agency. This problem typically occurs as a function of the confidence that product managers have in their ability to complete specific assignments. A clear identification of their role can avoid conflict and enhance the product planning and strategy process.

There are other problems inherent in the product manager organization.[11] Product managers are often encumbered by trivial tasks, such as corresponding with customers or salespersons. Not only are product managers often hindered by their lack of authority, but they are also not given the necessary support staff to carry out their duties. Furthermore typical product managers often have a short job tenure, as they may view the job as a stepping-stone to top management.

Recent research indicates that the role of the product manager may be changing to overcome some of these weaknesses.[12] Top management has recognized the need to participate more in the decision-making process to alleviate some of the authority problems that often arise in the product manager organization. The prestige of the product manager has been enhanced by providing the position with more authority and more clearly defined responsibilities that seem to have slowed job turnover. More emphasis is also being placed on marketing experience as a job requirement rather than education alone.

Product management is not the most effective organization for every company. Although it is still one of the most popular organizational approaches, particularly in complex multiproduct markets, it has been criticized for some of the obvious weaknesses that were discussed earlier in the chapter. With increased environmental pressures, as well as a greater concern for long-term planning, the product management system has been modified by some firms to achieve better marketing decisions in a changing environment. The following two sections discuss the impact of some of these factors.

Impact of Economic Environment on Product Management

The continued high inflation rate, as well as other economic and competitive pressures, has changed the product management system. The pressure of time constraints and day-to-day decision-making in the existing economic

---

[11]David J. Luck, "Interfaces of a Product Manager," *Journal of Marketing* 33, no. 4 (October, 1969): 35.
[12]Victor P. Buell, "The Changing Role of the Product Manager in Consumer Goods Companies," *Journal of Marketing* 39, no. 3 (July 1975): 3–11.

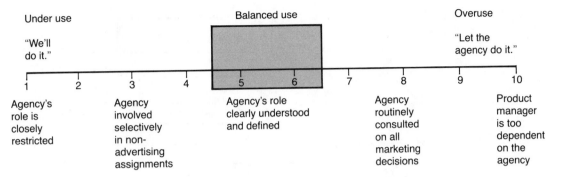

Under use | Balanced use | Overuse

"We'll do it." | | "Let the agency do it."

1 2 3 4 5 6 7 8 9 10

Agency's role is closely restricted | Agency involved selectively in non-advertising assignments | Agency's role clearly understood and defined | Agency routinely consulted on all marketing decisions | Product manager is too dependent on the agency

FIGURE 6–6
Balancing the use of an advertising agency

environment has made it increasingly difficult for the product manager to effectively manage and control a new product. As a result many firms are assigning the responsibility of new products to a new product manager who is experienced with the marketplace. In addition to experience, this new product manager should have the following traits:[13]

1. Should be a generalist and not oriented specifically to sales, research, or marketing management
2. Needs to have "clout" in order to persuade other people in the organization of the benefits of the new product
3. Should be an entrepreneur or risk-seeker, not a risk-avoider
4. Must be able to make judgments confidently with or without research data

The new product manager as an individual has provided important benefits to the product organization. With fewer people involved in the new product, fewer leaks occur to competitors in the marketplace. With the total commitment to the new product, there is no problem determining priorities, as there would be with a product manager. In addition, with this commitment, there is a greater chance for continuity in the management of the product. Once the new product has achieved some stability in the marketplace, the product may then be assigned to the product manager. The new product manager may at that point assume the responsibility for another new product in the organization.

Impact of Strategic Planning on Product Management

The increased role of strategic planning has also resulted in some changes in the responsibilities of the product manager. Typically the product manager is responsible for short-term market-product planning, whereas the strategic planner is responsible for long-term planning. The traditional product manager organization has been criticized for its short-term views, which limit effective control and management of return on investment and cash flow. Strategic management is less instinctive than product management and is

---

[13]*New Product Development* 1, no. 1, January 1981.

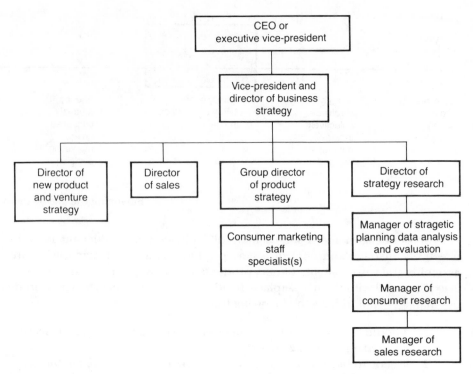

FIGURE 6–7
Strategic management organization structure
Source: Terry Haller, "An Organizational Structure to Help You in the '80's," *Advertising Age* (25 August 1980): 46.

regarded as more of an applied science. It integrates experience and knowledge from many different disciplines and capitalizes on the interaction of factors that affect long-term profitability. With this new emphasis on strategic management, an alternative organization structure to the traditional product management organization has been proposed. Figure 6–7 illustrates this proposed type of organization.[14]

In this new organization structure, marketing plays a significant role in the strategic planning process of the company. It also takes advantage of the skills of existing product managers and effectively assigns marketing responsibility as a staff function. The product manager's responsibility is expanded so that he or she takes a stronger role in the long-term impact of the product. In the past, product managers were typically concerned only with a brand or product over a short-time period. This new role would ensure that product planning and strategy is commensurate with the strategic planning of the organization.[15]

In Figure 6–7 the Director of New Product and Venture Strategy would be responsible for the development of new ideas and new markets. Again this

[14]Terry Haller, "An Organizational Structure to Help You in the 80's," *Advertising Age* (25 August 1980): 45–46.
[15]See "Brand Management—Boon or Boondoggle?" *Marketing and Media Decisions*, August 1980, 57–69, 113–14. Also Richard T. Hise and Patrick Kelly, "Product Management on Trial," 42, no. 4 *Journal of Marketing* (October 1978): 28–33.

person's responsibility will consider the long-term goals of the organization. The Director of Strategy Research would be responsible for marketing research and the direction of the generation of planning data, analysis, and evaluation of on-going business performance.

The new type of organization will enhance cash flow and return on investment because of its systematic concern for strategic options. Marketing budgets would be more effectively controlled and provide improved profit planning and risk reduction in long-term decision-making.

Venture Team

This type of organization is a new approach to new product development that seems better-suited to the design and development of products that do not necessarily fit into the on-going business of the firm.

A study of 98 venture managers for industrial and consumer products at large corporations revealed that a venture team has the following characteristics:

1. It is separate from the remainder of the organization.
2. Members are recruited from various functional areas, such as engineering, production, marketing, and finance.
3. Existing lines of authority in the permanent organization are not necessarily valid.
4. The venture team manager usually reports to the chief executive officer and is given authority to make major decisions.
5. The team is free of deadlines and remains together until the task is completed.
6. Freedom from time pressures fosters creativity and innovativeness.[16]

Figure 6–8 illustrates a venture team organization. Members of the teams (there may be more than one team operating at any time) are chosen from the various functional areas, with one person given the title of venture team manager. The venture team manager reports to the division head or some other upper-level administrator.

Advantages of the Venture Team Concept

The venture team concept of organization may be compared to new product departments, new product committees, or product managers. The biggest question in the three types of structure is who has the authority to make the final decision. The venture team tries to overcome this difficulty by providing leadership through the venture team manager who has a direct line with top management.[17]

The new product committee approach is the most temporary of all the organizational structures. Although the venture team consists of members from many functional areas, they are assigned to the team on a permanent basis until the task is completed.

---

[16]Richard M. Hill and James D. Hlavacek, "The Venture Team: A New Concept in Marketing Organization," *Journal of Marketing* 36, no. 3 (July 1972): 46.
[17]Hill and Hlavacek, "The Venture Team": 50.

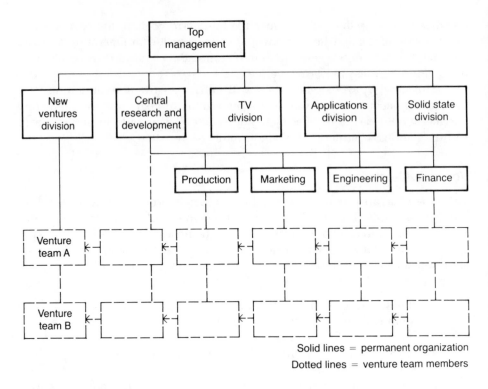

Solid lines = permanent organization
Dotted lines = venture team members

FIGURE 6-8

An example of the venture team organization

Source: Richard M. Hill and James D. Hlavacek, "The Venture Team: A New Concept in Marketing Organization," *Journal of Marketing* (July 1972): 47.

There are also disadvantages to venture team organizations. Interdepartmental conflict may often result, particularly when a department head feels
that one of the members of his or her staff is spending too much time on the
venture team instead of carrying out departmental responsibilities. The un-
willingness of departmental managers to cooperate may lead to conflict and
the weakening of the venture team approach.

ORGANIZATIONAL ALTERNATIVES NEED EXECUTIVE SUPPORT

One requirement is essential to the success of all four types of organizational
structures discussed here: executive support. Top management must realize
that their attitudes toward new product planning and development will be
reflected in the decisions made at each stage of the development process.
Negative attitudes will likely weaken new products, as well as stifle innovativeness.

WHICH ORGANIZATIONAL STRUCTURE?

The adoption of the right organizational structure depends largely on the size
of the firm, the type of market, the company objectives, and the availability of

expert personnel. The venture team structure may be more suited to large corporations interested in totally new products, rather than product extensions or "me-too" (imitation) products. Smaller firms that cannot afford full-time personnel for new product planning and development are likely to use the new product committee approach. This allows minimal commitment to new product development.

Firms with multiple product lines would probably find the product manager approach the most efficient structure, since it provides the firm with the advantage of a centralized coordinator of various departments involved in the planning and development process. The new product department is better-suited to those firms with a single product line and a traditional functional marketing organization.

A study by Grayson indicated that companies that maintain full-time new product executives produce 69 percent more new products than those with part-time executives. Firms with full-time new product executives produce 60.5 percent more new products than firms with no product executives.[18] These findings relate to firms that have sales of over $25 million annually and testify to the importance of a full-time commitment to new product development on the part of most large firms.

STAFFING DECISIONS

Regardless of the type of organization, certain qualities are desirable in the individual who manages the product planning group. Some companies staff their planning groups with former sales representatives.[19] These firms feel that the salesperson is best acquainted with the consumer and is more likely to know what their customers want. In addition many aspects of the product development process require the selling of ideas to others within the firm. Unfortunately the salesperson is often limited in other aspects necessary for product planning. Today most firms are staffing their product planning groups with individuals who have had experience with their firm in some marketing capacity, such as sales, advertising, or marketing research. This company-specific experience seemingly provides a solid understanding of corporate objectives, the product and its limitations, personnel, and the competition— all critical elements in the effective management of product planning and development responsibilities.

For large technically oriented firms, the manager of product planning and development should have a strong technical background in the particular market. Experience in marketing is becoming more necessary to the technical firm's leaders. For example in the chemical and instrumentation market, a recent study found that many product failures resulted from neglect of important marketing elements.[20]

---

[18]Robert A. Grayson, "If You Want New Products, You Better Organize to Get Them," *Proceedings of the American Marketing Association*, (June 1969): 78.
[19]William J. Constandse, "How to Launch New Products," *MSU Business Topics* (Winter 1971): 31.
[20]"Research Based Venture Companies—The Link Between Market and Technology," *Research Management* (May 1974): 18.

FROM PLANNING TO COMMERCIALIZATION

The new product planning and development organization often has difficulty in maintaining a balance of emphasis between existing products and products in various stages of development.[21] Why does the preoccupation of operating personnel with products already on the market cause difficulty in generating interest in new product plans? Personnel in the planning and development stages do not always participate in a product's commercialization. Once commercialization is reached, the planning personnel continue to plan and develop new products. The problem of transferring knowledge from planning to operations is frequently encountered but may be resolved by allowing development personnel to move into operations. This problem is not as great in the product manager organization because the product manager assumes responsibility for the product once it achieves commercialization. With committees, new product departments, and the venture team, this transfer of knowledge may be a more serious problem, since those groups are not likely to participate in the market introduction. This responsibility would be given to the marketing department.

The problems of knowledge transfer, role conflict, insufficient authority, and so on may all be minimized or eliminated completely if the firm and its employees maintain a positive attitude toward new product planning and development. A positive attitude will help to ensure the success of any of the four organizational structures presented.

THE RESEARCH AND DEVELOPMENT MARKETING INTERFACE

Individuals responsible for new product planning and development must interact with research and development in order to ensure the technical feasibility of any new product idea, as well as to determine the actual specifications required for consumer satisfaction. This interaction often breaks down for two reasons:

1. Lack of good communication during relevant stages in the new product evolution process
2. Value conflict between the research and development group and marketing personnel[22]

The first problem occurs because research and development has the difficult responsibility of explaining new technology used in new product development to the marketing staff. New designs, materials, and procedures that must be clearly communicated to those in the new product development organization are constantly being developed. It is also imperative that research and development groups be kept informed of the long-range goals (specifically, the strategic plan of the firm in the marketing of new products) so that effort is consistent with the objectives.

---

[21]Tiegjen, *Organizing Product Planning*, 66.
[22]Warren B. Brown and Lewis N. Goslin, "R&D Conflict: Manager Versus Scientist," *Business Topics* (Summer 1965): 75.

Value-conflict problems occur primarily because research and development employees may have little appreciation for problems in branding, packaging, distribution, pricing, and promotion. They must therefore be instilled with an awareness and understanding of how the problems faced by the marketing organization affect the success of any new product introduction.

Flexibility in managing the role of research and development seems to be the key to successful cooperation. Communication and understanding of the problems faced by each member of the team will help to minimize the conflicts of interest. Frequent meetings and careful scheduling will also aid in this interface.

CAREERS IN THE PRODUCT ORGANIZATION

Career opportunities in the area of product planning and development in an organization (particularly consumer goods) are generally restricted to MBAs or individuals with experience in sales, advertising, and/or market research. Some firms will hire students with undergraduate degrees for a training program that eventually leads to a position in product or brand management. However these opportunities are limited. The best training for a career in brand management is sales or advertising, and often individuals are able to attain such a position after spending a few years in one of these areas.

In the industrial goods market, product managers generally have a scientific or engineering degree and some technical knowledge of the product. Many of these individuals that obtain an MBA often become the leading candidates for product management jobs in the industrial market. The need for a particular degree or technical background, of course, depends on the nature of the product and market.

In addition to some of the background already mentioned, entry-level product or brand management positions require that the candidate spend some time in the field becoming totally familiar with the products, competition, market conditions and trends, and customer needs.

Although the sequence of actual positions varies, most entry-level positions are for marketing assistants, who not only spend time in the field but also work closely with past and present product and industry performance data needed by the product managers for developing market plans and strategies. After acquiring experience and familiarity with the product, market, and internal company policies, the marketing assistant is ready to move up to assistant product manager. The time needed to make this particular move varies with the company and individual but is typically about 12 months.

An assistant brand or product manager assumes some of the duties and responsibilities (mentioned earlier in this chapter) of the product manager. Assisting the brand manager in coordinating marketing activities and developing market plans is the prime responsibility at this stage, until the individual is ready to assume total responsibility for a specific product or brand. When this occurs the individual will be promoted to product or brand manager and will be assigned to a specific brand or product. Generally the first brand or product responsibility is with a reliable product. Product managers will often move laterally in the organization, changing from one brand to another. Each move generally includes a larger, more complicated market with a significantly higher budget responsibility.

The time required in any one position is difficult to predict, since so many different factors contribute to internal company personnel changes. However, in a consumer goods company, an individual should expect to become a brand or product manager in three to five years.

The next step after product or brand manager is group product manager, where responsibility is assumed for several products. Each of the product managers within the group are the responsibility of the group product manager.

The turnover in all of the positions mentioned above is fairly frequent. This implies that there are always opportunities in an organization to enter this particular career track. The position of product or brand manager and group product manager is felt to be excellent training for higher levels of management and as a vice-president of marketing. Because the product or brand manager has control over a large budget and a wide range of decision areas, he or she obtains a good picture of the overall operations of the firm. This knowledge and experience becomes invaluable to someone who is given a position in top management.

It is important to recognize that each industry and company has different requirements, procedures, and policies with regard to product management positions. Anyone interested in pursuing such a career should analyze and research each situation in order to fully understand the long-range opportunities.

SUMMARY

The new product planning and development process needs special attention from the firm and requires a separate organizational structure. Four organizational structures were presented:

1. New product department
2. New product committee
3. Product manager
4. Venture team

The choice of structure is a function of the size and resources of the firm, the company's objectives, the type of products, the industry in which the firm operates, and the availability of expert personnel.

For smaller firms with limited resources, the new product committee is most likely to be the most efficient organizational structure. The large, multidivisional firm is better-suited to the new product department organization, which provides centralization of decision-making across all divisions. The product manager approach appears to be most widely used by large, multiproduct firms. The responsibilities of planning and developing a particular product or group of products are managed by the product manager. If a large firm wishes to develop new products, it might want a venture team organization.

Once the planning and development decisions have been made, the firm must ensure the transfer of knowledge to those responsible for the commercialization stage. This may be accomplished by continuing the development personnel into operations decisions, as well.

In staffing the product planning and development groups, preference is usually given to those with experience in the relevant market as well as in marketing. Top management must maintain a positive attitude toward new product planning and development and show a willingness to support the organization responsible for these tasks.

SELECTED READINGS

DUNNE, PATRICK, and OBERHOUSE, SUSAN, eds. *Product Management: A Reader,* Chicago: American Marketing Association, 1980.
Provides a series of articles that relate the major environmental and economic factors affecting the role of the product manager. Includes articles on the complexities of the product manager's duties, growing risks, and the tools available to product managers. It has a section that discusses service and industrial marketing.

HALLER, TERRY. "An Organizational Structure to Help You in the 80's." *Advertising Age,* 25 August 1980, 45–46.
Provides valuable discussion of some of the issues related to successful brand management organizations. The author proposes an alternative organization structure that takes advantage of the skills of existing brand managers and effectively meets the needs of today's efforts by management to consider longer-range decisions through strategic planning.

HLAVECEK, J. D. "Toward More Successful Venture Management." *Journal of Marketing* 38, no. 4 (October 1974): 56–60.
Discusses a field study of 21 venture failures at 12 large multidivisional corporations. Numerous causal factors are identified, providing guidelines for more successful venture management.

HOPKINS, DAVID S. *Options in New-Product Organization.* New York: The Conference Board, Inc., 1974.
Reports the results, opinion, and views of 110 executives responsible for new product activities. The publication analyzes and compares the advantages and disadvantages of alternative organization approaches within different corporate environments.

KHANDWALL, P. N. *The Design of Organization.* New York: Harcourt Brace Jovanovich, Inc., 1977.
Provides a comprehensive view of organization structures in the firm. It is an ideal reference book for managers that are considering alternative organizational structures to effectively enhance the development of new products.

LORSCH, JAY W., and LAWRENCE, PAUL R. "Organizing for Product Innovation." *Harvard Business Review* (January–February 1965): 109–122.
One of the classic comprehensive articles that provides a comparison of various types of product organizations.

Management of The New Product Function. New York: N.Y. Association of National Advertisers, 1981.
Includes a variety of articles on current thinking on the management of the new product function. The chapters cover new ventures, "me too" products, obstacles to new product success, planning the new product function, staffing, and new idea evaluation and testing.

McDANIEL, CARL and GRAY, DAVID A., "The Product Manager." *California Management Review* 23 (Fall 1980): 87–94.

Examines product management from the viewpoint of the marketing manager, who is usually the immediate superior of the product manager. The study presented in this article found that the product management arrangement would become a more important part of future marketing organizations. Also found was disagreement about the authority and responsibility of the product manager. However there was agreement on the tasks and activities performed by the product manager.

SOUDER, W. E. "Disharmony Between R & D and Marketing." *Industrial Marketing Management* 10 (1981): 67–73.

Presents some potential solutions to research and development and marketing interface problems. The data and analyses are taken from a more comprehensive longitudinal data base on industrial innovations. Innovative management methods are then suggested to alleviate the problems between the research and development and marketing departments.

Planning Techniques for the
New Product and Its Mix

Regardless of the organization structure, each new product must be carefully defined in terms of its total marketing mix. While the product planning process itself will be introduced and expanded on in Chapters 9 through 11, it is important to consider the marketing mix early in the development process. Each of the factors in the mix must be carefully evaluated in an established analytical framework so that an optimal total product package with market appeal will result. Too often firms concentrate solely on the physical aspects of the product itself. While this may result in an excellent physical product, that product may have limited market appeal because it lacks the other elements in the mix.

This chapter discusses each of the elements in the mix, as well as various techniques for evaluating product and mix alternatives. The chapter concludes by discussing the role of a product information system for control and evaluation of the development and commercialization process.

INTEGRATING MARKETING MIX DECISIONS

A key ingredient in successfully developing and launching a new product is formulating a marketing strategy that contains the correct mix of product features, branding and packaging, price, distribution network, and promotion for the particular product/market situation. Establishing the best mix is not easy because of the various relationships that are possible between sales and mix levels, the initial level of satisfaction and subsequent purchase patterns of the consumer, and the potential and nature of competitive response. Additional complications can occur due to changes in business conditions, shifts in the economic environment, and changes in legal regulations.

Figure 7-1 illustrates the various aspects of developing the marketing plan for the new product. The starting point is the development of the various characteristics of the product. Decisions on the product's uniqueness, quality reliability, and durability must be made, taking into account consumer desires, contribution to the overall profitability and stability of the firm, and the competition. The length and coverage of product warranties (if any), guarantees, and servicing are often important additions.

Closely related to decisions regarding the characteristics of the new product is the price (as discussed in Chapter 16). Inflation, shortages of resources and materials, and consumer awareness have contributed to new product price decisions. Cost serves as the floor for all price decisions, but allocating costs is often a difficult process. Research and development costs, product and engineering costs, and administrative costs must be equitably allocated to the new product. Then the costs of selling, distribution, and promotion efforts should be delineated. The price of the new product must also take into consideration competitive reaction and consumer acceptance. The product has to be priced to meet the company's objectives yet not entice competitive reaction. At the same time, to stimulate sales, the price must appeal to the consumer.

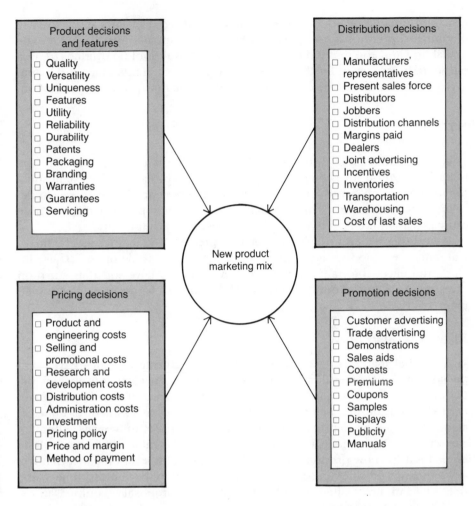

FIGURE 7–1
Designing the new product's marketing mix

The distribution system for the new product (thoroughly developed in Chapter 17) allows the firm to reach the government, industrial, or consumer markets. If a new product is very unusual, it is important to keep in mind the firm's corporate and product objectives when establishing the distribution strategy for the new product and when developing distribution systems for all products. Such factors as the nature of the market, margins required, available channel systems, and the degree of market exposure must be determined so that the new product will be available to meet demand. Specific decisions regarding the use of dealers, jobbers, distributors, sales force, or manufacturers' representatives must be made.

Promotion, which is the focal point of Chapter 18, is another area of the new product's marketing plan. Throughout the development process it is important to keep in mind that the choice of media and messages must be made in terms of the budget. Decisions regarding promotion require considerable interaction among the new product manager, the technical staff in the firm, and outside personnel. Aside from advertising, decisions about the use of contests, premiums, samples, displays, and manuals must be made. The availability and the method of using publicity should not be overlooked. It is particularly important for the firm to use a great deal of monetary and managerial resources to measure the effectiveness of the new product's promotion.

FINANCIAL CRITERIA

The new product mix must be established in terms of the profits and the relative return on investment that will occur. One procedure for establishing the marketing mix/profit relationship allows the new product manager to develop various sales and cost estimates for given new product mixes. These estimates are then used to determine the best new product mix.[1]

In designing and developing the new product and its mix, several financial criteria enable management to select the best possible option. These include the effect the new product will have on cash flow and the return on investment of the amount of income obtainable from the particular investment.

Effect on Cash Flow

Tight capital markets, decreasing profits, and high interest rates have compelled companies to evaluate much more carefully the cash flow characteristics of new product propositions. Accurate estimation of all monies spent and earned, given the mix being considered, requires an assessment of noncash elements—such as depreciation—as well as all opportunity costs. Although cash flow analysis ensures that sufficient working capital will be available to implement the new product mix, it is biased in that it favors the mix that will allow the new product to reach its break-even point in the shortest time. Using this criterion, firms often abandon new product proposals requiring an extended development and test period until liquidity problems have eased.

Return on Investment

Each possible new product mix should be evaluated in terms of its return on investment. This can be done through five basic approaches: payback method, average rate of return method, net present value method, internal rate of return method, and profitability index.[2]

The payback method is probably most frequently used to evaluate new product mix proposals. Management is always interested in knowing how long it will take to recover the initial investment. The payback period divides the total investment by the annual cash inflows for the recovery period. If the new

---

[1]Philip Kotler, "Marketing Mix Decisions for New Products," *Journal of Marketing Research* 1 (February 1964): 43–49. These same methods can be used to evaluate the profitability of new product alternatives throughout the product planning process discussed in Chapters 9 through 11.

[2]This discussion assumes that the expected cash flows are realized at the end of the time period being used (e.g., month, quarter, or year). The formulas used can be found in most financial management textbooks. For a good in-depth discussion, see James C. Van Horne, *Financial Management and Policy* (Englewood Cliffs, N.J.: Prentice-Hall, Inc., 1974), 70–77.

product mix had an initial investment of $90,000 with annual revenues of $25,000, its payback period would be figured as follows:

$$\text{Payback Period} = \frac{\text{Total Investment}}{\text{Annual Cash Inflows}} = \frac{90,000}{25,000} = 3.6 \text{ years}$$

A company should establish a minimum acceptable payback period against which all new product mixes are compared. If the payback period for the new product mix is less than the maximum, it is accepted. If not, the mix is rejected. When several new product mixes with payback periods that are less than the maximum are evaluated simultaneously, each payback period can also be compared with the projected life of the product. Although one mix may return its investment in a shorter period than others, that product may have such a short life that only a break-even situation will occur if that new product is introduced.

While the payback method gives management some insight into the liquidity and risk of the new product mix, the method has several shortcomings. It does not take into account interest income or other investment possibilities. It also neglects the size or the timing of the cash inflows during the recovery period. The major shortcoming of the payback method is that it does not consider any revenues occurring from the new product mix after the payback period is reached. In spite of these three deficiencies, the payback method is the most widely used method for evaluating new product mixes. This may be because the method is well understood and easy to use.

Another method used to evaluate new product mix alternatives is the average rate of return method. This accounting method indicates the ratio of the average annual profits (after taxes) to either the average investment or the total investment in the product mix. For example a new product mix with an expected product life of four years is projected to have annual profits of $4,000; with a total investment of $40,000, it would have a 10 percent (4,000/40,000) average rate of return on the total investment and a 40 percent (4,000/10,000) average rate of return on the average investment. As was the case with the payback method, a required rate of return is usually established by a firm so that each new product mix can be rated. Although this method is simple to use, it fails to take into account the timing of the cash inflows or the time value of money. It assumes benefits are the same regardless of whether they occur during the first year or last year of the product's life.

Three other methods of determining return on investment are variations of present value methods. These methods overcome some of the deficiencies of the payback and average rate of return methods. One of these, the net present value method, subtracts the present value of the total investment from the expected future cash inflows, which are discounted at the required rate of return.[3] The net present value of the product mix being considered is found through the following formula:

$$\text{Net present value} = \sum_{t=0}^{n} \frac{A_t}{(1 + k)^t}$$

---

[3] The required rate of return is an important part of this evaluation process. When evaluating various product mixes, it is important to consider the effect of the particular mix on the firm

where

$$A_t = \text{cash flow for period } t$$

$$k = \text{required rate of return}$$

If the product mix generates cash flows that, when discounted, are equal to or greater than zero, the mix is accepted. If the discounted cash flows are less than zero, the product mix is rejected. The product mix will be implemented if the present value of inflows is greater than the present value of outflows. For example if a firm wants a required rate of return of 10 percent after taxes on a product mix that will have cash inflows of $10,800 for each year of a five-year life, after an initial cash outlay of $20,000, the net present value is as follows:

$$
\begin{aligned}
\text{Net present value} = &-20{,}000 + \frac{10{,}800}{(1.10)} + \frac{10{,}800}{(1.10)^2} \\
&+ \frac{10{,}800}{(1.10)^3} + \frac{10{,}800}{(1.10)^4} + \frac{10{,}800}{(1.10)^5} \\
= &-20{,}000 + 40{,}940.64 \\
= &\ \$20{,}904.64
\end{aligned}
$$

Since this net present value is greater than zero, the new product mix should be adopted.[4]

The internal rate of return method is similar to the net present value method in evaluating alternative product mixes. In the net present value method, the cash flows and the required rate of return are given so that the net present value can be found. In the internal rate of return method, the cash flows are given; then the discount rate is found that will equate the present value of the cash inflows with that of the cash outflows. The resulting internal rate of return is compared with the required rate established by the firm to see if the new product mix should be adopted. If more than one product mix is being considered, then the internal rates of return can be used to rank the proposals in order of payback. The internal rate of return (r) is calculated as follows.

$$\sum_{t=0}^{n} \frac{A_t}{(1 + r)^t} = 0$$

---

as a whole. Its marginal contribution is dependent on the correlation of the mix with both existing projects as well as other new product mixes being considered. For simplicity we will assume that the firm has only one mix under consideration at a time, and all have the same degree of risk. For a discussion of the evaluation of risk, see Donald I. Tuttle and Robert H. Litzenberger, "Leverage, Diversification and Capital Market Effect on Risk-Adjusted Capital Budgeting Framework," *Journal of Finance* 23 (June 1968): 427–43; John Lintner, "The Valuation of Risk Assets and the Selection of Risky Investments in Stock Portfolios and Capital Budgets," *Review of Economics and Statistics* 47 (February 1965): 13–27; Robert H. Litzenberger and Alan P. Budd, "Corporate Investment Criteria and the Valuation of Risk Assets," *Journal of Financial and Quantitative Analysis* 5 (December 1970): 395–420; Mark E. Rubinstein, "A Mean-Variance Synthesis of Corporate Financial Theory," *Journal of Finance* 28 (March 1973): 167–82.

[4]An easier way to solve this problem would be to use a present value table. Look up the appropriate discount factor and multiply it by the annual cash inflow. For our problem, we would multiply the appropriate discount factor (3.7908) by the annual cash inflow (10,800). The resulting $40,940.64 has the cash outlay ($20,000) subtracted from it, giving $20,940.64.

where

$$A_t = \text{the cash flow for period } t$$

$$r = \text{internal rate of return}$$

When the initial cash outlay for the product mix occurs at time 0, the formula is as follows:

$$A_0 = \frac{A_1}{(1 + r)} + \frac{A_2}{(1 + r)^2} + \frac{A_3}{(1 + r)^3} + \cdots + \frac{A_n}{(1 + r)^n}$$

Using the previous example of product mix evaluation, this method yields the following results:

$$20,000 = \frac{10,800}{(1 + r)} + \frac{10,800}{(1 + r)^2} + \frac{10,800}{(1 = r)^3} + \frac{10,800}{(1 + r)^4} + \frac{10,800}{(1 + r)^5}$$

Solving for r gives a rate of return of approximately 45 percent. Since this high rate of return exceeds the required established rate of return of 10 percent, the product would be accepted.

The profitability index is the third present value method for evaluating the return on investment of various product mixes. This benefit/cost ratio (profitability index) of a product mix is actually the present value of future cash flows divided by the initial cash outlay:

$$\text{Profitability index} = \frac{\sum_{t=1}^{n} \frac{A_t}{(1 + k)^2}}{A_0}$$

where

$$A_t = \text{cash inflow in period } t$$

$$k = \text{required rate of return}$$

The profitability index for the same example problem considered for the present value methods is as follows:

$$\text{Profitability index} = \frac{\$20,940.64}{\$20,000} = 1.04$$

Since the profitability index is greater than 1, this product mix would be accepted. Again, as with the net present value method, various new product mix alternatives can be ranked according to the amount the indices exceed 1. While the profitability index and the net present value methods would yield similar results for any new product mix being considered, the net present value method is usually preferred in choosing between two alternatives since it expresses the expected economic contribution in absolute terms.

PROGRAM EVALUATION REVIEW TECHNIQUE

While timing is of course crucial during all aspects of the product planning and development process, it is especially important for successful commercialization of the new product. Sales and goodwill can be lost if the new product is not available at the time indicated by promotion. One useful method to achieve optimal coordination is the Program Evaluation Review Technique (PERT).[5] By identifying the critical operations in the commercialization process and producing status reports, the PERT technique effectively reduces or eliminates delays.

PERT requires the development of a detailed activity network depicting all important aspects of the commercialization process. By listing all activities that must be performed and their interrelationships, the network for commercialization can be derived, along with estimates of completion times. These estimates of the most likely completion times are prepared by individuals involved in completing the activity. In order to account for uncertainty about the time required to accomplish a particular activity, three time estimates for each activity are used:

$$t_e = \frac{t_o + 4t_m + t_p}{6}$$

t_e = expected time for completion

where

t_o = optimistic time for completion

t_m = most likely time for completion

t_p = pessimistic time for completion

An example of a PERT system for introducing a new product is shown in Figure 7–2. Six weeks (indicated in the figure by the number 6 in parentheses, located below the arrow concerning purchase equipment) was the time estimated to complete that activity. This figure is based on an optimistic time of four weeks, a most likely time of seven weeks, and a pessimistic time of 14 weeks:

$$t_e = \frac{4 + (4)\,(7) + 14}{6} = 6$$

In addition to the times in parentheses, each activity (task) is represented by a series of arrows that begin and end at points called events. These events are indicated by encircled numbers.

The most important path in the new product introduction network represents the sequence of activities that will take the longest time to complete. This critical path indicates the earliest possible time that the new product can

---

[5]For a good explanation of this technique, see: Joseph Newman, *Management and Applications of Decision Theory* (New York: Harper and Row Publishers, 1971), and Jerome D. Weist and Ferdinand K. Levy, *A Management Guide to PERT/CPM* (Englewood Cliffs, N.J.: Prentice-Hall, Inc., 1977).

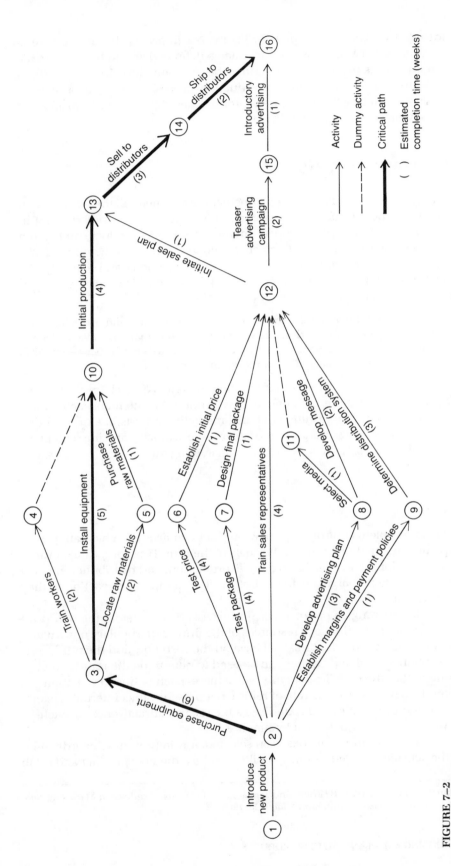

FIGURE 7–2
An introduction network for a new product

be introduced. The critical path in Figure 7–2 indicates that it will take 20 weeks to introduce the new product once a definite decision has been made. This path and its related time are critical because any delay in activities there will mean delays in the entire introduction process. PERT can be used to successfully control the entire new product development and introduction process, which is discussed in Chapter 9.

A NEW PRODUCT INFORMATION SYSTEM

As the new product mix is being planned and implemented, there is a need to establish a product information system. This system will provide relevant information so that the planned performance of the new product can be compared with actual results. Corrective actions can then be immediately undertaken to offset any deviations between projections and actuality. The basic characteristics of a product information system (PIS) are the same as those of a marketing information system.

> A structured, interacting complex of persons, machines, and procedures designed to generate an orderly flow of pertinent information, collected from both intra and extra-firm sources, for use as the basis for decision-making in specified responsibility areas of marketing management.[6]

In other words, a new product information system is a mechanism designed to gather and distribute data on the new product for the manager. By using this data base and the resulting detailed marketing reports, the new product manager can recognize trends and problems more quickly and implement any necessary changes in the new product's marketing mix.

SUMMARY

The main problem confronting a new product manager is orchestrating a marketing mix that will achieve the goals of the firm. This composite of various aspects of brands, packaging, price, promotion, and distribution must not only satisfy the target customer but ward off competitive reaction to the extent possible.

In formulating an optimal mix, it is necessary to consider the new product's impact on the financial resources of the firm. Of particular concern is the effect the new product mix will have on the cash flow and the return on investment. Several approaches can be used to determine the return on investment that the new product mix will achieve. Each of these evaluation methods (payback, average rate of return, net present value, internal rate of return, and profitability index) indicates the relative attractiveness of each new product mix under consideration.

Once the optimal mix has been selected, it is important to coordinate all the activities needed to successfully introduce the new product and estab-

---

6Samuel V. Smith, Richard H. Brien, and James E. Stafford, eds., *Readings in Marketing Information Systems* (Boston: Houghton Mifflin Co. 1968), 7.

lish a mechanism by which current information is available to management. PERT provides a means by which the critical path for successful introduction can be established. This network, combined with an established product information system, will greatly help the introduction and evaluation process.

SELECTED READINGS

AAKER, DAVID and SHANSBY, J. GARY. "Positioning Your Product." *Business Horizons* 25, no. 3 (May–June 1982): 56.
> Positioning strategy is crucial to successful introduction of new products. This article presents alternative positioning strategies and approaches, emphasizing key external and internal factors.

CRAWFORD, C. MERLE. "Defining the Charter for Product Innovation." *Sloan Management Review* 22, no. 1 (Fall 1980): 3–12.
> A study of the strategic plans of 125 firms resulted in the development of a new strategic concept, the Product Innovation Charter, which contains the target business areas by product type, end-user activity, technology, and intermediate or end-user group. Also included is the program of activities chosen to develop the target business areas.

DEMSETZ, HAROLD. "Barriers to Entry." *American Economic Review* 72, no. 1 (March 1982): 47–57.
> Bearing the risk of innovation has, in the past, been treated as an unproductive cost. However not all barriers can easily be tied to some purely objective measure of the cost of doing business, and the narrow view of these costs needs to be evaluated.

DUSENBERRY, WARREN. "CPMX for New Product Introductions." *Harvard Business Review* 45, no. 4 (July–August 1967): 124–139.
> Discusses the potential of CPM usage in product introduction, and concludes that, given problem constraints, CPM is a simple and inexpensive planning alternative.

GOULD, JAY M. "Managing in real-time." *Sales and Marketing Management* 126, no. 6 (27 April 1981): 22.
> The author is president of an information-retrieval system company that sells an on-line data base of marketing information to all types of companies. By requesting information based on SIC code, the data base produces share-of-market, line-of-business, and sales rankings for 450,000 establishments that account for 95 percent of all industrial sales and purchasing in seven key industries.

KEMP, DAVE. "Who Wants Your New Product? *Inc*, 3, no. 11 (November 1981): 161–2.
> Inadequate definition of new product niches is a serious problem for companies. This article recommends the identification of the product's most productive segment for focusing marketing dollars on lucrative prospects.

"The Marketers' Complete Guide to the 1980's." *Sales and Marketing Management* 123, no. 8 (10 December 1979): 27.
> The challenge of marketing in the 1980s will have a set of unique characteristics. Guide looks at the threats and opportunities posed by the economy, the changing population, and the marketplace itself.

McNEAL, JAMES U., and ZEREN, LINDA M. "Brand Name Selection for Consumer Products." *Business Topics*, 29, no. 2 (Spring 1981): 35–39.

Describes the results of a survey among 200 manufacturers of consumer products that sought answers to questions related to the brand name selection process. A discernible brand name selection process exists, and most firms consider the generation of an initial list of brand names as a critical marketing first step.

MURPHY, JOHN H., II, and AMUNDSEN, MARY S. "The Communications-Effectiveness of Comparative Advertising for a New Brand on Users of the Dominant Brand." *Journal of Advertising* 10, no. 1 (1981): 14–20.
Discusses an experiment to introduce a new brand of facial tissues. This experiment had three treatments—direct comparative copy appeal, brand X comparative copy appeal, and noncomparative appeal. The data indicates that the noncomparative advertising was the most effective for immediate and delayed claim recall and believability.

PARASARAMAN, A. "Hang on to the Marketing Concept." *Business Horizons,* 24, no. 5 (September–October, 1981): 38.
Determines that trends in domestic and international markets indicate that more, rather than less, emphasis should be placed on the marketing concept.

PEACOCK, JAMES. "Trends in Computing: Applications for the 1980's." *Fortune* 101, no. 10 (19 May 1980): 29.
Reviews the new modular systems with which CPM/PERT are competing.

"Selling With Your Back to the Wall." *Sales and Marketing Management* 124, no. 8 (9 June 1980): 35.
The economy of the 1980s will impact not only marketing of existing products, but the marketing of new products as well. The key factors affecting marketing and sales will be changes in the economy as a whole, rising inflation, competitive pricing, imperative cost reductions, and efforts to capitalize on the retreat of the competition.

SWAN, JOHN E., and RINK, DAVID R. "Fitting Market Strategy to Varying Product Life Cycles." *Business Horizons* 25, no. 1 (January–February 1982): 72.
The eleven different life cycles identified by this research have implications for marketing decisions. Strategy may be, in fact, a function of these product life cycles.

"Taming the Information Beast." *Advertising and Marketing Intelligence Service, Sales and Marketing Management* 124, no. 4 (17 March 1980): 13.
A new information system, Advertising and Marketing Intelligence Service (AMI), is now available to marketers. A joint venture formed by J. Walter Thompson and Times' Information Service, AMI is compatible with over 400 types of computer terminals and is an effort to resolve and network new product information.

WENSLEY, ROBIN. "Strategic Marketing: Betas, Boxes, or Basics." *Journal of Marketing* 45, no. 3 (Summer 1981): 173–182.
Criticizes both the financial and marketing approaches of resource allocation problems in multiproduct, multimarket firms. Contends that more emphasis is required on integrated, project-based assessment of market positions and financial returns.

WILLS, GORDON, and KENNEDY, SHERRILL. "How to Budget Marketing." *Management Today* (February 1982): 58.
States that the marketing budget, often set arbitrarily, should be developed with concrete objectives in mind. For this to be accomplished, marketing personnel should reassert their role in matching company resources to consumer needs.

ZIEGENHAGEN, M. E. "Thinking Like a Buy Sider: How to create marketing programs—not just pieces." *Industrial Marketing* 66, no. 10 (October 1981): 112. Fourth article in a continuing series. Develops the "full circle" approach to marketing communications. Emphasizes effectiveness and element coordination. A sound strategy, giving importance to early competitive positioning, will dictate a good marketing program.

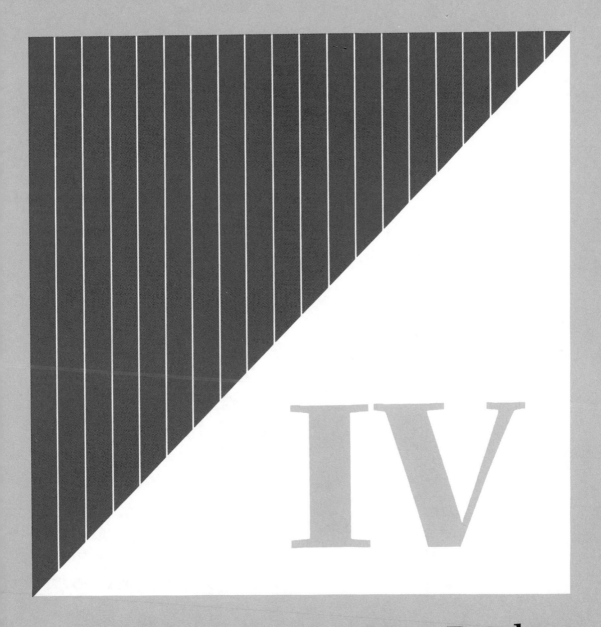

IV

New Product
Ideas: Obtainment
and Evaluation

Product Innovation

Olivetti, Italy's leading office equipment maker, will become the first major European company to introduce a small personal computer when it unwraps its 16-bit M20 model. Developed in Cubertino, California, at a cost of just under $10 million, the new office and home computer will be made in Italy and sold by Olivetti's dealers in Europe and through computer stores in the United States. The M20's base price of $3,000 can rise to $6,000, depending on additional peripherals and systems tailored to individual customers. While Olivetti is a "David," compared with the IBM and Apple personal computer "Goliaths," it hopes to take a successful swing at American leaders in the developing European market—which is estimated at $480 million for 1982.

Olivetti is looking hardest at business customers for its new computer, aiming for 10 percent of the world market for business-use personal computers. By 1983, when production of the M20 is fully underway, industry sources say that Olivetti is projecting annual sales of more than $300 million.

When Olivetti's computer hits the market, it will have to compete against IBM's personal computer and other makers of 16-bit computers, such as Digital Equipment, Data General, and Hewlett-Packard. Olivetti is banking on its solid European network of dealers and retail outlets for an edge in the competition.[1]

New York—Chase Manhattan Bank is taking the home bill-paying concept into a new dimension by offering the nation's financial institutions a computer home-pay service.

The system, known as Paymatic, has been test marketed to a sampling of customers of eight financial institutions, and, according to Chase vice-president Patrick J. Cunniff, the results of that test will be completed in June.

Paymatic is being marketed on a wholesale basis to financial institutions by Chase. The individual banks determine the rates to be charged and then market the system to customers.

Sources said the banks will pay between $200 and $300 per terminal and charge $1.25 to $2.25 per access. The banks will also probably charge a $6 monthly service fee.

Although the test is not yet complete, the service is now available to banks. However, Mr. Cunniff said, most banks are waiting for the test results before deciding on purchasing the service.

The terminal, which is a modified telephone, has a video display capable of screening 16 characters. A consumer can enter payment instructions into the computer by hitting certain codes. Company names are programmed into the phone and appear on the display terminal. The computer has the capability of storing 30 companies.

After a transaction is made, it moves to a clearing house, operated by Chase, before payment is made to a company.

Chase already offers its own customers a home bill-paying service through a regular telephone. Cuniff explains the new service as follows:

---

[1]"Italy's Olivetti Unwraps a Personal Computer," *Business Week* (5 April 1982): 42.

We are introducing the telephone terminal because we feel that it could be the next step and is going to allow us to operate in a cost-effective environment. That has not been the case with telephone bill-paying.[2]

The computer also has a feature that allows a consumer to prepay bills and store them until a certain release date. Chase can make payments electronically, but Mr. Cunniff believes that most companies prefer to be paid through the mail.

The computer was produced through the combined efforts of Los Angeles-based Telecredit and Santa Clara-based DMC Systems. Chase tested the system at banks in Connecticut, Mississippi, Maryland, Michigan, Texas, Pennsylvania, and Louisiana.

Mr. Cunniff said a decision on creating an advertising campaign for Paymatic will be predicated on the results of the test. Wells, Rich, Greene is Chase's agency.[3]

The two preceding scenarios indicate the output of the invention process in both the product and service section. This process can occur on an individual basis as well as in a corporate setting. But what actually is the invention process?

The starting point of the invention process is an inventor—a person who creates or produces something for the first time. While this is a unique position, in and of itself, the fruition of such effort lies in the successful sales of the end product. To have a product without a market is a tragic, costly proposition. It is tragic because the inventive genius could have been put to better use, and it is costly because time and money were wasted on a failure.

The area of product innovation has taken on an increasing level of importance, due to both the need for innovation to survive in highly competitive markets and the high cost of failure. This chapter addresses the topic of product innovation by first looking at the all-important, but little-understood, process of creative problem solving. Then the nature of the invention process is presented, followed by a discussion of the alternative sources of innovation. The chapter concludes with material on the role of inventors in the invention process.

CREATIVE PROBLEM-SOLVING TECHNIQUES

Creativity and problem solving are of course very generalized skills that have an area of mystique. Creativity, however, is an important part of every individual's composition. Unfortunately it declines with age and lack of use. In fact creativity has identifiable periods of decline. It declines when a person starts to attend school, becomes a teen, becomes 30, and becomes 50. In addition the latent creative potential of an individual can be stifled by perceptual, cultural, emotional, and organizational factors.

One way to unlock creativity and to stimulate the generation of creative ideas and innovation is to employ creative problem-solving techniques, which are indicated in Table 8–1 and discussed in the following pages.

---

[2]Robert Riassman, "Chase Offers Its Home-Pay," *Advertising Age* 53, no. 20 (10 May 1982): 8.
[3]Ibid., 8.

TABLE 8–1
A list of creative problem-solving techniques

Attribute listing	Collective notebook
Morphological analysis	Crawford slip-writing
Matrix charting	Bisociative thinking
Sequence-attribute/Modifications matrix (SAMM)	Heuristics
	Scientific method
Big-dream	Kepner-Tregoe method
Brainstorming	Forced relationships
Repeat	Buffalo method
Wildest idea	Synectics
Reverse brainstorming	Gordon method
Free association	Value analysis
Lateral thinking	Parameter analysis
Checklist method	

Attribute Listing

Attribute listing is a deliberate idea-finding technique requiring individuals to list the attributes of an item or problem and then look at each of them from a variety of viewpoints. Modification occurs in this process as each attribute suggests possible new uses to participants in the exercise.[4] In this way the method brings originally unrelated objects together to form a new combination that better satisfies a need.[5]

Since deriving a list of attributes is not an easy undertaking, the technique is best applied to specific problems or needs, rather than general ones. This keeps the scope of the endeavor within manageable levels. Usually the more familiar group members are with the product or problem, the more difficult it is for them to agree on attributes. Groups can assign attributes to areas outside their expertise much more easily than they can to problems within their expertise, due to the fact that too much familiarity with the problem prevents them from seeing it in a new light. If a problem is too familiar, a person is unable to be flexible or original in his or her ideas. The most productive group is generally one that is not extremely familiar with the problem or product whose attributes are being listed.

For example a group being asked to list attributes of a calculator might come up with the following: hand-held, plastic, LCD (Liquid Crystal Digit) display, reverse polish notation, and rechargeable battery. From this initial list, ideas and products could evolve.

As this example illustrates, probably the most important quality in this technique is that it "causes an immediate focus on the basic problem" and is chiefly an activity of individual analysis.[6]

Morphological Analysis

Morphological analysis attempts to look at all the possible combinations of variables in solving a problem. First the problem is stated and broadly defined;

---

[4]S. J. Parnes and H. F. Harding, eds., *A Source Book for Creative Thinking* (New York: Charles Scribner's Sons, 1962), 308.
[5]Ibid., 255.
[6]W. E. Souder and R. W. Ziegler, "A Review of Creativity and Problem-Solving Techniques," *Research Management* 20, no. 24 (July 1977): 36, Industrial Research Institute, Inc.

A_1	A_2	A_3		
B_1	B_2	B_3	B_4	B
C_1	C_2	C_3	C_4	
D_1	D_2	D_3		
E_1	E_2			

FIGURE 8–1

Example model for use in morphological analysis

Source: C. S. Whiting, *Creative Thinking* (New York: Reinhold Publishing Co., 1958). 63.

then all the conceivable solutions are listed in a model. Finally the suggested solutions are evaluated.[7]

The process is carried out through the aid of a model. Different dimensions are used, depending on the number of variables. A two-dimensional model is used for two variables, and a three-dimensional model is used for three variables. For more than three variables, the model in Figure 8–1 is used, where each letter is a major variable subdivided into various alternative forms. The difficulty with this technique, of course, is that it becomes unmanageable as the variables and the number of combinations increase. Since most problems do usually contain more than three major variables, morphological analysis is difficult to apply in most instances. Attribute listing and check lists are similar techniques, but are more practical and easier to control.

Matrix Charting

Matrix charting is a systematic way to search for new opportunities by listing elements along both axes of a chart and then asking basic questions regarding these elements. The answers are then recorded in the relevant boxes of the matrix. The basic questions used to elicit creative responses are as follows: What can it be used for? Where can it be used? Who can use it? When can it be used? How can it be used?[8] The matrix includes as many boxes as the ideas or elements present. Since there is no universal way to create a matrix, the variables may often have to be rearranged several times to find the best order.

Since answers may often be numerous, space should be provided for recording. The process can be conducted by a single individual or by pairs or teams of individuals.

Matrix charting can be used to devise new options for packaging (product/package), for production processes (product/process), or for market segments (product/market). As the example in Figure 8–2 indicates, one of the major advantages in matrix charting is viewing a standard issue from a new perspective.

Sequence-Attribute Modification Matrix (SAMM)

The sequence-attribute modifications matrix (SAMM)—which is more effective in group settings—identifies areas to be evaluated but does not provide actual

---

[7]The process involved in using morphological analysis can be found in C. S. Whiting, *Creative Thinking* (New York: Reinhold Publishing Co., 1958), 63, and Souder and Ziegler, "Review of Techniques," 38.

[8]Charles H. Clark, *Idea Management: How to Motivate Creativity and Innovation* (New York: AMACOM, 1980), 47.

FIGURE 8–2
Example of matrix charting

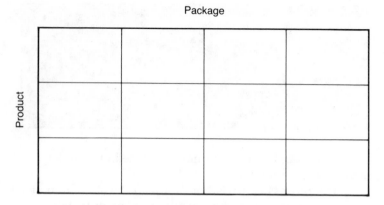

Package

Product

solutions. This approach centers around developing a matrix that indicates on the left-hand side a sequence of activities and on the right-hand side possible modifications to the process. Each activity is evaluated and necessary modifications checked off. A sample of the technique for the steel industry is indicated in Table 8–2.

Big-Dream Approach

The big-dream approach requires the individual to produce as grandiose a dream about a problem as possible; that is, the individual is to think *big*. Then

TABLE 8–2
Illustration of SAMM technique

| | | | | | Modifications | | | | | |
| | | | | | | | | | | |

Sequence/Attribute Description	Item Number	Eliminate A	Substitute B	Rearrange C	Combine D	Reverse E	Enlarge F	Reduce G	Modify H	Separate I
Heat steel slag to pliable state	1							X		
Transfer from heating furnace	2									
Position rolls to desired setting	3			X	X					
Pass slab through rolls (elongate)	4			X	X					
Check slag gauge	5					X				
Shear slab to desired length	6	X								
Transfer sheared product	7							X		

Key: X = possible priority items that can be modified, combined, etc.
Source: Souder and Zeigler, "Review of Techniques," 38.

the individual investigates every possible subject related to the big-dream idea developed. Once this step is completed, a secondary big-dream idea is similarly researched and so on until the dream is molded into a workable form.[9]

Brainstorming

The brainstorming technique is probably the most well known and widely used creative problem-solving technique. It is an unstructured process for generating—through spontaneous contributions of participants—all possible ideas about a problem within a limited timeframe. While the technique can be used by individuals, it is primarily a group process. Some people think this technique is chaotic, but, if it is practiced correctly, it provides an organized generation process.

Because brainstorming is so well known, its all-important planning process is often disregarded. Groups sit down and merely start brainstorming. When there is no planning, a chaotic session frequently results. For a good brainstorming session to occur, a problem statement that is neither too broad (which would diversify ideas too greatly so that no one area may successfully emerge) nor too narrow (which would tend to confine responses) needs to be carefully prepared.[10]

Once the problem statement is prepared, group members are chosen. The ideal size of a group is between 6 to 12 people. It is important to the group process that a wide variety of knowledge be represented by group members and that persons with supervisor-subordinate relationships on the job are not in the same group. If group members are experts in the field and are therefore dealing with an extremely familiar issue, the range of new ideas that result is usually severely restricted. Having superior-subordinate relationships present in the group also usually represses the creativity. After the group is formed the method for recording ideas needs to be selected. Clark suggests that the leader write *numbered* ideas on large pieces of newsprint and display them in concert with a tape recording of the session.[11] It is better not to use a tape recorder alone because the displaying process is a critical part of the ideation process.

Once the planning stage has been carefully completed, the six states of brainstorming can begin:

1. State the problem and discuss it for familiarity and background.
2. Restate the problem.
3. Write down the restatement responses for all to see.
4. Start a warm-up session to get participants freewheeling, laughing, and generally in the mood for brainstorming.
5. Conduct the actual brainstorming session.
6. Come up with the "wildest idea."

Participants in the brainstorming session must never criticize or evaluate or analyze while the session is occurring. All ideas, no matter how illogical, be-

---

[9]For a discussion of this approach, see M. O. Edwards, "Solving Problems Creatively," *Journal of Systems Management* 17, no. 1 (January–February 1966): 16–24.
[10]For a discussion of this aspect, see Clark, *Idea Management*, 42.
[11]Ibid., 43.

come important contributions. Any judgment is completely suspended during the session.

After the close of the session it is important to communicate to each participant a complete list of the generated ideas as soon as possible. It is the leader's responsibility to organize the list into related areas as this tends to motivate suggestions for implementation from the group members. It also provides the necessary closure to the activity.

The brainstorming technique has been successfully used in many business applications since its inception in 1938. For example, Corning Glass found new uses for glass in autos, General Electric developed value engineering, and B. F. Goodrich devised work-simplification techniques, all as a result of brainstorming sessions.

Repeat Technique

The repeat technique is actually a variation of brainstorming and is used when the same idea is mentioned identically in the second brainstorming session. When this occurs (and it must be an exact replication, word for word), the leader refers to the first idea and asks the participant if the idea was what was meant. If it was, the session continues. If it was not, the participant is asked to make another suggestion.[12]

The technique puts some burden on the leader to police the ideas and interrupt the process, then to start it up again. Nevertheless the repeat technique is useful in clarifying and maximizing ideas.

Wildest Idea

Another approach used during a brainstorming session is what is called the wildest-idea technique. It is utilized toward the end of the brainstorming session when new ideas become more and more infrequent. The leader asks the participants to look at the lists of ideas generated earlier and to choose from among them the wildest ones. Then the leader records those suggestions on a single sheet entitled *wildest idea*.[13] Some of the suggestions draw laughter and are dropped as still being quite unrealistic. However some may elicit still more ideas.

This iterative process recurs for each of the wildest ideas until a new subset of ideas is finally developed. Often many of these wildest ideas are actually quite reasonable and may even lead to the problem solution sought at the outset of the brainstorming session. Should a possible solution emerge, the group then proceeds to the evaluative steps in brainstorming.

Reverse Brainstorming

Reverse brainstorming, also called *The Devil's Advocate technique*, is similar to brainstorming, except that criticism is allowed. In fact the technique deliberately tries to find fault by asking the question, "In how many ways can this idea fail?" Because participants are critical of others' suggestions, the leader

---

[12]For a discussion of this method, see J. Geoffrey Rawlinson, *Creative Thinking and Brainstorming* (New York: John Wiley & Sons, Inc., 1981), 52.
[13]Rawlinson, *Creative Thinking*, 53.

must be careful to keep the group's mood optimistic. Often reverse brain-storming is used prior to brainstorming or other creative techniques to initiate serious efforts at innovative thinking.[14] The process most often involves the identification of everything wrong with the issue at hand followed by a discussion of ways to overcome the flaws identified.

Free Association

One of the simplest techniques in creative problem solving, which can be used in groups or by individuals, is free association. The idea of this technique is to stimulate the thinking process to elicit an entirely new slant to a problem. To begin the process, a word or phrase related to the problem is written down, then another, and another. Each new word adds something fresh to the on-going thought-processes, thereby creating a chain of ideas. Before beginning the technique, a decision should be made on whether to establish a time limit. Since ideas build on one another, the process could continue for quite a long period of time.

Free association is best used in instances where no structure is required and no specific output desired. For example, it could be used in an advertising agency to develop a slogan for a product.

Lateral Thinking Process

Edward DeBono describes the creative problem-solving arena as necessitating lateral or divergent thinking. DeBono's contention is that much of an individual's lifetime is spent, not in creating ideas, but in logically arriving at solutions via step-by-step analytical or vertical processes. As shown in Table 8–3, lateral thinking is the creative partner of analysis and includes several distinguishing characteristics in direct opposition to vertical analysis.

Three specific stages provide the basis for conducting lateral thinking sessions: (1) background, in which an idea arises from the nature of a sequential patterning system, (2) process, in which old ideas are abandoned and new ideas provoked, (3) method, in which attitude (awareness), techniques, and skills are practiced without any judgment of ideas occurring.[15]

Lateral thinking views creativity as unique and quite separate from analytical reasoning. The two types of thinking (convergent and divergent) are, in fact, quite dissimilar: logic and imagination are separate vehicles, ones that managers can effectively combine to their advantage. They can be used to pinpoint problems and analytically deduce a few solutions. Then divergent thinking can be employed to stimulate change or new ideas. Finally the ideas can be analytically evaluated to derive solutions.

Vertical or analytical thinking places parameters around problems, while lateral or divergent thinking searches beyond known boundaries to generate change. As such, lateral thinking is a powerful creative problem-solving technique.

---

[14]For a discussion of this technique, see Rawlinson, *Creative Thinking*, 124, 126, and Souder and Ziegler, *"Review of Techniques,"* 35.
[15]This method is discussed in Edward Debono, *Lateral Thinking for Management: A Handbook for Creativity* (New York: American Management Association, 1971), 4–13.

TABLE 8–3
Differences in vertical and lateral thinking techniques

Vertical	Lateral
Solution	A far-reaching exploration
Definition/description	*Makes* something happen
Narrowly confined by rules	Open-ended
Develops ideas	Generates ideas
Makes one choice, decision, or action	Elicits possibilities for restructuring an idea for change
Is an orderly, predictable process	Characterized by discontinuity

Source: Edward DeBono, *Lateral Thinking for Management: A Handbook for Creativity* (New York: American Management Association, 1971). 4–13.

Checklist Method

In the checklist method, a list, composed of related issues or suggestions, is prepared and a problem is analyzed. While sometimes the list is presented in statement format, more often it is a series of questions.

The participant(s) uses the list as a way to guide the direction (not necessarily the content) or the ideas. The checklist is intended primarily as an aide to force the user to concentrate on specific areas. Checklists may be specialized or generalized and of any length. A specialized checklist, often included in business journals, aids in the development of new ideas and is a reminder of the essential steps in a process. A typical example might be "Ten Suggestions on How to Sell Products to Clients." However generalized checklists are usually much more valuable in the creation of original ideas.[16] Osborn developed a general list using nine ways of changing an existing idea. These include the following questions, which are further amplified in Table 8–4:

- "Put to other uses?
- Adapt?
- Modify?
- Magnify?
- Minify?
- Substitute?
- Rearrange?
- Reverse?
- Combine?"

The checklist serves to simplify the problem by drawing imaginary boundaries around it through the questions themselves. However, in so doing, the checklist may also restrain the creative-thinking process and actually limit the kind and variety of ideas that evolve. The checklist method usually works best when used on familiar problems.

[16]For a discussion of generalized versus specialized checklists, see Whiting, *Creative Thinking*, 60–61.

TABLE 8–4
Osborn's checklist

Put to Other Uses?   New ways to use as is? Other uses if
modified?

Adapt?   What else is like this? What other ideas does this suggest? Does past offer parallel? What could I copy? Whom could
I emulate?

Modify?   New twist? Change meaning, color, motion, odor, form,
shape? Other changes?

Magnify?   What to add? More time? Greater frequency? Stronger?
Larger? Thicker? Extra value? Plus ingredient? Duplicate? Multiply? Exaggerate?

Minify?   What to substitute? Smaller? Condensed? Miniature?
Lower? Shorter? Lighter? Omit? Streamline? Split up?
Understated?

Substitute?   Who else instead? What else instead? Other ingredient? Other material? Other process? Other power? Other place?
Other approach? Other tone of voice?

Rearrange?   Interchange components? Other pattern? Other lay-
out? Other sequence? Transpose cause and effect? Change
pace? Change schedule?

Reverse?   Transpose positive and negative? How about oppo-
sites? Turn it backward? Turn it upside down? Reverse roles?
Change shoes? Turn tables? Turn other cheek?

Combine?   How about a blend, an alloy, an assortment, an en-
semble? Combine units? Combine purposes? Combine appeals?
Combine ideas?

Source: Alex F. Osborn, *Applied Imagination* (New York: The Scribner Book Companies,
Inc., 1957), 318.

Collective Notebook Method

In the collective notebook (CNB) method of problem solving, a notebook is
carefully prepared that includes a statement of the problem. Ample blank
pages are also provided, as well as any background data or graphics pertinent
to the problem. The preparatory stage should be planned and executed with
great care. Any information that might assist in reaching a creative solution to
the problem is given to participants.

Each participant receives an identical copy of the prepared notebook
and agrees to review the problem, think about its solution, and record one to
several times a day, for at least one month, his or her thoughts and ideas. At
the end of the month each individual prepares a list of the best ideas, suggestions for future thoughts and research, and other new ideas unrelated to the
original problem.[17]

Notebooks are then given to a central coordinator who synthesizes the
data and summarizes all the material. The summary becomes the topic of a
final creative discussion by the group.

[17]For a thorough discussion of the collective notebook method, see J. W. Haefele, *Creativity and
Innovation* (New York: Van Nostrand Reinhold Co., Inc., 1962), 152.

This method of creative problem solving reduces tension and pressure to a very low level, since participants develop their own notebook responses and record these notations at their convenience. A participant can enter *no idea today,* for example, without fear of penalty. Participants are generally very committed to making the daily entries over a month's time and usually find the process enjoyable. The *modus operandi* promotes a positive experience for participants, especially since there are no quality judgments about suggestions. Each participant can see all the other notebooks at the end of the recording period.

The method lends itself to groups of varying sizes—including very large ones—and to problems with wide scopes. It is heavily dependent, however, on the synthesizing ability of the coordinator, who has the rather large task of bringing the multiple suggestions into focus for a final discussion that leads to suggesting a simple problem solution.

Crawford Slip-Writing

The Crawford slip-writing technique is used in group problem-solving sessions when a proliferation of ideas is sought.[18] The group leader announces the problem in a "how to" format. Then, small pads of paper (about 3 × 5 inches) are given to each participant for writing down ideas or solutions to the problem. As each participant thinks of a new idea, he or she writes it down— one idea per slip of paper. A specific time period is designated and, at its conclusion, all the slips of paper are collected by the leader and processed outside the group. An edited list is then made available to the participants.

The success of the slip-writing technique, like so many of the other techniques, depends on careful preparation by the leader—particularly in his or her statement of the format. If possible the question should be distributed before the actual meeting so participants can think about ideas and discuss them.

Bisociative Thinking

Bisociative thinking consists of combining existing information in a new way. Its results are solutions that existed all along but were not recognized because no one had thought to combine the information in the right ways. An example of bisociation is evident in the development of the printing press. Gutenberg attended a wine harvest where, upon seeing the wine press, he "discovered" the idea of a printing press for evenly reproduced copies. While the wine press was not new, the specific application of the process of printing was.[19]

The word for this process was coined by Arthur Koestler who describes it in terms of two planes or matrices.[20] The information does not provide a solution on the first or second plane until the individual experiences information from both matrices at the same time and establishes a relationship between them. The relationship is called *dissociation* and, interestingly, is

---

[18]This technique is discussed in Clark, *Idea Management,* 36–38.
[19]Rawlinson, *Creative Thinking,* 10.
[20]For a discussion of this method, see Arthur Koestler, *The Act of Creation* (New York: Mac-Millan, 1964), 178–211.

often accompanied by a release of tension—laughter, sighing—stemming from the awareness of the new insight.

Heuristics

Heuristics is a method of independent problem solving. It relies solely on the individual's propensity to indulge in the art of discovery—not through a rigid scientific method, which is the antithesis, but through progression of thoughts, insights, and learning. Heuristics is dependent on self-motivation to delve into any facet of a problem from any angle. Furthermore the reasoning involved is not thought of as final but rather as provisional.[21]

Heuristics is probably used more in business applications than imagined simply because business managers most often settle for a plausible estimate at the outcome of a decision rather than on any assured certainty. While they may use standard and often complicated methods to forecast outcomes, managers also use their own investigation of parameters, some of them outside the boundary of logical thought. A creative product manager might use heuristics for example, to investigate all possible ways to stimulate consumer purchase behavior.

While heuristics has many specific applications in creative problem solving, one approach is called the heuristic ideation technique (HIT).[22] The process involves locating all relevant concepts that could be associated with a given product area and generating a set of all possible combinations of ideas. Each concept is broken down into others and may yield so many ideas that it may turn out to be costly to reduce these to a workable quantity. The advantages and disadvantages of the method are indicated in Table 8–5.

Scientific Method

The scientific method, widely used in various fields of inquiry, consists of principles and processes required in any investigation. These include steps for developing rules for concept formation, followed by conducting observations and experiments, and finally validating the hypothesis. In business applications, the method is modified into what has become a traditional approach: (1) defining the problem (2) analyzing the problem (3) gathering and analyzing data (4) developing and testing potential solutions and (5) choosing the best solution.

The scientific method or its modification can be usefully applied to any problem-solving situation to instill clarity and a sense of organization. Because of its inherent orderliness and step-by-step methodology, a solution emerges slowly, but assuredly. Completing one step in a thorough manner automatically pushes the individual or group onward to the next step. While the method does not allow for random ideas or for unsubstantiated results, nevertheless creativity may be aptly utilized in the stage calling for the development of potential solutions.

The method may be used in almost any business problem situation where the quality of the solutions—not the quantity—is considered of primary

---

[21]G. Polya, *How to Solve It* (Princeton, N.J.: Princeton University Press, 1945), 113.

[22]See Edward M. Tauber, "HIT: Heuristic Ideation Technique," *Journal of Marketing* 36, no. 1 (January, 1972): 58–70.

TABLE 8–5
Advantages and disadvantages
of the HIT process

HIT advantages are that it:

Is systematic

Allows management to review a great number of alternatives

Is flexible

Can be used by individuals

Requires no specialized training

Its disadvantages are that it:

Relies on management to specify what is defined as "relevant"

Does not guarantee an optimal solution

Can only deliver a combination of ideas

May elicit unfeasible ideas because they are not screened as they are brought up

Source: Edward M. Tauber, "HIT: Heuristic Ideation Technique," *Journal of Marketing* 36, no. 1 (January 1972): 60.

importance. An example is a company deciding whether to grow through diversification, mergers, or new products. The deliberate, careful scientific method acts as a basis for sound decisions.

Kepner-Tregoe Method

The Kepner-Tregoe method is a creative problem-solving technique for analyzing problems in a group setting.[23] While basically a training method for developing decision-makers, it can be easily applied in the business environment. Since the method isolates or finds a problem and then decides what should be done about it, it is similar to the modified scientific method.

Using the steps shown in Figure 8–3, the final decision is made after several iterations. The method lends itself very well to made-to-order situations, with each member of the group receiving an assignment. For example one individual may role-play the production manager, and another the financial manager, and so on. After the fourth step, the process is repeated, and each member is given a new role to play. In this way a participant gets to view the problem from different dimensions.

The method works best when a course leader, who is not an active participant, monitors the group's conversations. Following the making of the final decision, the group analyzes, under the leader's guidance, the information and assumptions that went into the decision. The participants also evaluate how the decision-making process could have been improved.

---

[23]For a discussion of this method and its application to the business environment, see Charles Kepner and Benjamin Tregoe, *The Rational Manager* (New York: McGraw-Hill Book Co., 1965), 39–56; and Charles Kepner and Benjamin Tregoe, "Developing Decision Makers," *Harvard Business Review* 38, no. 5 (September–October 1960): 115–124.

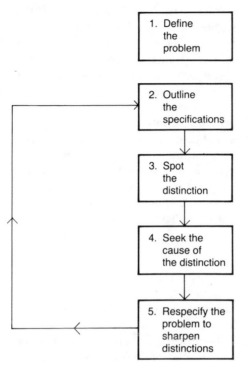

FIGURE 8–3

Steps in the Kepner-Tregoe method

Source: Perrin Stryker, "Can You Analyze This Problem?" *Harvard Business Review* 43, no. 3 (May–June 1965): 73–78, and "How To Analyze That Problem," *Harvard Business Review* 43, no. 4 (July–August 1965): 99–110.

Forced Relationships

Forced relationships is a creative problem-solving technique used by groups to develop a new object. The technique asks questions about objects, or ideas, in an effort to determine what new object or idea would be produced if new combinations were made. The objective is to come up with a totally new concept through a five-step process:[24]

1. Isolate the elements of the problem.
2. Find the relationships among these elements.
3. Record the relationships in an organized fashion.
4. Analyze the record of relationships to find ideas or patterns.
5. Develop new ideas from these patterns.

Buffalo Method

The Buffalo method teaches creative problem solving through the use of an instruction manual, a teaching guide, and a student workbook.[25] These resources teach and improve creative problem solving through practice.

---

[24]Rawlinson, *Creative Thinking*, 52–59.
[25]For a discussion of the Buffalo method, see Alex F. Osborn, *Applied Imagination* (New York: Charles Scribner's Sons, 1957), 86, and Sidney J. Parnes and Harold F. Harding, *A Source Book for Creative Thinking* (New York: Charles Scribner's Sons, 1962), 307–23.

This process is divided into four steps. The first step is fact-finding, which involves problem definition and gathering and analyzing data. The second step is idea-finding, which involves idea production and idea development. The final step is finding the solution, which includes the evaluation and adoption of one outcome. Throughout the process, students alternate between judicial and creative thinking in various informal settings.

The Buffalo method emphasizes the importance of creativity and imagination, not as belonging to an elite minority, but as inherent in all walks of life. The method focuses on demonstrations and discussions, as well as the practicing of various creative-thinking techniques, including brainstorming, the deferred judgment principle, attribute listing, forced relationships, and the checklist procedure. The elements of taking notes, setting deadlines and quotas for the production of ideas, and setting aside special times and places for deliberate idea-production are important ingredients in successfully implementing the method.

Synectics

Synectics is a creative process that forces participants to consciously apply, through analogy, preconscious mechanisms to solve problems. In this method a group works through two steps. The first step is to make the strange familiar. This involves basic analysis that enables the group, through the use of generalizations or models, to understand the intricacies of the problem presented. The analysis serves to put the problem into a readily acceptable or familiar perspective, thereby eliminating the strangeness.

Once the strangeness is eliminated, participants engage in the second step, which is a reversal of the first step—making the familiar strange, or seeking novel solutions. Gordon suggests four mechanisms of analogy for accomplishing this: personal, direct, symbolic, and fantasy.[26] Each type of analogy is used to speculate about the success of the proposed solution. Group participants verbally exchange information, using any of these types of analogy, to discuss solutions to the problem.

The group's activities bring together different and apparently irrelevant elements, which is the Greed definition of synectics. This joining of elements in synectics theory is based on the belief that the creative, emotional component in problem-solving behavior is more important than the intellectual—that the irrational is more critical than the rational. The probability of success with synectics is increased when group members are conscious of emotional, irrational elements in arriving at solutions to problems.

This process yielded the following results when it was applied to inventing a more serviceable roof. The analysis first uncovered the possibility of having a white, sun-deflecting roof in summer and a black, heat-absorbing roof in winter. Then the exchanged analogies included comparison with several kinds of animal that change color. After discussing the chemical process of animal color change in depth, the information was applied to possible use in roofing materials.[27]

---

[26]For a thorough discussion and application of this method, see W. J. Gordon, *Synectics: The Development of Creative Capacity* (New York: Harper & Row Publishers, Inc., 1961), 37–53.
[27]Ibid., 55.

Gordon Method

The Gordon method, unlike many other creative problem-solving techniques, begins with group members not knowing the exact nature of the problem. This is based on the premise that when members are aware of the actual problem, the solution is clouded by preconceived ideas and habit patterns.[28]

The Gordon method depends heavily on a group leader. The leader is the only one at the three-hour-minimum session who is aware of the actual problem. A session begins with the leader mentioning an associated concept that is quite general in nature. For example, in deciding the best way to park cars in an overcrowded area, the leader might begin the session by telling the group to think about storing things.

The group begins by conceiving and expressing a number of ideas. Then a concept is developed, followed by related concepts. For instance, in the previous example, the efficient use of space may evolve in different forms from storing things in circles to standing them on end. During this process it is the leader's responsibility to guide these discussions from the general to the specific, since he or she is the only person aware of the real problem.

Ultimately the leader guides the discussion so that the ideas are close to solving the actual problem. At this point, the leader reveals the actual problem, enabling the group to make suggestions for implementation or refinement of the final idea.

The group is expected to create ideas freely without constraints in methodologies, number or kind of ideas, or physical configurations. It is the responsibility of the leader to keep the discussion moving toward a reasonable solution. This includes not only guiding the discussion pertinent to the solution but also using techniques to optimize the efficiency of the group. This frequently would entail dealing with group members' fatigue, since the exercise is quite tiring. Several techniques to minimize or dispel fatigue include the interjection of facts into the free-association stage, humor, or physical room arrangements.

Value Analysis

Another creative problem technique—value analysis—is a method for suggesting ways of doing something that will maximize its value to the firm.[29] This method uses a question such as "Can this part be of lesser quality, since it isn't a critical area for problems?" To implement a value analysis procedure, a firm has to establish regularly scheduled meetings. Once the role and importance of value analysis is established, the topic selected is discussed and recorded by the meeting chairman. The ideas indicated in this discussion are then evaluated and refined until an agreed-upon idea needs outside input or is ready for action.

Frequent use of value analysis occurs at General Electric. The company has developed various keys to value, including creative thinking, removing

---

[28]This method is discussed in Haefele, *Creativity*, 145–47; Parnes and Harding, *Source Book*, 265–75; and Souder and Ziegler, "Review of Techniques," 34–42.
[29]For a discussion of value analysis and its application at General Electric, see "A Study on Applied Value Analysis," *Purchasing* 46 (8 June 1959): 63–65 and "The 20 Keys to Value," *Purchasing* 46 (8 June 1959): 66–67.

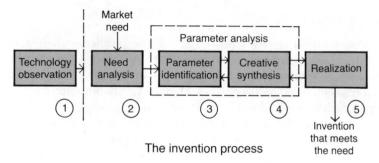

FIGURE 8-4

Illustration of parameter analysis

Source: Yao Tzu Li, David G. Jansson, and Ernest G. Cravalho, *Technological Innovation in Education and Industry* (New York: Van Nostrand Reinhold Company, 1980). 142.

obstacles, bringing in new information, obtaining information from the best sources, spending the company's money as if it was yours, working on specifics rather than generalities, and using standards.

Parameter Analysis

The final method useful in creative problem solving and developing inventions is parameter analysis.[30] Parameter analysis is a matching process involving two aspects—parameter identification and creative synthesis (see Figure 8-4). In step one (parameter identification), variables in the situation are analyzed to determine which are most important to the problem's solution. These variables become the focus of the investigation, and the other variables are set aside. For example, in focusing on the problem of energy loss through windows, several features are of prime importance: heat retention in the winter, heat reflection in the summer, light preservation, ease of installation, and aesthetics. These features are needed for a window insulation to be successful on the market. Once the primary issues have been identified, secondary issues are isolated and evaluated in terms of their impact on the problem.

After the primary issues have been identified, the relationships between parameters that most effectively describe the underlying issues should also be examined. For example, if we have decided that heat retention in the winter is a very important attribute, we must then analyze the trade-offs between this attribute and the costs and aesthetic appeal.

Following an evaluation of the parameters and relationships, the development of the solution(s)—the second aspect of parameter analysis—occurs. This step combines creative synthesis with a total understanding of the problem so that a unique product can be developed. This process begins by focusing on the critical issues and the technology involved. In the example of the window insulation, it may be decided that some heat retention should be sacrificed so that a more aesthetically appealing window insulation can be developed for the consumer market. A more heat-retentive, costly, less-appealing shade could be appropriate for the industrial market.

---

[30]The procedure for parameter analysis is thoroughly discussed in Yao Tzu Li, David G. Jansson, and Ernest G. Cravalho, *Technological Innovation in Education and Industry* (New York: Reinhold Publishing Company, 1980), 26–49, 277–86.

TABLE 8-6
An inventor's profile: some dominant personal characteristics

On *self-assessment* of 13 specific personal characteristics, an inventor is one who

- Has initiative and seeks personal responsibility
- Is a driver with a high level of energy
- Is a persistent problem-solver
- Is a moderate risk-taker
- Deals well with failure
- Uses feedback to take corrective action
- Has a high degree of long-term involvement

On levels of *satisfaction,* the activities that rate the highest for giving a feeling of success are

- Coming up with an innovative idea or solution
- Discovering new theoretical insights or solutions
- Working on a problem that contributes to the general well-being of the nation or society
- Inventing a product that is successfully licensed

On other measures, inventors

- Are able to tolerate uncertainty and ambiguity and are able to work with scant information
- Are generally self-confident but are more self-confident in their area of expertise
- Are willing to invent in an unfamiliar area
- Are highly achievement-oriented

Source: Robert D. Hisrich, *"Inventors: A New Investment Opportunity for Venture Capital?"* (unpublished working paper, March, 1982).

INVENTORS AND THE INVENTION PROCESS

An inventor is a person who creates or produces something for the first time. What kind of person is an inventor? Is an inventor a good source of new product ideas? What are an inventor's personal characteristics? Are they as unique as the invention itself?

The inventor's profile shown in Table 8–6 includes the following characteristics:

- A very high level of satisfaction in being an inventor
- An ability to tolerate ambiguity and uncertainty
- A very high level of self-confidence, especially in one's own area of expertise
- A willingness to take risks
- A high premium placed on being an achiever[31]

These characteristics indicate that at least some inventors are not significantly different from their entrepreneur counterparts and, as such, are good

---

[31]Robert D. Hisrich, *"Inventors: A New Investment Opportunity for Venture Capital?"* (Unpublished working paper, March 1982).

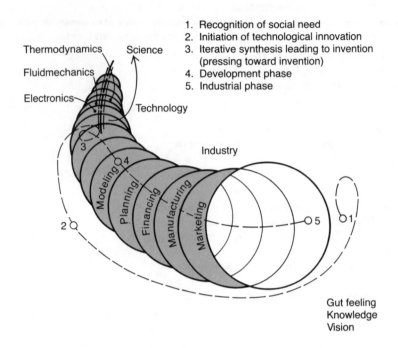

1. Recognition of social need
2. Initiation of technological innovation
3. Iterative synthesis leading to invention (pressing toward invention)
4. Development phase
5. Industrial phase

FIGURE 8–5

Innovation process

Source: Yao Tzu Li, David G. Jansson, and Ernest G. Cravalho, *Technological Innovation in Education and Industry,* (New York: Van Nostrand Reinhold Company, 1980). 27.

candidates for new product ideas. From a company's standpoint, ideas from inventors mean a far lower risk and a shorter packback period than the development of an entirely new product from its inception.

An inventor, whether he or she is an independent consultant or part of the firm, goes through the innovation process indicated in Figure 8–5. The process, represented here as a cornucopia—the traditional symbol of abundance—begins with knowledge and proceeds along the traditional disciplines when a technological innovation is involved. The critical point is when technology interacts with a need, which leads to an innovation being marketed. Several problems occur at this juncture.[32] Innovation is difficult in many large corporations because of the lack of incentives to take risks and the emphasis on short-term profits. The small inventor, on the other hand, does not have the knowledge, ability, or financing to take the invention to market. For example, even though there is a venture capital pool of $4 billion available to the inventor, little if any of it is for seed capital ($50,000 to $150,000 may be needed). The final problem occurs in the area of technology transfer. Despite the many attempts and high level of interest, there is surprisingly little technology transfer taking place. The fallout from much scientific research rarely makes its way to the market. Few inventors and patents of universities and government agencies ever become actual products available on the market.

---

[32]These problems are discussed in Li, Jansson, and Cravalho, *Technological Innovation in Education,* 27–28.

In spite of all the difficulties, there are methods for making the innovation process more effective. One of these is to make sure that new product ideas are not lost. Tips for developing and remembering ideas are indicated in Table 8–7. These include making sure pencil and paper are always within easy reach for recording the ideas; immediately writing down and sketching each idea; continually projecting into the future, and not focusing on the negatives of the idea too early in the innovation process.

Another effective mechanism for increasing the results of the innovation process is to employ the market evaluation methodology indicated in Figure 8–6. This process is expanded on in Chapters 9 and 10, but it is important at

---

TABLE 8–7
Tips for developing new product ideas

Good ideas are hard to come by, and remembering them isn't easy, either. Thomas Edison kept a pad and pencil next to his bed and in the bathroom. He said that "ideas are like lightning flashes: moments of brilliance in the darkness of thought."

Edison is gone, but the pattern of new product (new thought) creation goes on. Howard Zunger, a Denver-based business consultant and student of thought processes, is continuing the study. "Everyone suffers from a loss of creative thought," Zunger said. "Normal people suffer because they aren't used to the pattern. Creative people suffer because their minds are too crowded with such matters."

He offers these 10 tips for those who have trouble remembering new product ideas.

- Keep pencil and paper handy.
- Don't ever think that an idea is so good that you will remember it.
- Write it down and make a sketch.
- Stop whatever you are doing and concentrate on the idea. When brain circuits are clicking, keep them going.
- New ideas are especially difficult to remember because there is no base to which you can return.
- New ideas are risk ideas. It isn't natural to take risks. The mind will erase the idea, if you aren't careful.
- Don't examine the "why nots" in the early stages of an idea. Keep the creative—positive—process in motion. (Your superiors and inferiors will tell you "why not" in due time.)
- Keep projecting into the future. Write down the entire project: markets, colors, and styles. Go into great detail. You can cross out the bad points later.
- Cool off. Review your ideas the next day, but have good notes. It is not a time to try to remember, only a time to review the total project.
- Creative thought is a matter of skill, training, and practice.

Note: This story is reprinted by permission from the December, 1981, issue of "New Product Development," a newsletter published and copyrighted by New Product Development, P.O. Box 1309, Point Pleasant, NJ 08742

Source: "Keep Brain Circuits Clicking When Hit By New Product Idea," *Marketing News* (5 February 1982): 10.

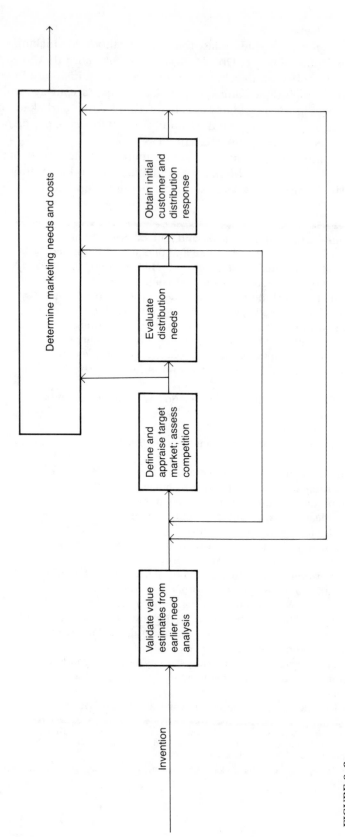

FIGURE 8-6
Market evaluation methodology

TABLE 8–8
Determining new product
success

- Were the company's last two products successful?
- Has the product been in development for a year?
- Does the company make a similar product?
- Does the company sell to a related customer market?
- What was the source of the product idea?
- Will the product be test-marketed?
- Can the company afford advertising?
- Will a recognized brand name be on the product?
- Can the decision to buy it be made by only one person?
- Is the product to be made in fewer than five versions?
- Will the product need service and repair?
- Will the product be on the market for more than ten years?

this point to know that the innovation process must be closely aligned with the market. There is nothing worse than having a good innovation without a market. This danger can be minimized by focusing on the market early in the innovation process by defining and appraising the target market and assessing competition. This often takes the form of a determination of the market needs and associated costs (see Fig. 8–6). One critical area frequently overlooked in the assessment process is the distribution system and its requirements and costs. Frequently modifications in the innovation or its package need to occur before the product is acceptable to the distribution system. One new dog-treat product, for example, was not carried by the supermarkets until it was introduced in a package size consistent with shelf-height requirements of the pet food aisle and other dog-treat products.

A final method for maximizing the innovation process is employing the criteria for determining new product success, as indicated in Table 8–8. This includes asking such questions as "Does the company make a similar product?" "Does the company sell to a related market?" "Can the decision to buy the innovation be made by only one person or a committee?" "Will the product have a few or many versions?" "Will the product need service and repair?" By assessing each innovation in this way, necessary modifications can be made early in the development process. When management understands and employs creative problem-solving techniques and develops methods for maximizing the innovation process, product innovation will be a much more successful endeavor.

SUMMARY

An area of particular importance for successful product planning and development is product innovation. Product innovation requires creativity on the part of the individual and company. Since creativity declines with both age and lack of use, it is often beneficial to use one of the many possible creative problem-solving techniques to unlock and stimulate creativity and innovation. These techniques include the more commonly used techniques of attribute listing, big dream approach, brainstorming, reverse brainstorming, free association, collective notebook method, checklist method, and parameter analysis. The end result of using one of these or a lesser used technique is a unique product or service that should be further evaluated and refined in the product planning and development process.

SELECTED READINGS

DeBONO, EDWARD. *Lateral Thinking for Management: A Handbook of Creativity.* New York: American Management Association, 1971.
Describes lateral and vertical thinking. Lateral thinking is emphasized and its concepts detailed. Chapters on judgment, criticism, change, and discontinuity include useful summary on the methodology eschewed. Written by one of Britain's foremost scholars in creative problem-solving.

GORDON, W. J. J. *Synectics: The Development of Creative Capacity.* New York: Harper & Row Publishers, Inc., 1961.
Objective is to develop an operational concept of human creativity based solely on synectics as an approach. Applies the synectic method of joining together different and apparently irrelevant elements. Includes chapters on synectics history, operational mechanisms, its role in industry, and the place held by play and irrelevance. Completes the discussion with a description of social applications.

HAEFELE, J. W. *Creativity and Innovation.* New York: Reinhold Publishing Co., 1962.
Discusses nature and theories of creativity as they apply to business situations. Gives a review of the literature, introduces new methods, illustrates ways in which managers can enhance creativity, and discusses group and individual aids. Includes chapters on theory, process, and related issues.

OSBORN, ALEX F. *Applied Imagination.* Rev. ed. New York: The Scribner Book Companies, Inc., 1957.
Aimed at illustrating how individuals can be trained to use creative talent through courses in creative problem-solving. Includes chapters on forms of imagination, factors that limit creativity, development of creative processes, and individual ideation vs. team collaboration.

PARNES, SIDNEY J. and HARDING, HAROLD F. *A Source Book for Creative Thinking.* New York: The Scribner Book Companies, Inc., 1962.
A collection of 29 articles and addresses plus 75 research summaries that examine from various viewpoints the area of creative thinking. Incorporates sections on the need for creative education, the concept of creative talent, research on the identification and development of creativity, operational procedures for problem-solving, and descriptions of several educational programs for the development of creative problem-solving ability.

POLYA, GYORGY. *How to Solve It.* Princeton, N.J.: Princeton University Press, 1945.
Directed toward teaching how to solve problems mathematically. It includes a lengthy and detailed section on heuristics, which is in dictionary format. While mathematical in nature, it is of interest to those concerned with invention and discovery.

SOUDER, WILLIAM E. and ZIEGLER, ROBERT W. "A Review of Creativity and Problem Solving Techniques." *Research Management* (July 1977): 34–42.
Gives short descriptions of over 22 different creative problem-solving techniques including, for example, brainstorming, synectics, attribute listing, free

association, collective notebook, Kepner-Tregoe, heuristics, and problem inventory analysis. Includes an extensive reference list of cited literature.

STRYKER, PERRIN. "Can You Analyze This Problem?" *Harvard Business Review* 43, no. 3 (May–June 1965): 73–78.
Provides examples of the Kepner-Tregoe method of problem analysis. They deal with an examination of a real-life situation using the K-T training technique as it was applied to a group of managers.

TAUBER, EDWARD M. "HIT: Heuristic Ideation Technique." *Journal of Marketing* 36, no. 1 (January 1972): 58–70.
Describes in detail the heuristic ideation technique (HIT)—that is, generating all relevant concepts and combination of ideas that can be associated with a product. Discusses the suggesting of an idea and its subsequent breakdown into parts; sometimes so many suggestions are received that it becomes difficult and costly to reduce them to manageable quantities. Relates HIT to morphological analysis.

WHITING, CHARLES S. *Creative Thinking.* New York: Reinhold Publishing Co., 1958.
Examines creativity in a common-sense, objective manner. Looks especially at how creative training methods can provide an organization with creative and communications' benefits. Focuses on operational techniques, including group discussion methods and analytical and mechanical ways to produce new ideas.

9

Introduction to the Product Planning and Development Process

A s we have stressed throughout this book, a constant flow of new products—or at least improvements in existing products—is necessary for a firm to continue to profit. The need for new profits, coupled with the high rate of failure of new products, requires a continuous search for sources of new product ideas, as well as for new product evaluation at all stages in the development process. The techniques of creative problem solving discussed in Chapter 8 provide methodologies for product innovation. This chapter discusses the major stages in obtaining and evaluating new products, sources of new product ideas, and the role of models in new product evaluation and planning. In addition, it provides some specific new product models and examines their usefulness.

THE PRODUCT PLANNING AND DEVELOPMENT PROCESS

Although the product planning and development process varies from industry to industry, as well as from firm to firm within a given industry, its activities generally follow the pattern indicated in Figure 9–1. The process can be divided into five major stages: idea stage, concept stage, product development stage, test marketing stage, and commercialization.

In the idea stage, suggestions for new products are obtained from all possible sources—all available devices for generating new product ideas should be employed. Frequently one of the creative problem-solving techniques discussed in Chapter 8 can be used to develop marketable ideas. These ideas should be carefully screened to determine which are good enough to require a more detailed investigation. It is important for the company to establish objectives and define growth areas to provide a basis for this analysis.[1]

Ideas passing the initial screening enter the concept stage, where they are developed into more elaborate product concepts. In evaluating the product concept, each company should keep in mind its own strengths and weaknesses, as well as the needs of potential buyers. A tentative business plan consisting of product features and a marketing program should then be developed and a sample of potential buyers should be presented with the concepts for evaluation. This presentation can be verbal or pictorial; however, the latter will generally produce more accurate results.

Once the concept for the new product has been approved, it is developed into a prototype and tested. This is the product development stage, in which the technical and economic aspects of the potential new product are explored by assigning specifications for the development process to members of research and development. Unless excessive capital expenditures make it impossible, laboratory-tested products should be created at this point. These products can then be produced on a pilot run basis, which will allow for production control and product testing. The products are then evaluated by in-use consumer testing, which will determine whether the potential new product has features superior to products currently available.

---

[1]The planning and evaluation process in the idea, concept, and product development stages is discussed in Chapter 4; the activities in the test marketing stage are presented in Chapter 5.

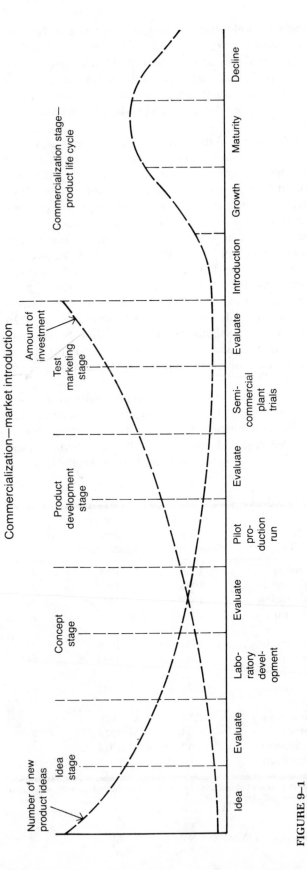

FIGURE 9–1
The product planning and development process

The results of the product development stage form the basis of the final marketing plan for the new product. In some cases a market test is undertaken to increase the certainty of successful commercialization. This last step in the evaluation process—the test marketing stage—provides actual sales results indicating the acceptance level of consumers. Of course positive results cannot guarantee success, but they do indicate that the new product can be marketed profitably on a larger basis.

The feasibility of a market test and the extent of the evaluation in each stage of the product planning and development process depends to a great extent on the product, market, and competitive situation confronting the firm. These factors must be evaluated in terms of the time elapsing between idea generation and commercialization and the costs involved. Time and cost are then weighed against the high costs of unsuccessful commercialization. Even though great technological advances may occur, they are not usually quickly transformed into finalized new products. As Table 9–1 indicates, the average time elapsing between conception and new product realization for the ten sample products investigated was 19.2 years. This time difference varied from a low of six years for the video tape recorder to a high of 32 years for the heart pacemaker. It does not appear that this time period is becoming shorter. In fact the government is so concerned about the length of time before new products are commercially available once an idea has been generated that some federal funding has been committed to research in the area of technology transfer.

The long time period in the product planning and development process must also be carefully weighed against the costs associated in the evaluation process and the costs of commercialization. Even though eliminating 90 percent of available ideas, the idea and concept stages still take very little time and money—only about 10 percent of the total time and expenditures involved

TABLE 9–1
Duration of the innovative process for ten innovations

Innovation	Year of First Conception	Year of First Realization	Duration (Years)
Heart pacemaker	1928	1960	32
Hybrid corn	1908	1933	25
Hybrid small grains	1937	1956	19
Green revolution wheat	1950	1966	16
Electrophotography	1937	1959	22
Input-output economic analysis	1936	1964	28
Organophosphorus insecticides	1934	1947	13
Oral contraceptives	1951	1960	9
Magnetic ferrites	1933	1955	22
Video tape recorder	1950	1956	6
Average duration			19.2

Source: Battelle/Columbus Laboratories Study, report to the National Science Foundation (Washington, D.C., 1973).

in the entire product planning and development process. The product development stage requires about 30 percent of the total expenditures and 40 percent of the total time. On the other hand, the test marketing stage takes 12 percent of the total money and 20 percent of the time. These time and cost percentages must be carefully weighed against those associated with the commercialization stage, in which about 50 percent of the total expenditures occur in only 23 percent of the total time.[2] Regardless of the time and cost tradeoffs it makes, a company still has to actively develop sources for new product ideas so that the product planning and development process can continually lead to commercialization.

SOURCES OF NEW PRODUCT IDEAS

There are many possible sources of new product ideas. The major ones are customers, competition, distribution channels, federal government, research and development, the company sales force, upper-level management, and employee suggestions.[3]

Customers

Companies are paying more and more attention to what should be the focal point of new products—their customers. This can take the form of monitoring ideas received on an informal basis or formally arranging for customers to have an opportunity to express their opinions. One company that has an informal format for monitoring customer ideas is General Foods.[4] General Foods receives about 80,000 letters from consumers annually. At one point, a consistent complaint occurred regarding the size and shape of the cereal box. Generally, the complaint was that the cereal boxes were too tall to stand upright on standard kitchen shelves and tipped over easily. These complaints led to the development of a new, more compact package.

Kimberly-Clark Corporation, on the other hand, formally monitors customer opinions. Groups of consumers meet and discuss the relative advantages and disadvantages of the company's products. From one series of discussions, it became apparent that some consumers felt the tissue papers should be larger. This sparked the development and eventual introduction of *Man Size Kleenex*.[5]

Competition

Companies should also establish a formal procedure for monitoring the new product activities of their competitors. This procedure should have two fac-

---

[2]These time and cost estimates are presented in *Management of New Products* (New York: Booz, Allen and Hamilton, 1968), 10.
[3]There are many comprehensive listings for sources of new product ideas. See, for example, Peter Hilton, *Handbook of New Product Development* (Englewood Cliffs, N.J.: Prentice-Hall, Inc., 1961), 9–16; and U.S. Chamber of Commerce, *Developing and Selling New Products* (Washington, D.C.: U.S. GPO, 1955), 3–14.
[4]"Helpful Consumers," *Wall Street Journal* 2, June 1965, 1.
[5]Ibid., 1.

ets. Information concerning forthcoming new products should be obtained from trade sources via the company's sales force. Also, when a competitive product is introduced, the sales performance should be carefully monitored, either by the company's sales force or through an established, outside research service.

Distribution Channels

Members of the distribution channels can be excellent sources for new product ideas. While channel members can come up with suggestions for a completely new product due to their familiarity with the needs of the market, a company can also achieve new product additions by analyzing its distribution system. In other words, a company can add new products in order to generate more merchandise for the stores handling its products.

A case in point occurred in the greeting card industry. The growth in this industry had been decreasing to such an extent that in the 1960s Hallmark, Gibson, American Greetings, Rust Craft, and Norcross expanded their product lines into related merchandise.[6] Hallmark introduced candles, paper plates, napkins, books, and costume jewelry. Each of these new products could be sold through the same outlets as the greeting cards. Hallmark successfully expanded its entire product line by taking advantage of the established distribution system.

Federal Government

The federal government can be helpful with new product ideas in two ways. First, the files of the Patent Office contain numerous new product possibilities. Although these possibilities, in themselves, may not be feasible product introductions, a patent can frequently be a stimulus that leads to other new product ideas. There are several government agencies and publications that can also be helpful. For example the *Official Gazette* published by the United States Patent Office weekly summarizes every patent granted and lists any patents available for license or sale.

The Government Patents Board publishes lists of abstracts of thousands of government-owned patents. An example is the *Government-Owned Inventories Available for License*. There are several government agencies, such as the Office of Technical Services, that assist manufacturers in obtaining specific product information.

Secondly, new product ideas can come from government regulations. For example the Occupational Safety and Health Act (OSHA), which was an attempt to eliminate unsafe working conditions in industry, mandated first-aid kits in business establishments. The act stipulated that every business establishment with three or more people must have a first-aid kit that contained specific items, depending on the industry. (The weatherproofed first-aid kit needed for a construction company, for example, was different than the one needed by a company manufacturing facial cream.) In response to OSHA, both established and newly formed manufacturers marketed a wide variety of first-

---

[6]"Hallmark Tries Out the Jewelry Trade," *Business Week* (3 November 1975): 29.

aid kits. For example, one newly formed company—R&H Safety Sales—was successful in filling the need for companies not in the construction industry.

Research and Development

Many new product ideas originate in the company's own research and development department. Studies show that this department is the source of more new product ideas than any other. In one survey, Booz, Allen, and Hamilton found that 88 percent of new product ideas came from within the companies themselves; of this, research and development contributed 60 percent. Realizing the potential of basic research, many companies have large research and development budgets.

One industry where technology from research and development has increased at an accelerated rate is the telephone industry. Innovations have made the 200,000 private brand exchanges throughout the nation almost obsolete. The new computerized exchanges not only make the office telephone more versatile and efficient, but also allow cost control.[7]

When considering research and development as a source of new products, it is important to differentiate various types of research. This allows research money to be allocated and the correct expectation of the probability of new product ideas to be established. Fundamental research is done purely for the sake of knowledge. Applied research consists of locating, classifying, and interpreting the results of fundamental research so that they can be used when problems arise. The last type of research—developmental research—is done to generate new product ideas, as well as adjust and find new users for present products. Since all three types of research are important for most companies, the research and development should be evaluated by classification so that this important source for new products can be better managed and controlled.

Company Sales Force

Given their grass-roots experience and constant contact with consumers, the company sales force is an excellent source of new product ideas. This is especially the case for industrial products, where the salesperson is in contact with the actual end-user. By observing trends and usage patterns and listening to customer complaints, the salesperson can often formulate a new product— or at least provide the direction for research and development to develop one. For example one sales manager of a small company observed the growth in the use of metallic films at Christmas. This led the company to develop metallic papers used in candy wrappings and ice cream bags.

Top Management

Top management can be another good source of new product ideas and should at least define general areas to explore for new products. These areas should be related to the strengths and weaknesses of the company and its product line. A company should search for new products that are complemen-

---

[7]"Technology Changes the Office Telephone," *Business Week* (19 January 1975): 42–44.

tary to its existing products. A cake-mix manufacturer could add frosting mixes to the product line, or a company with highly seasonal sales could introduce products with more stable sales. For example toy manufacturers could produce adult plastic products or toys.

Top management can provide new product ideas by examining existing products of the firm, reviewing abandoned projects, and considering the possibilities of by-products. When examining existing products, management should review such factors as the design, materials, packaging, and channels of distribution.

Similarly, by evaluating the raw materials, production facilities, or technical process the firm already uses, top management may get ideas for new products. For example, IBM and Digital Equipment Corporation modified the design of large computers to introduce smaller business and personal computers. Also, Anderson Little carries in their retail stores men's accessory products manufactured by other companies to complement their own line of men's suits.

Management should also check the by-products of the production process for potential product ideas. This was successfully done by a small manufacturer of wooden pallets. In response to an increased demand by landscapers, the company placed announcements in newspapers and local gardening magazines that it was a source of wood chips. This eventually became a substantive source of revenue for the firm.

Employee Suggestions

Another excellent internal source for new product ideas is employee suggestions. The importance of salespeople and top management as idea generators has already been discussed. Other employees can also be a very valuable source. In fact, three-fourths of the top 500 firms have formal employee suggestion procedures.

In order for this to be a viable source of new product ideas, certain things must be established. First the company needs to establish specific goals for the number of suggestions desired. A company must also help an employee to develop a feasible idea. Such incentives as cash, contests, or corporate recognition and praise are important for employees to actively think about potential products. Finally management must review all ideas from employees promptly and monitor the entire process. A corporation with a well-managed, formal employee suggestion system can gain many new product ideas.

METHODS FOR GENERATING PRODUCT IDEAS

Several methods have been developed to generate better ideas. While some of these have been discussed in the context of creative problem solving in the previous chapter, this is such an important factor that they will be discussed in terms of idea generation as well. These methods include focus groups, attribute listing, forced relationships, brainstorming, reverse brainstorming, and problem inventory analysis. Because good new product ideas come from diverse sources, these methods should only be viewed as mechanisms by which new product ideas can sometimes be more easily generated.

Focus Groups

Focus group interviews have been used in many aspects of marketing research since the 1950s. A focus group interview consists of a moderator leading a group of people through an open, in-depth discussion. This is much different from a group interview in which the moderator simply asks questions to solicit responses from participants. In the focus group, the moderator focuses the discussion of the group on the new product area in either a directive or a nondirective manner.

In addition to generating a new idea, the focus group is an excellent means for the initial screening of ideas and concepts. Recently several procedures have been developed so that more quantitative analyses can be used in interpreting the focus group results. With increased use, and the development of more procedures for evaluation and quantification of the results, the focus group is becoming an increasingly valuable method for generating new product ideas.[8]

Attribute Listing

This technique consists of listing the existing attributes of a product idea or area; these attributes are then modified until a new combination of attributes emerges that will improve the product idea or area. A small company manufacturing pallets used for shipping or moving a product on a conveyor along an assembly line wanted to devise a better product. It listed the attributes that defined the existing pallets, such as wood composition, rectangular runners, and accessibility from two sides by a fork lift. It then examined each attribute for any change(s) that would improve the product. For example the wood composition could be changed to plastic, resulting in a cheaper price; and the rectangular wooden runners could be replaced by cups, making the pallets easier to store and allowing them to be accessed from all four sides for easier pickup. Through attribute listing, this small company achieved an idea for a much-improved product.

The following questions are often used in attribute listing to generate new product ideas:

- Put to other uses?   New ways to use as is? Other uses if modified?
- Adapt?   What else is this like? What other idea does this suggest? Does past offer parallel? What could I copy? Whom could I emulate?

---

[8]For a more in-depth presentation on focus group interviews in general and quantitative applications, see "Conference Focuses on Focus Groups: Guidelines, Reports, and 'the Magic Plaque,'" *Marketing News*, 24 October 1975, 1; Alvin J. Rosenstein, "Quantitative—Yes Quantitative—Applications for the Focus Group," *Marketing News*, 21 May 1976, 8; Keith K. Cox, James B. Higginbotham, and John Burton, "Application of Focus Group Interviews in Marketing," *Journal of Marketing*, 40, no. 1 (January 1976): 77–80; and Robert D. Hisrich and Michael P. Peters, "Focus Groups: An Innovative Marketing Research Technique," *Hospital and Health Services Administration* 27, no. 4 (July/August 1982), 8–21.

■ Modify?	Change meaning, color, motion, sound, odor, form, shape? Other changes?
■ Magnify?	What to add? More time? Greater frequency? Stronger? Higher? Longer? Thicker? Extra Value? Plus ingredient? Duplicate? Multiply? Exaggerate?
■ Minify?	What to subtract? Smaller? Condensed? Miniature? Lower? Shorter? Lighter? Omit? Streamline? Split up? Understate?
■ Substitute?	Who else instead? What else instead? Other ingredient? Other material? Other process? Other power? Other place? Other approach? Other tone of voice?
■ Rearrange?	Interchange components? Other pattern? Other layout? Other sequence? Transpose cause and effect? Change pace? Change schedule?
■ Reverse?	Transpose positive and negative? How about opposites? Turn it backwards? Turn it upside down? Reverse roles? Change shoes? Turn tables? Turn other cheek?
■ Combine?	How about a blend, an alloy, an assortment, an ensemble? Combine units? Combine purposes? Combine appeals? Combine ideas?[9]

The one major drawback to attribute listing is that it focuses on the product at hand. It cannot be used in all new product situations. It may even stifle imaginative thinking to some extent. Yet, as in the case of the pallet manufacturer, it is often a useful method for developing a new product idea.

Forced Relationships

A third method for generating new product ideas is forced relationships.[10] In this technique many new ideas are first listed. Then, as the name implies, the new product ideas are considered in pairs. By considering one idea in relation to every other, new ideas are often generated. Even though this technique is not in wide use, it is a good, systematic procedure to see whether there are any new products that stem from a combination of existing products. These new products would then naturally fit into the existing product line and management expertise.

Brainstorming

This technique evolves from the belief that people can be stimulated to greater creativity by meeting with others and participating in organized group experiences. Top management of a company meet frequently in small groups of between six and ten to generate new product ideas. Many of the results are absurd and have no basis for consideration for development, but this method often produces a large number of ideas. This is especially true when the

---

[9]Alex F. Osborn, *Applied Imagination*, 3rd ed. (New York: The Scribner Book Companies, Inc., 1963), 286–7.
[10]For an application of this technique, developed by Charles S. Whiting, ibid., 213–14.

meetings, lasting about an hour, focus on a specific area. There are four rules to be followed for most effective use of management brainstorming:

1. No criticism. Negative judgments must be withheld until later.
2. Freewheeling is encouraged. The wilder the idea, the better; it is easier to tame down than to think up.
3. Quantity is wanted. The greater the number of ideas, the more the likelihood of useful ideas.
4. Combinations and improvements are sought. In addition to contributing ideas of their own, participants should suggest how ideas of others can be used to produce still another idea.[11]

It is also particularly important that no participant be an expert in the field or have any other trait that would stifle the group. In keeping with this notion, meetings should be play- rather than work-oriented sessions.

Reverse Brainstorming

This approach is a modification of brainstorming. The objective of reverse brainstorming is to take a particular product, such as a dishwasher, and generate a list of its shortcomings. This list of negative attributes then provides the direction for discussion on new products and product improvements. The general advantages and disadvantages of this approach are similar to the brainstorming technique. The major limitation is that the approach focuses on the problems of a product as perceived by management. These may or may not be important problems for consumers.

Problem Inventory Analysis

This is a recently developed method of generating new product ideas that uses consumers in a manner analogous to focus groups. However, instead of being asked to generate new product ideas themselves, consumers are provided with a list of problems from a general product category, such as food, clothing, or cosmetics. They are then asked to identify and discuss products that have this particular problem. This method is often more effective than a focus group because it is much easier for consumers to relate known products to suggested problems and arrive at a new product idea than to generate an entirely new product idea with minimal guidance. This approach is also an excellent way to test a new product idea.

An application of this approach in the food industry appears in Table 9–2. One of the most difficult aspects of this approach is developing an exhaustive list of all the problems. In food we see that such attributes as weight, taste, appearance, and cost affect a consumer's decision. Once a list of problems is developed, participating consumers can usually indicate products associated with each problem.

Results from product inventory analysis must be carefully evaluated as they may not represent new product opportunities. For example, in a study of a food product, 49 percent of the respondents mentioned that a package of cereal did not fit on the shelf well. Yet General Foods' introduction of a small,

---

[11]Ibid., 156.

TABLE 9–2
Problem inventory analysis

Physiological	Sensory	Activities	Buying Usage	Psychological/Social
A. Weight · fattening · empty calories	A. Taste · bitter · bland · salty	A. Meal planning · forget · get tired of it	A. Portability · eat away from home · take lunch	A. Serve to company · would not serve to guests · too much last minute preparation
B. Hunger · filling · still hungry after eating	B. Appearance · color · unappetizing · shape	B. Storage · run out · package would not fit	B. Portions · not enough in package · creates leftovers	B. Eating alone · too much effort to cook for oneself · depressing when prepared for just one
C. Thirst · does not quench · makes one thirsty	C. Consistency/texture · tough · dry · greasy	C. Preparation · too much trouble · too many pots and pans · never turns out	C. Availability · out of season · not in supermarket	C. Self-image · made by a lazy cook · not served by a good mother
D. Health · indigestion · bad for teeth · keeps one awake · acidity		D. Cooking · burns · sticks	D. Spoilage · gets moldy · goes sour	
		E. Cleaning · makes a mess in oven · smells in refrigerator	E. Cost · expensive · takes expensive ingredients	

Source: Edward M. Tauber, "Discovering New Product Opportunities with Problem Inventory Analysis," *Journal of Marketing* (January 1975), 69. Reprinted from *Journal of Marketing*, published by the American Marketing Association.

compact cereal box was not successful. It appears that the perceived problem of package size actually had very little effect on actual purchasing behavior. Problem inventory analysis should be used primarily to identify product ideas for further investigation. These ideas should then be studied in depth to determine their importance to consumers.

QUALITATIVE SCREENING CRITERIA

The new product planning and developing process requires continual monitoring of all products in the product line, as well as changes in the social and competitive environments. Companies marketing both consumer and industrial products often rely more on executive judgment than anything else in evaluating individual product proposals and determining their marketability. Although judgment is probably the single-most important element in the new product planning and evaluation process, it is better to facilitate that judgment by systematizing and quantifying as many of the decision elements as possible.

It is particularly important to continually rate the product and evaluate the market and financial criteria of each of the potential new products. This rating is most effective when a standard form is used. Table 9–3 illustrates a form that can be generally applied in any company or product situation. Each qualitative screening criterion is rated from superior to poor and is assigned a standard (weight) depending on its importance in successful commercialization. The three major criteria are then further delineated into specific characteristics relevant to the company situation. For example specific product criteria can include product uniqueness, use of existing facilities and skills, impact of changing economic conditions, and availability of raw materials. Once the forms have been filled out by selected company managers, the combined ratings should give an indication of the probable market success of the proposal being evaluated.

The need to carefully screen each new product proposal quantifying the evaluation procedure to the extent possible has led to an increasing interest in establishing and using models for analyzing and predicting sales and profits for each new product proposal. Models are also used in evaluating variations of the introductory marketing plan for commercialization. In addition to their use in evaluation, marketing models can contribute to the marketing orientation of the firm, thereby improving the company's new product performance.

TYPES OF MARKETING MODELS

There has been a growing interest in using models for evaluating and marketing new products. A model is basically a representation or abstraction of a real-world system and is usually a logical representation of a problem. Since a model is an abstraction, it is not a perfect representation of the real-world phenomonen. Therefore the model is always easier to understand and manipulate than the real world phenomenon represented. In designing and implementing a model for new product planning and evaluation, a balance must be achieved between the completeness of the model (a measure of its validity) and its utility.

TABLE 9–3
Qualitative screening criteria

Criteria	Standard or Weight	Superior	Excellent	Good	Fair	Poor	No Opinion
Product criteria							
Product uniqueness							
Use of existing facilities and skills							
Patent position							
Servicing requirements							
Technical feasibility							
Technical know-how							
Legal considerations							
Organizational support							
Seasonality							
Impact of changing economic conditions							
Availability of raw materials							

TABLE 9–3
(continued)

Criteria	Standard or Weight	Superior	Excellent	Good	Fair	Poor	No Opinion
Market criteria							
Market size							
Market growth potential							
Customer need							
Effect on existing product line							
Distribution requirements							
Market life							
Competitive advantage							
Financial criteria							
Cost of entry							
Profit contribution							
Effect on cash flow							
Payback							
Return on investment							

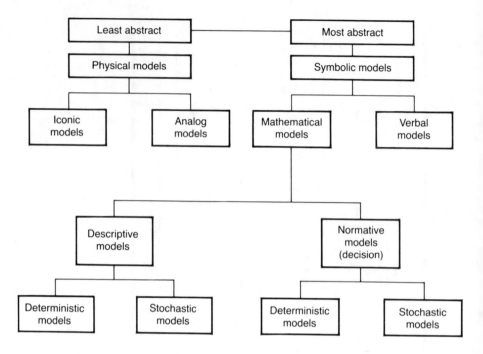

FIGURE 9–2

Types of marketing models

Source: Adapted from James H. Donnelly and John M. Ivancevich, *Analysis for Marketing Decisions* (Homewood, Ill.: Richard D. Irwin, Inc., 1970), 16, 18.

There are several ways to classify models. Perhaps one of the earliest classification schemes developed is based on the degree of abstraction of the models.[12] As indicated in Figure 9–2, the least-abstract models are physical ones—models that give the appearance of the actual system they represent. Although these models are easy to describe and to observe, they are very difficult to manipulate and are therefore not generally used in new product analysis and prediction. Depending on their appearance and behavior, physical models can be classified into two groups: (*a*) iconic, which appear like, but do not behave like, the system being represented and (*b*) analog, which behave like, but do not appear like, the system being represented. An example of an iconic model for a new product would be alternative layouts for a new store. An organizational chart of a new company would be an example of an analog model.

Figure 9–2 indicates two types of symbolic models. Symbolic models are the most abstract and are called *mathematical models* and *verbals*. While a verbal model is a written description of the system, a mathematical model is a representation of a detailed system in equation form. A mathematical model often uses symbols to describe the same variables and relationships described in the verbal model. DEMON, SPECS, and SPRINTER are examples of new product mathematical models that are described later in this chapter.

---

[12]James H. Donnelly and John M. Ivancevich, *Analysis for Marketing Decisions* (Homewood, Ill.: Richard D. Irwin, Inc., 1970), 13–27.

Mathematical models can be further classified as either descriptive or normative. A descriptive model describes a particular phenomenon under study, such as the proposed distribution system for the new product. This model is used to provide an understanding of the process; it does not make any judgment concerning its applicability, nor does it attempt to determine the best alternative. On the other hand a descriptive model becomes a normative or decision model when it is used to evaluate several new product decision alternatives. A normative model is constructed to enable the company to select, from several alternatives, the best one for the new product under consideration.

Finally both descriptive and normative models can be either deterministic or stochastic in nature. When a new product model is deterministic, no chance comes into play. All of the factors surrounding the new product are considered to be exact and have determinate quantities. Although assuming that all factors are deterministic simplifies the model-building process, this is usually not an accurate portrayal of the reality of new product introduction. Therefore a more realistic new product model is a stochastic one. In a stochastic model conditions of uncertainty regarding the successful introduction of the new product are introduced. The uncertainty reflects data on real occurrences to the extent possible.

New Product Model Development

Any manager involved in the new product development process can either develop or aid in the development of a new product model. This would involve helping to identify the constructs and relationships involved in the planning and marketing process. There are four primary steps in developing a new product model:[13]

1. Define and formulate the problem.
2. Construct the model.
3. Test the model and develop controls.
4. Implement the solution.

As was the case in creative problem solving, probably one of the most difficult tasks in developing a new product model is to define and formulate the problem in a meaningful way. This process lays the foundation for all subsequent steps. If the problem is not succinctly defined and formulated, then the model will be of little value in aiding in the planning, evaluation, or marketing of the new product. One important process in problem definition is making the complex problem of a new product as manageable as possible. This simplification process should be done to expedite model building without eliminating any important variables. Three methods of simplifying the problem can be employed. The first way is to assume certainty. In some new products it may be feasible to assume that some, if not all, the variables are deterministic in nature; that is, chance does not come into play. Even if this assumption cannot be employed in the final new product model, it can be made at first to ease model development.

---

[13]This discussion is based in part on material found in Donnelly and Ivancevich, *Marketing Decisions*, 21–27.

Simplifying relationships is another way to reduce the complexity of the problem in model building. The relationships can usually be assumed to be less complex than they actually are in the new product situation being modeled. For example, linearity can be assumed (even if only on an initial basis) when the relationships among the variables may be actually nonlinear.

A final way to make the problem more manageable is to isolate the operations involved in the new product planning, evaluation, and introduction processes. Each part of the problem being modeled should be isolated and broken down into its smallest components. These can later be combined in the final model if needed. In our experience with building models in the new product area, it has always been easier to combine elements than to break them down at a later date once the model has been developed.

Once the problem has been delineated, then the model must be constructed. A model is basically composed of constructs and relationships. There are three types of constructs: inputs, intervening, and output. Input constructs are those controllable and uncontrollable elements that are the basic components of the model. These would include such things as advertising expenditures, new product, price, and competitive prices. Intervening constructs are those that relate the input constructs and often describe the state of a component. Consumer attitudes, stage in the family life cycle, and store traffic are examples of intervening constructs. Output constructs are factors such as sales, costs, demand, or profit—the results of the new product model.

Aside from constructs, relationships are the other basic part of every new product model. Basically these specify how the constructs are related to each other. Through these relationships, the model reflects the new product marketing process.

After construction, the model must be tested before it is applied to the firm's new product operation. From this test, the effectiveness of the model can be determined on a small scale. If necessary the model can then be altered before final implementation. At the same time the necessary controls for the new product model must be established. Since changes are constantly occurring in the dynamic environment surrounding new product planning and introduction, tight controls are necessary to ensure that the external environment and the variables and relationships in the model have not dramatically altered, thereby rendering the new product model results inappropriate.

The final step in new product model development is implementation of the model in the company's operations. This is often a more difficult task than it would seem. Some managers have apprehensions about the usefulness of any models. This apprehension can be eased to some extent by making sure the managers understand the objectives, assumptions, functions, and limitations of the model. In particular, the advantages gained by implementing the model and the model's solution should be pointed out.

SPECIFIC NEW PRODUCT MODELS

Several new product models have been developed, including DEMON, SPRINTER, STEAM, depth of repeat forecasting, NEWS, SPECS, PROBE, and

NEWPROD. The following paragraphs discuss each of these in terms of composition and output.

DEMON

Decision Mapping via Optimum GO NO Networks (DEMON) is a basic new product model built on the premise that product decisions are made in an information system in which the choice is either full-scale product development (GO), investigate more (ON), or reject the new product (NO).[14]

As indicated in Figure 9–3, a GO decision means that the company is committed to the new product and can engage in either product or market testing before full-scale introduction. When the ON decision is reached, more information on the new product is needed before a GO or NO decision can be made. When this is the case, the process is repeated. When a NO decision occurs, the new product process is discontinued.

In the DEMON model, three controllable marketing variables affect the number of consumers trying the new product: advertising, sales promotion, and distribution. The profits for the new product are obtained by estimating the relationship between these variables and sales over the estimated life of the product less the cost of sales. Uncertainties are incorporated into the model by establishing confidence limits for the best yearly sales estimates.

The DEMON model can provide good insight into the process of consumer acceptance of new products by optimizing the marketing variables of the new product situations in which the inputs can be quantified.

SPRINTER

Specifications of Profits with Interaction under Trial and Error Response (SPRINTER) is characterized by the classic GO-NO-GO decision through the use of an information network.[15] It is composed of three primary elements—demands, cost, and profit. The model develops an equation that describes the relationships between these elements, which maximizes the expected value of the total discounted differential of profit and uncertainty. This equation is solved through a computer simulation based on a trial-and-error technique that maximizes profits under various scenarios of price, advertising, distribution, and competition.

The inputs of the model consist of various price levels, advertising expenses, distribution efforts, and life cycle. Through use of four submodels (demand, cost, differential profit, and uncertainty), the output of GO (accept

---

[14]For a thorough presentation of DEMON see David B. Lerner, "DEMON New Product Planning: A Case History," *New Directions in Marketing*, Proceedings of the American Marketing Association, 48th National Conference, June 1965 (Chicago, Ill.: American Marketing Association, 1965), 489–508; A. Charles et al., "DEMON; Decision Mapping via Optimum GO NO-GO Networks—A Model for Marketing New Products," *Management Science* 12, no. 11 (July 1966): 865–88; and "DEMON: MARK II: External Equations Approach to New Product Marketing," *System Research Report* no. 110 (Evanston, Ill.: Northwestern University Press, 1965).

[15]The SPRINTER model is presented in Glen L. Urban, "SPRINTER: A Tool for New Product Decision Makers," *Industrial Management Review* 8, no. 2 (Spring 1967): 43–54; and Glen L. Urban, "SPRINTER MOD III: A Model for the Analysis of New Frequently Purchased Consumer Products," *Operations Research* (September 1970): 805–54.

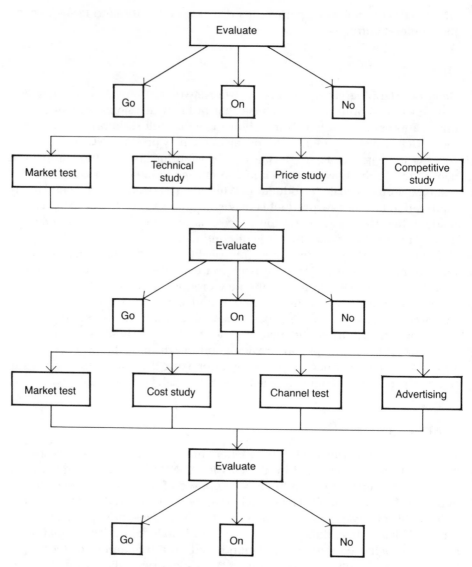

FIGURE 9–3

Decision networks for new product evaluation in the DEMON model

Source: David B. Montgomery and Glen L. Urban, *Management Science in Marketing*. Englewood Cliffs, New Jersey. (Reprinted by permission of Prentice-Hall, Inc., 1969): 312.

the new product) or NO GO (reject the new product) is based on the probability of achieving the company's target rate of return.

Three basic versions of SPRINTER are MOD I, II, and III. SPRINTER MOD I describes the acceptance process of a new consumer product through trial, repeats, and loyalty. SPRINTER MOD II is a more detailed version that includes awareness levels and links price and advertising to the diffusion process. Sales and profits forecasts are generated for alternative strategies of the new product and its competitive products. Additional controllable variables, as well as word-of-mouth communication, are included in SPRINTER MOD III.

In providing a view of the consumer acceptance process of new products, SPRINTER has three major advantages. First, it considers the impact of a competitive environment, as well as interactions of the new product with existing products. Second, the model optimizes the marketing variables of the new product used in inputs. Finally, it allows the product manager to ask what-if questions and vary the constraints. A major disadvantage of the model is that it must have quantitative inputs for which it can be very expensive to collect data.

STEAM

STEAM is a depth-of-trial-class model for new convenience products.[16] It describes the propensity of consumers to enter various product classes. This is called depth-of-trial class. Generally it is a dynamic model that presumes population heterogeneity. The model uses such factors as average product use and time since last purchase. Future purchase patterns of individual households are simulated; then the results are projected to the total market. Test market data on trials and repeat purchases of households in a consumer panel serve as inputs. From this, the model produces not only the depth-of-trial-class data but simulates future sales data as well.

The STEAM model can be advantageously used for long-range forecasts of frequently purchased products. However it does require consumer panel data on a household basis and does not take into account any promotion efforts employed after the introduction of the new product.

"Depth of Repeat" Forecasting

When making early forecasts of demand for new consumer products, a good approach is to use a depth of repeat forecasting model.[17] This model, similar to one of the earliest-developed formal sales projection models, plots the values of sales of the new product over time and determines which extension of the sales curve is most likely to occur.[18] The inputs for this model are data from consumer diary panels. The model then determines whether the sales of the new product will grow, level out, or decay based on the initial curve.

NEWS

The New Product Early Warning System (NEWS) model, developed at Battan, Barton, Durstine, and Osborn, Inc., predicts the performance of a new product

---

[16]See William F. Massy, "Forecasting the Demand for New Convenience Products." *Journal of Marketing Research* 6, no. 4 (November 1969): 405–12; William F. Massy, "Stochastic Models for Monitoring New Product Adoptions," in *Applications of the Sciences in Marketing Management*, eds. Frank Bass, Charles King, and Edgar Pessemier (New York: John Wiley & Sons, Inc. 1968), 85–111; and William F. Massy, David B. Montgomery, and Donald G. Morrison, *Stochastic Models of Buying Behavior* (Cambridge, Mass.: The MIT Press, 1970), 326–61.

[17]This approach is discussed in Gerald J. Eskin, "Dynamic Forecasts of New Product Demand Using a Depth of Repeat Model," *Journal of Marketing Research* 10, no. 2 (May 1973): 115–29.

[18]For a discussion of this early sales projection model, see Louis A. Fourt and Joseph W. Woodlock, "Early Prediction of Market Success for New Grocery Products," *Journal of Marketing* 25, no. 2 (October 1960): 31–38.

in a test market. The inputs for the model are such factors as amount of advertising, number of consumers exposed and number of exposures, brand awareness, product use rate, and probability of a repeat purchase. From these inputs, the model predicts a trial level for the new product, market share by purchase cycle, and short-term sales.

SPECS

The Strategic Planning, Evaluation, and Control (SPECS) model incorporates the Ayer New Product model for use in planning the marketing of new consumer products.[19] This model allows alternative assumptions and strategies to be evaluated prior to the actual test marketing of the new product. By putting in various levels of product, media, promotional weight, distribution, price and costs, the model forecasts probable market reaction, profit-and-loss projections, discounted cash flow, and the payback. This model is unique in that it can provide valuable information as early as the concept stage of a new product. At this stage, the inputs are, of course, subjective and are derived from norms from previously marketed new products. When data on the new product concept itself are available, these can be used as inputs to produce more product-specific results.

The model differs from many of the previously discussed new product models in that its output is not a GO or NO GO decision, but rather various forecasts for the new product. The one limitation is that the model can be successfully implemented only when the company has had previous experience with a similar new product.

NEWPROD

NEWPROD is a model for consumer nondurables that predicts market share for the first year of national market introduction.[20] It does this by simulating the number of potential buyers who are at various stages in the adoption process. As was the case with many of the new product models discussed, NEWPROD allows the marketing manager to determine the sensitivity to input variables because various combinations of key decision variables can be used. The model incorporates key internal and external variables influencing the new product's market, such as advertising expenditures, number of samples and coupons distributed, and awareness, trial, and repeat-purchase rates. These variables, based on actual data from the test market and subjective estimates from the new product manager, are inputs used to predict the market share primarily through regression analysis.

---

[19]This model is described in Henry J. Claycamp and Lucien E. Liddy, "Prediction of New Product Performance: An Analytical Approach," *Journal of Marketing Research* 6, no. 4 (November 1969): 414–20.

[20]See Gert Assmus, "NEWPROD: The Design and Implementation of a New Product Model," *Journal of Marketing* 39 (January 1975): 16–23.

The NEWPROD model can be used to predict new product success before any expensive test marketing is undertaken. Another advantage is that only those variables considered critical to the new product are used in determining market share. In addition, such data are easily obtained, facilitating the measurement of the effectiveness of alternative decision strategies.

USES AND LIMITATIONS OF NEW PRODUCT MODELS

One purpose of any new product model is to predict or provide a better understanding of the new product planning and marketing system. Relevant constructs (both internal and external) can be delineated and the nature of the relationships between them can be specified through the use of models. Models can be used to forecast probable sales of the new product and indicate which of the decision alternatives would be best suited to the specific product market context.

A good new product model also serves as an information organizer. For example any information regarding new product sales under certain competitive reactions can be summarized in an orderly fashion. This often aids in relevant discussion of various decisions regarding the new product. By providing a system for organizing information, the new product model guides the marketing research undertaken as it identifies areas in need of information.[21]

Although a new product model can aid in new product planning, evaluation, and marketing, it may not, in all instances, bring more precision to the solution. The usefulness of the new product model is dependent upon the logic with which the model was constructed, as well as the accuracy of the information put in. Four rules can make the use of a model more effective:

- Making the methodology explicit and easily communicated, with all assumptions clearly spelled out
- Basing the methodology on solid theory and testing
- Basing the model on hard, accurate information
- Validating the model on the company's own brands, since no reliance on outside case histories is likely to be convincing
- Better and more productive methodologies should continually be sought to avoid too narrow or inbred an approach and to encourage probes into creative new areas.[22]

By implementing these four rules, the new product manager can construct a model, which at the lowest level, can make the thinking and decision-making process more rigorous. Through the model various possible decisions can be evaluated with the best decision being selected and implemented.

---

[21]For a discussion of this aspect of model building from a general viewpoint, see David A. Aaker and Charles B. Weinberg, "Interactive Marketing Models," *Journal of Marketing* 39, no. 4 (October 1975): 16–23; and David B. Montgomery and Charles B. Weinberg, "Model Marketing Phenomena: A Managerial Perspective," *Journal of Contemporary Business* 2, no. 4 (Autumn 1973): 17–43.

[22]Michael J. Naples, "First Understand Assumptions Behind Models," *Marketing News*, 19 November 1976, 4.

SUMMARY

The product planning and development process provides the basis for new product development. As a proposed new product idea moves through the major stages of the process—idea, concept, product development, test marketing, and commercialization—various factors must be constantly evaluated and monitored.

The ability of the firm to obtain and evaluate new product ideas is fundamental to its continuing success. Sources, such as consumers, distribution channels, government laws, and internal company resources, should be continuously monitored and screened for potential new products. To better use these sources, several methods for generating new product ideas can be employed, including focus groups, attribute listing, forced relationships, management brainstorming, reverse brainstorming, and problem inventory analysis.

The decision on whether a new product should be further developed or which particular marketing variable should be implemented in the product's introduction is often aided through the use of a new product model. New product models vary in their degree of abstraction. Physical models are the least abstract and symbolic models are the most abstract. Regardless of the type, the new product model is built on a definition of the problem, construction of the model, establishment of parameters, tests of solution, and implementation.

There are many new product models, such as DEMON, SPRINTER, STEAM, depth of repeat forecasting, SPECS, PROBE, and NEWPROD. Each has certain parameters and outputs that make its usefulness dependent in part on the new product marketing conditions.

Regardless of the new product model employed, caution must be used in implementing the solution. Although a new product model is particularly useful for testing various hypotheses concerning the new product, it does not necessarily result in a precise solution. Yet, by presenting and giving an evaluation of the various alternatives, the new product model can greatly aid in the product planning and development process.

SELECTED READINGS

BUCKNELL, ROGER W. "The Product Timing 'Window.' " *Industrial Marketing* 67, no. 5 (May 1982): 62.
Explains that successful new product development takes time and requires continued monitoring and identification of marketplace needs. Provides three case histories.

ENGLE, P. H. "Six Sacred Cows of New Product Development." *Marketing Times* 12, no. 13 (July–August 1982): 28–31.
Targets six problem areas in new product development as being classic product failure areas to be avoided. Gives guidelines for circumventing these "sacred cows."

GREENWALD, GEORGE. "Seven Steps Toward New Product Success." *Advertising Age*, 27 April 1981, 52.
Proposes seven steps for new product introduction, asserting that the new

product process "renews itself." Marketers should learn from companies with a record of successful new product introduction.

LAWTON, LEIGH. "The Impact of the Marketing Concept on New Product Planning." *Journal of Marketing* 44, no. 1 (Winter 1980): 19–25.
Reveals that little, if any, empirical research has been done to examine the impact of the adoption of the marketing concept on new product planning. The purpose of this article is to describe a research study done to fill this void.

"The Next Product Revolution Will Be . . . Selling Solutions." *Sales and Marketing Management* 126, no. 1 (12 January 1981): 38.
Contends that new product consultants will flourish as consumption options increase and new products are developed to fill those options.

PILDITCH, JAMES. "Product Planning—We're Still Getting It Upside Down." *Long Range Planning* 14, no. 5 (October 1981): 20–21.
Looks at historical approach to new product development in Britain and suggests reason for that country's slow competitive development. "Pure technological innovation" should take second place to the needs of consumers. Describes four phases for new product development in an attempt to put innovation into its "fruitful context."

RABINO, SAMUEL and MASKOWITZ, HOWARD. "Optimizing the Product Development Process: Strategical Implications for New Entrants." *Sloan Management Review* 21, no. 3 (Spring 1980): 45–51.
Describes product optimization, an efficient approach to product development. Two major factors of importance to the product managers are shortened lead time and determination of factors enhancing acceptability. Discusses applications with respect to marketing strategy that focuses on brand market share development and consumer loyalty.

RAICHE, ARTHUR J. "Marketing has to Initiate Innovation Flow." *Industry Week.* 211, no. 3 (2 November 1981): 15.
Traces possible stages for new product development, which is crucial for the health of the organization.

LaPLACA, PETER J. "Regulatory, Communications Energy, R&D, and Strategic Concerns on Market Planning for the 1980's." *Marketing Strategies for a Tough Environment*, American Marketing Association Proceedings, Series no. 45 (Chicago, Ill.: American Marketing Assoc., 1980), 3–6.
States that planning must come before the product introduction, and, with a tighter market, more marketers are turning toward planning research on regulatory, communications, energy, and product issues to beat the competition.

"What It Takes to Spur Innovation." *Business Week,* no. 2643, (30 June 1980): 122.
Makes the case that freeing industry to develop new products will be a key action on the part of government and the industries themselves. The country needs a push to innovation to bring it out of the "technological recession" of the 1970s.

10

Evaluating New Product Ideas

A s indicated in Chapter 9, the product planning and development process requires a well-defined system of evaluation. At each stage of this process, critical decisions on whether to proceed with the product or drop it have to be made, in spite of uncertainty and insufficient data. In order to help minimize this uncertainty, general evaluation criteria must be established. This chapter discusses these criteria and examines various screening criteria and evaluation checklists for analyzing new product proposals. It concludes with a discussion of all aspects of new product research.

GENERAL EVALUATION CRITERIA

As a new product evolves, management should establish criteria for evaluating it. These criteria should be broad enough and yet quantitative enough to carefully screen the product at all stages of development. Criteria should be developed to access market opportunity, competition, the marketing system, financial factors, production factors, legal implications, and the impact on company image.[1]

Market Opportunity

First a new or current need for the product idea must exist. The evaluation of adequate market demand is by far the most important criterion of a proposed product. Assessment of the market opportunity and size should account for such factors as the characteristics and attitudes of consumers or industries that may buy the product, the size of this potential market in dollars and units, the nature of the market in respect to its stage in the life cycle (growing or declining), and the share of the market the product could reasonably capture. This determination is more important and revealing on a unit basis than a dollar basis, as price can then be treated as a separate variable.

Competition

The current competing producers, prices, and marketing policies should be evaluated, particularly in their effect on the target market share of the proposed product. The new product should be able to compete successfully with products already on the market by having features that will meet or overcome current and anticipated competition. Consideration should be given to the ease with which either a present competing product could be improved or a new, strongly competitive item could be marketed. The new product should have some unique differential advantage. All types of competitive products should be evaluated, not just those that are in direct competition.

---

[1]For an example of a new product evaluation system, see John T. O'Meara, Jr., "Selecting Profitable Products," *Harvard Business Review* 39, no. 1 (January–February 1961): 83–89; for an evaluation matrix that quantifies some qualitative considerations involved in selecting new products, see Barry M. Richman, "A Rating Scale for Product Innovation," *Business Horizons* 5, no. 2 (Summer 1962): 37–44.

Marketing System

It is imperative that the new product be compatible with existing management capabilities and marketing strategies. The firm should be able to fully use its marketing experience and other expertise in this new product effort. For example General Electric would have a far less difficult time adding a new kitchen appliance to its line than Procter & Gamble would. While at times the general marketing relationship may be difficult to distinguish, the new product should fit the existing marketing structure to the extent possible. Several factors should be considered: the degree to which the ability and time of the present sales force can be transferred to the new product, the ability to sell the new product through the company's established channels of distribution, and ability to "piggy-back" advertising and promotion required to introduce the new product.

Financial Factors

The proposed product should fit into the company's financial structure. To this end, financial managers should estimate the manufacturing cost per unit, sales and advertising expense per unit, amount of capital required, and amount of inventory required. These figures, combined with the price consumers would be willing to pay, will give an estimate of the time necessary for the product to reach the break-even point. The long-term profit outlook for the product should also be determined.

Production Factors

Along with the financial criteria, production or operation managers should determine how compatible the new product's production requirements are with existing plant, machinery, and manpower. If the new product idea cannot be integrated into existing manufacturing processes, not only will the idea be less favorably received by management, but new plant and production costs—as well as plant space—must be determined if the new product is to be manufactured efficiently. Of course all required materials should be available in sufficient quantity. When the production process is new, even greater controls must be established in order for the new product to be available on a timely basis once commercialization commences. There are few things more costly than having customers wait, ready to purchase but unable to get the product.

Legal Implications

Before any new product idea receives further research, all legal requirements should be delineated and met. Any patentable features should be registered, the trademark protected, and all pertinent regulations adhered to. Patents provide a firm with a strong market position and an additional incentive to fund research and development activities. RCA Corporation's original patent in color television allowed the company to develop a strong market position by inhibiting competitive entry and growth. Similar protection of market entrance can be accomplished by registering a distinguishing trademark with the U.S. Patent Office. Once registered, the company has exclusive use of the trademark for the product or product lines.

The new product should comply with, and indeed add to, the company's overall image and positioning; it should not be contrary to the company's established self-concept. For example a firm normally manufacturing high-priced, high-quality, prestige merchandise should exercise great caution in adding a low-priced, low-margin item to its line.

EARLY-STAGE TESTING OF NEW PRODUCTS

Aside from establishing criteria for each new product idea, management should be concerned with formally evaluating an idea throughout its evolution. From the marketing perspective, three pretest marketing stages can be delineated in the product's evolutionary process: the idea stage, the concept stage, and the product development stage. It is important in each stage to establish criteria for the decision to stop or continue the development process. Whether a formal research study is conducted in each stage depends on the firm's market and financial strengths, the competitive environment, the newness of the product, and the risks involved. Some products are evaluated in each of the three pretest marketing stages; others are not. The cost of the test market, as well as the cost of the product's failing, must be weighed against a longer time span before introduction, as was discussed in Chapter 9.

Even though significant costs might be involved, care must be taken against too little evaluation being done. The probability is against a new product's surviving—approximately 70 percent of test marketed products are never expanded nationally. Of the ten most common reasons for new product failure indicated in Table 10–1, four can be directly attributed to lack of proper evaluation in the idea and concept stages. These reasons for failure—poor communication across marketing/research and development interface, passive idea generation, misuse of marketing research, and lack of a product-abandonment program—can be at least partially prevented with proper evaluation procedures.

TABLE 10–1
Ten most common reasons for a new product to fail idea and concept testing

Incomplete overall strategy
Inadequately communicated strategy
Poor communications across marketing/research and development
Passive IDEA Generation
Wasteful extremes in financial evaluation
Misuse of marketing research
Poor use of tracking concepts
Delusion/self-deception/over-confidence
Organizational confusion
Lack of a "product-abandonment" program

Source: "Common Mistakes in Product Development," *Management Review* 66, no. 6 (June 1977): 61.

Idea Stage

Promising new product ideas should be identified and impractical ones eliminated, allowing maximum use of the company's resources. One method employed in this stage is the systematic market evaluation checklist. This is based on the premise that consumers buy ideas rather than physical products.[2] In this method each new product idea is expressed in terms of its chief values, merits, and benefits. Consumers are presented with these clusters of new product values to determine which, if any, new product alternatives should be pursued and which should be discarded. A company can quickly test many new product idea alternatives with this evaluation method. Promising new product ideas can be developed more easily without wasting resources on ideas not compatible with the market's values.

Regardless of whether the market evaluation checklist is used, it is equally important for the company to determine the need for the new product as well as its value to the company. If there is no need for the suggested product, then its development should not be continued. This should also be the case if the product idea does not have any benefit or value to the firm. In order to effectively determine the need for a new product, it is helpful to define systematically the potential needs of the market in terms of the kind, timing, satisfaction alternatives, benefits and risks, future expectations, price-versus-product performance features and market structure, size, and economic conditions. These factors should be evaluated not only in terms of the characteristics of the potential new product, but also in terms of the new product's competitive strength relative to each factor.

There are two general methods for initially assessing the need for a new product on an overall basis: the Standard Industrial Classification System (SIC), and the input/output method. The SIC code method is very appropriate for an initial appraisal of the need for industrial products. Standard industrial classifications, which are the means by which the federal government classifies manufacturing industries, are based on the product produced or operation performed. Each industry is assigned a two-digit, three-digit, and, where needed for further breakdown, a four-digit code. There are 82 two-digit industry groupings, such as 01 Agricultural Production Crops, 23 Apparel and Other Textile Products, 50 Wholesale Trade Durable Goods, 57 Furniture and Home Furnishing Stores, 62 Security, Commodity Brokers and Services, 70 Amusement and Recreation Services, and 94 Administration of Human Resources. Each two-digit group is further broken down into three- and four-digit groups, depending on the industry grouping being considered. For example the three-digit groups for 72 Personal Services are as follows: 721 Laundry, Cleaning and Garment Services; 722 Photographic Studios, Portrait; 723 Beauty Shops; 724 Barber Shops; 725 Shoe Repair and Hat Cleaning Shops; 726 Funeral Services and Crematories; and 729 Miscellaneous Personal Services. Where needed each three-digit group is further refined. The 721 Laundry, Cleaning and Garment Services includes such categories as 7211 Power Laundries, family and commercial; 7214 Diaper Service; 7215 Coin-Operated Laundries and Cleaning; and 7217 Carpet and Upholstery Cleaning.

---

[2]An example of the use and importance of this process is given in Louis Gedimen, "How to Screen New Product Ideas More Effectively," *Printer's Ink* 291, no. 4 (27 August 1965): 63–64.

To determine the primary market demand using the SIC method, it is necessary to first delineate all potential customers that have a need for the product or service being considered. Once the groups have been selected, the appropriate base for the demand determination must be established and the published material on the industry groups obtained from the *Census of Manufacturers.*[3] Then the primary demand can be determined from the relationship established.

Consider the primary demand estimation problem facing the Kekaka Corporation, a small, Chicago-area manufacturer of a grill-cleaning compound that cleans hot working grills better than any commercial product available. Although the cleaning compound would be of interest to homemakers, the firm will look into this market at a later date because of its limited resources. The more easily accessible market is commercial restaurants. The SIC code for this market is 58 Eating and Drinking Places. This two-digit category is composed of 5812 Eating Places and 5813 Drinking Places. Since the company is not sure whether it has the capability of initially marketing the product on a nationwide basis, it obtains the information for this SIC code from the *Census of Manufacturers* (specifically the *Census of Retail Trade*) on both a national and state basis. These figures are indicated in Table 10–2.

Kekaka's management believes that due to the characteristics of the product, eating places would be the most likely prospect. As indicated in Table 10–2, there are 253,136 eating places in the United States and 13,634 in the state of Illinois. Since the company believes that each would use approximately one gallon of the grill cleaner every other week, this market represents an annual potential primary demand of 6,636,536 gallons nationally and 354,484 gallons in Illinois. It appears that there is indeed a viable primary market for the grill cleaner on both a national and a state basis. From this point, Kekaka's corporation can further analyze the market by evaluating the number of products presently on the market, the strength of the competition, and the market's growth rate before finally deciding on market entrance.

Another technique for evaluating the overall market for a particular product or product category is the inut/output method. A table can be used to determine the number and size of the transactions occurring within specific sectors of the total economy. On a macro basis this table provides a summary of all exchanges between each industry grouping, as well as between all industries and the final consumer. Although the total input/output structure of the United States economy is given on an 85-industry category basis, a sample is presented in Table 10–3.

The table reveals that household furniture sold more than four-fifths of its output to final markets and would therefore be strongly affected by any changes occurring in these markets. On the other hand, wooden containers sold almost all of its output to intermediate customers and would therefore only be indirectly affected by changes in the final markets.

The relative primary demand among industries can be derived from input/output tables by allocating the proportion of total sales of an industry to each particular industry segment. For example, of the $12,905 million of

---

[3]U.S. Department of Commerce, Bureau of the Census, *Census of Manufacturers* (Washington: GPO, 1980).

TABLE 10–2
Determining market potential using SIC code method

1972 SIC CODE	Kind of Business	Number	Sales ($1,000)	Operated by Unincorporated Businesses		Establishments with Payroll		
				Sole Proprietorships (number)	Partnerships (number)	Number	Sales ($1,000)	Paid Employees for Week Including March 12 (number)
58	Total United States	359,524	36,867,707	164,023	34,778	287,250	35,047,577	2,634,457
5812	Eating places	253,136	30,385,361	108,159	23,986	208,899	29,312,731	2,317,425
5812	Restaurants and lunchrooms					112,656	16,652,826	1,353,843
5812	Social caterers					3,944	663,046	51,592
5812	Cafeterias					8,162	1,587,166	127,399
5812	Refreshment places					72,850	8,537,626	634,813
5812	Contract feeding					5,836	1,515,755	122,008
5812	Ice cream, frozen custard stands					5,451	356,312	27,770
5813	Drinking places (alcoholic beverages)	106,388	6,482,346	55,864	10,792	78,351	5,734,846	317,032
58	Total Illnois	21,388	2,254,889	10,904	2,065	16,110	2,106,897	157,412
5812	Eating places	13,634	1,790,811	6,022	1,450	11,278	1,736,663	139,036
5812	Restaurants and lunchrooms					6,105	1,006,134	82,711
5812	Cafeterias					383	75,724	6,528
5812	Refreshment places					3,798	485,001	36,458
5812	Other eating places					992	169,804	13,339
5812	Drinking places (alcoholic beverages)	7,754	464,078	4,882	615	4,832	370,234	18,376

Source: U.S. Dept. of Commerce, Bureau of the Census. *Census of Retail Trade Summary and Area Statistics.* (Washington, D.C.: GPO, 1972).

TABLE 10-3
Determining market potential using input/output method (figures in millions of dollars)

Industry	Lumber and wood products, except containers	Wooden containers	Household furniture	Other furniture and fixtures	Paper and allied products, except containers	Paperboard containers and boxes	Personal consumption expenditures	Net inventory change	Total federal government purchases	Total state and local government purchases	Total
Lumber and wood products, except containers	3,492	198	702	170	1,186	13	259	121	30	4	12,905
Wooden containers	33	17	2	1	1			3	24	28	543
Household furniture	29	4	78	55	1	1	3,861	49	56	28	5,122
Other furniture and fixtures	8	2	16	70	1	1	174	41	89	297	2,822
Paper and allied products, except containers	67	2	8	7	2,683	2,444	1,502	228	116	179	16,733
Paperboard containers and boxes	24	3	65	34	338	109	73	39	34	21	6,031

Source: This is an abbreviated table to give an indication of the complete tables found in U.S. Department of Commerce *Survey of Current Business* (Washington, D.C.: GPO, 1974), Vol. 54, no. 2.

lumber and wood products (except for containers), 1.5 percent were sold to the wooden container industry, 5.2 percent to the household furniture industry, and 9.2 percent to the paper and allied products (excluding containers) industry. This suggests that future new product efforts of the lumber and wood products group should be oriented toward the latter industry group, as it provided the largest share of previous industry business.

Input/output tables allow an initial evaluation of possible market demand. Although these are only overall estimates of the market potential, they can provide a method for determining which markets for the new product need further analysis. Once the markets have been delineated, then the appropriate market research can be undertaken to define the exact nature of the market and its value to the firm. By comparing these, the opportunities the idea affords can be appraised.

The need for each overall market delineated by the SIC code method or input/output method should focus on type of need, its timing, the users involved with trying the product, importance of controllable marketing variables, the overall market structure, and the characteristics of the market indicated in Table 10–4. Each of these should be evaluated in terms of the characteristics of the new idea being considered and the aspects and capabilities of present methods for satisfying the particular need. This analysis will indicate the extent of the opportunity available.

In determining the worth of the new product to the firm, various considerations must be made. Financial scheduling, such as cash outflow, cash inflow, contribution to profit, and return on investment, should be synchronized with other product ideas as well as investment alternatives (see Table 10–5). In this determination, the dollar amount of each of the considerations important to the new product idea should be as definite as possible. However these dollar amounts should be estimated for each item, regardless of accuracy, so that a quantitative evaluation can be made. These figures can then be revised as better information becomes available as the product continues to be developed.

Concept Stage

After a new product has been identified in the idea stage as having potential, it should be further developed and refined through the interaction with consumers. In the concept stage, the refined product idea is tested to determine consumer acceptance, without necessarily incurring the costs of manufacturing the physical product; that is, initial reactions to the concept are obtained from potential customers or members of the distribution channels, if appropriate.[4] One method used to measure consumer acceptance is the conversational interview, in which selected respondents are exposed to statements that reflect the physical characteristics and attributes of the product idea. Where competing products exist, these statements can be used to compare the primary features of existing products. Favorable as well as unfavorable product features can be uncovered from analyzing the consumers' response. Favorable features can then be incorporated into the product, avoiding the

---

[4]For a discussion of the importance and implementation of concept testing, see A. R. Kroeger, "Test Marketing: The Concept and How It Is Changing," *Media Scope* (December 1966): 63–68.

TABLE 10–4
Determining need for idea in
idea stage

Type of Need
  · established
  · emerging
  · time span

Timing of Need
  · frequency
  · duration
  · occurrence
  · variation

Trial Risks
  · value
  · customer tastes and preferences
  · buying behavior pattern
  · norms
  · expectations

Controllable Marketing Elements
  · price sensitivity
  · performance criteria
  · promotion needs
  · information needs
  · personal contact needs
  · support service needs
  · distribution requirements

Market Structure
  · geographic constraints
  · purchasing influences
  · purchasing process and requirements
  · degree of concentration
  · buying process
  · funds available

Market Size and Characteristics
  · size
  · trends
  · growth
  · general economic conditions

difficulties and costs in changing product prototypes. It should be remembered that the concept is designed to determine consumer interest in the basic idea of the product by indicating whether the product idea should be developed further. It is not a substitute for product testing. By discovering any major deficiencies in the product idea, the concept test results can direct research and development to culminate in a more marketable product.

Features, price, and promotion must be evaluated to determine the viability of the concept. These aspects should be considered for both the concept being studied, as well as for any major competing products. In this way, any deficiencies or benefits can be noted. By pointing out any major deficiencies

TABLE 10–5
Determining product's worth
to firm at idea stage

Cash Outflow Over Time
- · research and development costs
- · marketing costs
- · capital equipment costs
- · inventory costs
- · interest costs

Cash Inflow Over Time
- · sales of idea
- · sales of ancillary products
- · increased sales of present products
- · salvage value

Profit
- · from idea
- · from effects on existing products
- · tax implications

Relative Return
- · on investment (ROI)
- · on shareholders equity (ROE)
- · on assets (ROA)
- · on sales
- · cost of capital
- · present value

Compared to Other Opportunities
- · other new ideas
- · possible acquisitions
- · real estate
- · securities

in the product concept, the test results can direct any research and development to present a more marketable product.

The features that should be included in this evaluation are indicated in Table 10–6. Each applicable feature should be evaluated in terms of the new concept versus features of competitive products. The relative advantages should then be noted. For example how does the new concept compare with competitive products in terms of quality and reliability? Is the concept superior or deficient compared to products currently available in the market? What does this mean in terms of market opportunity for the firm? Similar evaluations should be done on each of the remaining apsects of features of the product, price, promotion, and distribution.

The idea and concept stages are pretesting stages. It is in these stages that interest in a product idea can be determined and the idea refined without incurring the direct and indirect costs of actually manufacturing prototypes. The actual evaluation needed is determined by the nature of the new product and the time available for its testing. For example a technically complex product that is difficult to describe to consumers would not be as suitable for pretesting as a product idea that is less complex. As is discussed in Chapter

TABLE 10–6
Competitive evaluation of a new concept

Product Features
- quality
- reliability
- durability
- utility
- versatility
- options available
- special features
- unique aspects
- warranty
- guarantee
- support services

Price Features
- cost
- price
- terms of sales
- special allowances
- shipping costs

Distribution Features
- product availability
- special allowances
- distribution options available
- margins

Promotion Features
- advertising
- sales promotion
- allowances
- publicity
- news releases
- trade shows
- coupons
- packaging
- quality of sales personnel
- number of sales personnel
- compensation system

13, the complexity of an innovation is inversely proportional to the rate of adoption.

Product Development Stage

Product development is the last stage of product evaluation, and is followed by the firm's dropping, test marketing, or commercializing the product. It is at this stage that consumer reaction to the physical product is determined. One tool frequently used in product development is the consumer panel, where a group of potential consumers is given product samples to use. Participants keep a record of when they use the product and comment on its virtues and deficiencies.

There are several problems in using consumer panels. For example consumers tend to react favorably to all products they test. Moreover they often make erroneous comparisons between the new product and one previously used (i.e., they incorrectly recall the attributes of the previously used product). The extent of this problem is, as yet, undetermined. These difficulties have led to the use of other testing procedures. One that is commonly used is to give a panel of potential customers a sample of the product to test and one or more competitive products simultaneously. For example, one test product may be already on the market, whereas the other test product is new; both products can also be new with some significant variation between them. Then, one of several methods, such as multiple brand comparisons, risk analysis, level of repeat purchases, or intensity of preference analysis, is used to determine consumer preference.

To measure consumer preference, the manufacturer's corporate or brand name must be eliminated by identifying each product in an equally obscure way. Care must be taken so that the packages do not bias the results. For example, a test product packaged in a white container may be preferred over one in a yellow container—not because the first product has outstanding attributes, but merely because panel members are more accustomed to products of this type being in a white package.

Consumers selected for panels to test a new product in the development stage should represent a segment of the market that has a buying interest in the product being tested. Then a paired comparison or monadic test can be used.

Paired comparison tests are generally of two types: side-by-side tests or staggered tests. Each type normally uses two unidentified products—a manufacturer's potential product and a leading competitive counterpart already on the market.[5] In the side-by-side test the two products are judged simultaneously, whereas in the staggered test each product is judged separately in a short time. In using the side-by-side test, care must be taken not to overevaluate the results. Small product differences, which may be of little (if any) importance to the actual marketing of the product, can often be magnified. The staggered test, on the other hand, is more like the usual consumer use situation, because a consumer generally tries one brand after another before selecting the preferred product. However the results in a staggered test can be seriously affected by the sequence in which the products are tested. For example, in a staggered test comparing two grades of women's stockings, the product tested second was preferred to that tested first among the women in the panel. When the women were asked to rate the stockings on product attributes, such as resiliency and durability, the stocking worn first tended to be regarded more favorably.[6]

In the monadic test, only one product is evaluated by the panel. Therefore it is necessary when comparing two different products to employ two separate panels—one for each product. In a monadic test, various scales are

---

[5]The comparison with a product already being marketed makes it possible to use the test results to forecast probable sales for the new product. These are problems associated with using these results in this manner, however, as discussed in Chapter 14.

[6]Allan Greenberg, "Paired Comparisons in Consumer Product Tests," *Journal of Marketing* 22, no. 4 (April 1958): 411–14.

used to derive rating criteria for the proposed new product. The ratings are then compared on either a combined or separate basis, with ratings given to a competitive product by a different consumer panel. By allowing enumeration of various product attributes without specific comparison with any other product, the monadic test gives less restrictive results in a shorter period of time.[7]

When testing a new product in the product development stage with either the paired comparison or monadic method the consumer panel must be selected very carefully. It should be of sufficient size and its members should have characteristics representative of potential users. For example, if the innovator and early adopter (identified in Chapter 9) are not represented, the results may be misleading because their initial response to a new product is so critical. Results can also be affected by the length of product use and the methodology of the test. A test should be conducted over a sufficient period and under conditions as normal as possible to reflect true use patterns. The test should be designed to test product characteristics and to show the competitive strengths of the product.

RESEARCH DESIGN

SUMMARIZE IN APPENDIX

When research is conducted in the idea, concept, or product development stages, it is very important that the group of consumers to be studied is defined correctly. A subset of consumers should be selected so that the total potential market is represented. Once the group has been defined, several techniques can be employed to select members who will represent this population. The actual research methodology employed for a given new product depends on cost and time limitations, as well as on how much a wrong decision will cost the firm.

IMPORTANT EFFECT ON FAILURE

Determination of Population

The first step in using consumers in testing the new product is to define the population to be studied. All individuals, companies, or stores possessing characteristics relevant to the problem must be enumerated. This process is often one of the most difficult tasks in evaluating a new product because a variety of factors must be considered. For example it might be important for the population from which the sample will be selected to be differentiated according to the product use; that is, when the new product is eventually marketed, its purchasers will be composed of new purchasers as well as brand-switchers. The number of new purchasers that includes former users of a previously purchased brand can give a clear indication of the market share the new product may attain. In addition, by discerning the reasons for switching, it is possible to categorize buyers according to the degree of their brand-switching tendency. This tendency can differentiate consumers who

---

[7]For a thorough discussion of the use and comparative results of paired comparison and monadic tests, see Jack Platten, "Where to Test and for What?" *Printer's Ink*, 287, no. 9 (29 May 1964): 118–19, and Jean Caul and Shirley Raymond, "Principles of Consumer Testing," *Journal of the Society of Cosmetic Chemists* (May 1965): 763–76.

may be expected to continue to buy the new product from those who will switch brands once another alternative is available. Other factors to consider are large-volume versus low-volume consumers, the length of time needed to establish repeat purchasing, and the degree of innovativeness of the consumers purchasing the product.

In light of these factors, a list of the members of the defined population should be accumulated. This list would ideally include the name, address, telephone number, and any appropriate characteristics of each consumer in the population. Sometimes this is a comparatively easy task. For example, if a company is interested in surveying automobile owners, a complete registration list can be obtained from the state department of motor vehicles or a commercial list broker, such as R. L. Polk and Company.

Specialized lists for certain product categories are often available from list brokers, trade associations, government agencies, or the records of the firm. If there are no lists available, the company can compile a list from tract maps and a directory that indexes telephone subscribes by city and street. Although this list would contain only the address and telephone number of each resident in an area, the company can be reasonably sure that a population is at least segmented by economic and ethnic characteristics.

Selection of Sample

Once the population has been defined, the best way to represent it within the company's time and cost constraints must be determined. Figure 10–1 indicates two different categories of sampling techniques that can be used to collect data from that portion of the population (the sample). Both categories—probability and nonprobability sampling—can be used to provide accurate information about the entire population. The goal for the company is to obtain as much information as possible about the population with each unit of cost.

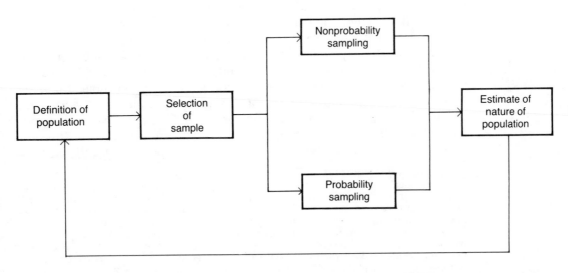

FIGURE 10–1
Designing a test market sample

The first step is to select the sample, either by a probability or nonprobability method. A nonprobability method is one in which the company uses subjective judgment to decide the size and composition of the sample. This is generally a simpler method than probability sampling because it relies on the judgment of a member of the company or of an outside expert or on the use of a previously used list of consumers. The accuracy of the new product test results depends on the soundness of the judgment or mailing list.

Judgment samples are often used during the exploratory phase of researching the new product, when the company is attempting to get an early indication of problems or wants to pretest a questionnaire for clarity and reliability of responses. In other words, the company is not trying to gather data that can be statistically evaluated or generalized to the entire population. Sometimes it is possible to use a judgment sample to select consumers who are truly representative of the population. This is especially true in the case of industrial products. For example, if a company is testing a new product that could be used only in certain types of automobiles, a judgment sample could be as reliable and representative as a probability sample. Often judgment and probability samples are combined. This is done, for example, when a portion of the population, such as the residents of Indianapolis, is selected using management's judgment, and a probability sample is drawn from this population. In this case statistical testing methods may be used to generalize the sample results to the entire city, but not to the entire population. (Additional cities would be needed to generalize to the entire population.)

Statistical techniques cannot be used to evaluate the results for nonprobability sampling. Therefore probability methods are generally used to achieve objectivity in the sample. In a probability sample, every consumer has a known probability of being selected. A variety of sampling techniques can be used, including simple random sampling, stratified sampling, cluster sampling, multistage sampling, or systematic sampling.[8]

Simple random sampling means that each consumer has the same chance of being in the sample. A company would have a simple random sample if the defined population contains 20,000 consumers and 300 are randomly selected. But how can the members of the 300 be selected so that each consumer in the population of 20,000 has an equal chance of being in the sample? This may seem to be a very simple process, but unless great care is taken, some subjective bias will occur. An easy way to avoid subjective biases in the selection of samples is to use a table of random numbers. By using any of the available standard random number tables, the sample can be obtained without any bias.[9] Care must be taken in using these tables in new product research to distinguish between sampling with replacement and sampling

---

[8]For a thorough presentation of the methodology of these and other sampling techniques, see Morris H. Hanse, William N. Hurwitz, and William G. Madow, *Sampling Survey Methods and Theory, Vol. 1* (New York: John Wiley & Sons, Inc., 1966), 110–332, or William G. Cochran, *Sampling Techniques* (New York: John Wiley & Sons, Inc., 1963), 11–30, 65–110, 160–87, 215–67.
[9]Although there are many random number tables available, the following three are perhaps the best known and most convenient to use: M. G. Kendall and B. B. Smith, *Tables of Random Sampling Numbers* (New York: Cambridge University Press, 1954), *Table of 105,000 Random Decimal Digits* (Washington, D.C.: Interstate Commerce Commission, Bureau of Transport, Economics and Statistics, 949), Rand Corporation, *A Million Digits* (New York: The Free Press, 1955).

without replacement. When sampling the population of 20,000 consumers with a sample of 300 without replacement, the probability that any single consumer will be selected is 1 in 20,000. Because one consumer is selected and removed from the total population of 20,000 (sampling without replacement), the probability of selection of the second consumer will be 1 out of 19,999, and so forth. When sampling with replacement, the selected consumers are not removed. Thus, even though a consumer has already been selected, he or she still has a chance of being selected for the sample a second or even a third time. The probability of every selection remains at 1/20,000. Sample selection for new product research is generally done without replacement.

The first step in the sample survey is to project from the sample the parameters of the population. Specifically, the mean $(\overline{X})$, the standard deviation (S), and sometimes the proportion $(\hat{p})$ of the sample must be determined. Each of these characteristics is used to project its counterpart in the population. In other words, the sample mean $(\overline{X})$ is used as an estimate of the population mean (u); the sample standard deviation (S) is used to estimate the true standard deviation of the population (σ); and the sample proportion $(\hat{p})$ is used as an estimate of the population proportion (P).

Any size sample can be obtained by using simple random sampling, but it might not be as efficient as other techniques in terms of time and cost. For example, if there are certain known demographic or economic characteristics that must appear in the sample, then the stratified sampling technique may be superior.

There are two factors that distinguish one sampling problem from another in simple random sampling: the information available on the universe, and the type of data that will be the basis for the sample determination. The information available on the universe allows more options in the sampling methodology employed. Of particular importance is whether the standard deviation (S) of the universe is known.

The level of information upon which the sample is based also affects the sample. Since the types of information possible are discussed in the last part of this chapter, the key differentiating factor in sample design is whether more than two responses are desired. When more than two responses are desired, the sampling methodology becomes slightly more cumbersome. One sample situation will illustrate these differences.

T.L.R. is a small firm about to market a new product—a novel alarm clock that stops ringing when its upper part, a thickly covered ball, is thrown against the wall. When the ball hits the wall, the shut-off mechanism is triggered. The ball can then be replaced on the base of the clock for the next alarm situation.

Before investing the effort in tooling management, T.L.R. wants an indication of market acceptance. They know that one exclusive mail order catalogue, with a mailing list of 2,000, sold a similar novel item to 84 percent of its catalogue recipients. T.L.R. management wants to determine how many members of this group need to be contacted regarding their opinion of the novel clock concept so that the company can be assured of the sample results at a 95 percent confidence level and 4 percent error level.

This sampling situation requests the lowest level of data—dichotomous data, which has only two possible answers. In addition no information is known about the variation that is occurring in the universe. This sampling situation can be solved using the following formula:

$$n = \frac{\dfrac{z^2\, p(1-p)}{d^2}}{1 + \dfrac{\dfrac{z^2\, p(1-p)}{d^2}}{N}}$$

where: n = sample size required
z = table value associated with the desired confidence level
p = probability of a response
d = error level desired before actually undertaking the research
N = population size

In the problem at hand z = 1.96, p = .84, d = .04, and n = 2,000. Therefore the sample size needed by T.L.R. is as follows:

$$n = \frac{\dfrac{(1.96)^2\,(.84)\,(.16)}{(.04)^2}}{1 + \dfrac{\dfrac{(1.96)^2\,(.84)\,(.16)}{(.04)^2}}{2,000}} = 279$$

If T.L.R. management polls 279 individuals from the catalogue's mailing list, they will be sure that evaluations of the proposed clock will accurately reflect the total mailing list at a 95 percent confidence level and 4 percent error level.

In stratified sampling, the defined population is divided into layers called *strata* so that elements within each layer are as similar as possible (see Fig. 10–2). For example one company test marketing a new chili product thought it necessary to ensure that the sample contained final consumers who had been using frozen chili, canned chili, and homemade chili. Therefore the population was divided into these three strata, and final consumers were selected from each stratum and given the new chili product to try. This example shows the population is stratified on the basis of characteristics related to the information desired. The more similar the elements within each stratum, the more efficient the sample. Stratified sampling can be used when elements in each stratum are not similar, but it will not be as efficient in terms of time and cost.

The new chili product example shows that the critical factor in stratified sampling is judgment—breaking the units of the population into strata so that the members within each stratum are as similar as possible.

Because the producer of the new product will not be able to stratify based on the characteristics being measured—that is, on how well consumers like the new product or how much they will purchase—a characteristic that is highly correlated to the characteristic to be measured is used. If many potential characteristics for stratification are available, regression analysis

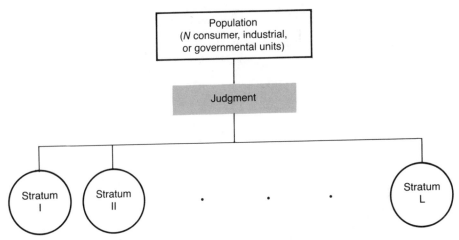

FIGURE 10–2
Stratified sampling in new product research

should be used to determine which characteristic is most closely related.[10] For example one case involved establishing a new, statewide preventive maintenance system for Bell Telephone. The best sampling procedure was needed to determine the conditions of telephone poles in the state so that those in the worst condition could be repaired or replaced. This procedure would avoid disrupting service, as well as expensive repairs if the pole were to break. The only information available was the year poles were installed, their height, and their location. Which of these factors was most closely correlated to the condition of the poles?

Obviously the statement that the condition of a pole is a function of its height is not reasonable at all. The condition of a pole as a function of its geographic location could be a possibility, especially if the region has wide differences in temperatures or if some poles were more apt to be involved in automobile accidents. However the best single prediction of the condition of a pole is the year it was installed. The condition of a pole should be excellent when it is new, and it should deteriorate over the years. This example demonstrates the importance of judgment in stratified sampling. Without a good determination of the best stratification criteria, stratified sampling has no more value in terms of time and money than simple random sampling.

Once the strata have been delineated, a procedure similar to that used in simple random sampling is used to determine the mean, standard deviation, interval estimates, and sample size. As was the case with simple random sampling, either variable or percentage information (or both) on the new product can be the basis for determinations.

Although the company may be primarily interested in segmenting the population by relevant characteristics, such as the type of product presently used, if such information is not available or if there is a great distance between the units, cluster sampling may be more efficient (see Fig. 10–3). In cluster

[10]Although the use of regression analysis in determining the best variable for stratification is beyond the scope of this book, a thorough explanation can be found in N. R. Draper and H. Smith, *Applied Regression Analysis* (New York: John Wiley & Sons, Inc., 1966).

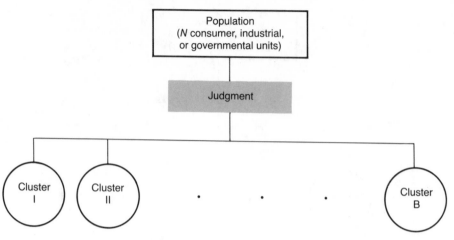

FIGURE 10–3
Cluster sampling in new product research

sampling, the population is grouped into clusters based upon physical location. For example one company wanted to test industrial consumers' response to a newly developed valve. Because the potential customers were located throughout the United States, they were first grouped into eight clusters, each representing a group of industrial customers. A sample was then drawn from each group. In this way the company was assured of obtaining results from various regions of the country, each with its different climatic conditions. Cluster sampling is most efficient when the consumers within each cluster are as different as possible and the means of each cluster as similar as possible.

The use of cluster sampling in new product research has several advantages. First, precise information on the nature of the consumer's attitude can be obtained without sampling the entire area; therefore less time would be used and lower travel costs would be incurred. Second, information is needed only on the primary sampling units (the clusters). Because the units within each cluster must only be identified in the first stage sampling, a substantial cost savings is achieved if it is very expensive to obtain information on the population. Finally, cluster sampling is generally faster and cheaper than simple random sampling for a given sample size.

These advantages must be balanced against several disadvantages. For example the precision of cluster sampling is much lower than that of simple random sampling for a given sample size. Although it can often be alleviated by using a larger sample at no extra cost, this problem should still be considered when choosing a sampling technique. Another major drawback of cluster sampling is that it is very difficult to determine the sample sizes required before instituting the research. Even though these disadvantages are troublesome, cluster sampling is frequently used in new product research, especially for new industrial products where the customers are geographically dispersed.

A fourth technique that can be used in new product research is multistage sampling, where several sampling techniques are used in stages. For example if a company testing a new valve also wanted to stratify the companies in eight geographical clusters by firm size before selecting the sample, a multistage sampling technique would be used. This technique allows the testing company to balance the costs of obtaining the sample with the reliability desired for the information on the new product.

If the selection process itself is costly or if there is little information available concerning the population, then systematic sampling may be best. The mechanics of taking a systematic sample are not at all complex. If the defined population (N) consists of consumers in some order, and if the desired sample size in n units, then every N/n unit in the population is chosen after the first consumer is chosen at random. For example one company had a list of 20,000 potential customers for its newly designed projector fading unit (a device for fading slide pictures in and out). The only information known about these customers was their name, address, and telephone number. A sample size of 400 was desired, so the 28th customer from the first 50 customers on the list was chosen because 20,000 divided by 400 gives groups of 50; the 28th is randomly determined. This customer was the first customer in the sample, the second customer was the 78th $(28 + 50 = 78)$, and the third was the 128th $(78 + 50 = 128)$. In systematic sampling, the order of the elements can either have no effect, or they can increase or decrease the reliability of the sample.

Regardless of the sampling technique used to select the sample for researching the new product, care must be taken so that inaccurate information is not collected. Consumers have the tendency to give positive ratings to new products, especially if they believe that using the product enhances their status. In addition, outside distractions may affect the response, as well as bias in the phrasing or sequencing of the questions. To overcome these difficulties, proper scaling and questionnaire design is imperative.[11]

QUESTIONNAIRE DESIGN FOR NEW PRODUCTS

In designing the questionnaire for any new product research, the type of questions used depends upon the type of information needed, the method of data collection, the coding and tabulation requirements, and precautions necessary to avoid influencing the respondent. Types of questions available for use, a sample question, and some advantages and disadvantages in the use of each are presented in Table 10–7. A dichotomous question is one in which the respondent can give only one of two possible answers. Although such questions are generally easy to ask and easy to answer, they often yield biased answers. Respondents may be far more positive in their answers than in their purchasing behavior.

One way to avoid arbitrary answers to dichotomous questions is to use multiple-choice questions. These questions are still easy to ask, answer, and tabulate. However one must be sure that the choices presented are all-encompassing.

---

[11]An example of some of these difficulties is given in Robert N. Reitter, "Product Testing in Segmented Markets," *Journal of Marketing Research* 6, no. 2 (May 1969): 179–84.

TABLE 10–7
Questionnaire design for new product research

Types of Questions	Example	Advantages	Disadvantages
Dichotomous	Do you usually like to try new products? • Yes • No	1. Easy to answer 2. Can be used to screen before asking further questions. 3. Easy to tabulate 4. Provides definite answer	1. Forces a choice 2. Provides no detailed information
Multiple-Choice	Which of the following four packages to you like? • Package A • Package B • Package C • Package D	1. Usually avoids forcing an arbitrary choice 2. Easy to answer 3. Easy to tabulate.	1. Choices may not be all-encompassing 2. Choices may not be clearly distinctive
Preference	Which of these products do you most prefer? • Brand A • Brand B • Brand C • Brand D	1. Gives information on preference 2. Easy to respond	1. Preference may not reflect purchase choice 2. Choices may present some confusion
Rating	On a scale from 1 to 9 (where 1 represents "did not like at all" and 9 represents "liked it very much") indicate your overall	1. Gives important information on relative feelings about various product attributes 2. Does not force an	1. Distinctions on scale may not be clear to respondent 2. Provides scale gradations that may not be consistent

Preference questions are very similar to multiple-choice questions. Preference questions yield information on a respondent's view of which of the alternative products, packages, or prices is most preferred. Even though the preference indicated may not reflect the actual purchase choice, valuable information is obtained on what consumers like best among the options presented.

An important tool for obtaining opinions about various attributes of the new product is the rating question. In a rating scale question, respondents are asked to indicate on a scale their feelings about various product features. Care must be taken to ensure that the distinctions on the scale are very clear and are consistent with the respondents' knowledge. This technique produces a wide range of responses on product features. If many features are used, the ratings of one attribute can be compared with those of another attribute to obtain the relative importance of each new product attribute.

TABLE 10–7
(continued)

MENTION

Types of Questions	Example	Advantages	Disadvantages
	feelings about the new product by circling the number that corresponds to your feeling: Did not like at all 1 2 3 4 Like very much 5 6 7 8 9.	arbitrary choice 3. Provides a wide range of responses for comparative purposes.	with knowledge of respondent
Ranking	Rank in order from 1 to 5 (where 1 is the best and 5 the worst) your opinion of the following products: · Product A · Product B · Product C · Product D · Product E	1. Provides valuable information on relative consumer opinions on products or attributes 2. Provides a definite answer 3. Yields information quickly	1. Is probably the most confusing type of question for the consumer to answer 2. Provides no information on how good the best product is 3. Provides no information on relative differences between ranks of products
Open-ended	Why do you buy that particular product?	1. Does not bias response with established answers 2. Provides a wide range of information 3. Provides information of more depth	1. Interpretation of answers requires skill and may vary among interpreters 2. Difficult to tabulate

If the testing firm wishes to find the order of preference of various new product alternatives, then the ranking question is used. In this technique the respondent indicates the order of preference of the alternatives given. Although this provides sound information on relative consumer opinions on products or attributes, it has been the most confusing of all types of questioning in our experience. Significant instructions must be provided to help consumers answer this type of question carefully and thoroughly. No more than five alternatives should be presented at one time, and the alternatives must be consistent with respondents' knowledge. Even with these precautions, other questions should be included in the new product research to provide information in case respondents become confused by the ranking questions.

The last type of question useful in product research is the open-ended question. Although responses to open-ended questions are often difficult to tabulate and are sometimes not definitive, they do provide information without restricting respondents to an established answer. Open-ended questions are especially useful when used in conjunction with other, previously discussed questioning techniques.

SUMMARY

The ability of the firm to obtain and evaluate a new product idea is fundamental to its continuing success. In the new product process, it is very important to establish evaluation criteria that include all-important aspects relevant to new product success. New products should be evaluated in terms of market opportunity and the marketing, financial, and production capability of the firm. Specific evaluation stages (the idea stage, concept stage, and product development stage) are useful in formally evaluating the product as it evolves.

In any type of sampling, one must be careful in defining the population, selecting the sample, and designing the questionnaire. So that time and cost constraints can be met, a firm can use simple random, stratified, cluster, multistage, or systematic sampling to select consumers. Effective evaluation of the new product is needed as the idea evolves, so it will be test marketed and commercialized only if warranted.

SELECTED READINGS

BREENE, TIM. "Why Product Managers Fail." *Management Today*, March 1980, 91.
Pinpoints the reason for lack of dynamism among product managers of consumer products in the United Kingdom as over-reliance on marketing problems at the expense of the "big picture," lack of creativity, incentives that don't motivate the product managers, and lack of broad experience among new managers breaking into a new organizational system.

FORBIS, JOHN L. and MEHTA, NITIN T. "Value-Based Strategies for Industrial Products." *Business Horizons* 24, no. 3 (May–June 1981): 32–42.
By using product life cycle costing, develops a model for comparing competing industrial products. The Economic Value to the Customer (EVC) concept includes price, start-up costs, and competitive costs. The author reviews a descriptive case and a procedure for implementing the EVC approach in the sale of industrial products.

FREIDMAN, HERSHEY H. "Futuristics: Reducing Marketing Myopia." *Business Horizons* 23, no. 4 (August 1980): 17.
Argues that one reason for product failures is the inability or refusal of a company to forecast effectively. Describes seven techniques for prediction: trend extrapolation, trend impact analysis, barometric forecasts, Delphi technique, cross-impact analysis, computer simulation, and scenario writing.

HILLS, GERALD E. "Evaluating New Ventures: A Concept Testing Methodology." *Journal of Small Business Management* 19, no. 4 (October 1981): 29.
Working from the premise that entrepreneurial start-ups often fail because owners take too little time to test market new products, outlines methods to test new product acceptance and illustrates new product concept testing.

LAWTON LEIGH, and PARASARAMAN, A. "So You Want Your New Product Planning to be Effective." *Business Horizons* 23, no. 6 (October 1980): 29–34.

Surveys 107 large United States manufacturers to determine the impact of product source ideas on the nature and success of new products, and the role marketing research plays in this process. Concludes that no marketing research is done for over half of the new products put on the market today.

MAHAJAN, VIJAY, PETERSON, ROBERT A., JAIN, ARUN K., and MALHOTRA, NARESH. "A New Product Growth Model With a Dynamic Market Potential." *Long Range Planning* 12, no. 4 (August 1979): 51.
Presents a new product growth model for durables. Its purpose is to improve product growth predictions, relaxing the assumption of constant market potential.

MASSEY, ANNE. "Why a New Product Fails." *Marketing* 6 (29 July 1981): 23–24.
Explores several reasons for new product failure, using an actual business case to illustrate each concept.

THODES, WAYNE L., JR. " 'Crystal' Ball Helps Locate Design Problems." *Infosystems* 28, no. 3 (March 1981): 28.
Describes "Crystal," a program developed for evaluation of new product potential, takes new product design feature inputs and analyzes the potential output problems and/or successes.

TIGERT, DOUGLAS, and FARIVAR, BEHROOZ. "The Bass New Product Growth Model: A Sensitivity Analysis for a High Technology Product." *Journal of Marketing* 45, no. 4 (Fall 1981): 81–90.
Objective of this analysis was the development of a forecasting equation to aid in production scheduling and market development. The Bass Model forces a disciplined approach to estimating market potential.

"Using S&MM's Surveys to Define Tomorrow Today." *Sales and Marketing Management* no. 8 (7 December 1981): 59.
Describes the use of Survey of Buying Power features (Buying Power Index and Effective Buying Income) in locating test markets, establishing sales potential, and projecting demographic data in county-by-county markets.

WILTON, PETER, and PESSEMIER, EDGAR. "Forecasting the Ultimate Acceptance of an Innovation: The Effects of Information." *Journal of Consumer Research* 8, no. 2 (September 1981): 162.
Develops a unique method for forecasting an innovation's ultimate acceptance prior to introduction and lists practical considerations for the decision-maker. The key steps in marketing new products include measuring state of knowledge, advancing knowledge to the state normally encountered prior to an adoption decision, measuring the probability of adoption, predicting purchase choices of these individuals, and validation against external data.

11

Test Marketing the New Product

T

est marketing is a stage that many new products go through in the evolutionary process toward commercialization. The decision to test a new product should, of course, never be made routinely. Test marketing is costly and time-consuming, and the same information can often be obtained more efficiently through other research techniques. Frequently the information is not needed at all.

This chapter describes the test marketing and discusses how to determine whether test marketing is really needed. It also explains factors that affect test marketing and discusses problems and alternatives.

NATURE OF TEST MARKETING

Test marketing a new product has two primary objectives: to provide a "laboratory" where management can experiment with various options of the new product and its mix, determine any problems with the product in the marketplace, and obtain feedback vital to successful new product launching; and to forecast national sales by projecting sales results from the test market areas. In other words the purpose can be to test only the physical product or the entire marketing mix (the package, promotion, price, and distribution channels). As a single evaluation procedure, test marketing gives the consumer the opportunity to express product preferences in the actual market environment.

Test marketing has a very broad scope and many interacting factors (see Fig. 11–1). When the product is tested in an actual market situation, the interdependence of the variables of the marketing mix can be evaluated and determined. Still the decision to test market must be evaluated in terms of the amount of risk involved.

There are two risks inherent in test marketing: actual quantifiable costs, and losses that cannot be quantified. Actual quantifiable costs include the costs of the plant and machinery needed to manufacture the new product, the cost of advertising to introduce the new product, and the cost of salesforce time. Equally important to the firm, but not always recognized, are the other nonquantifiable costs, such as the loss of retail shelf space of other company products to the product under test; the loss of time and funds that could have been used for other company products; and, most importantly, the negative impact on the company's reputation at both the consumer and distribution levels if the new product should fail. Investment in a test market allows these risks to be reduced while an accurate forecast of national sales is obtained. However it should be noted that test marketing can only minimize company losses; it cannot maximize profits.

Four main factors should be evaluated before deciding whether to test the market.[1] First, the company should compare the costs and risks of product failure with the profit and probability of success. If the costs and risks of product failure are low, then commercialization and national launching can

---

[1]These factors are discussed in detail in N. D. Cadbury, "How, When and Where to Test Market," *Harvard Business Review* 53, no. 3 (May–June 1975): 96–105.

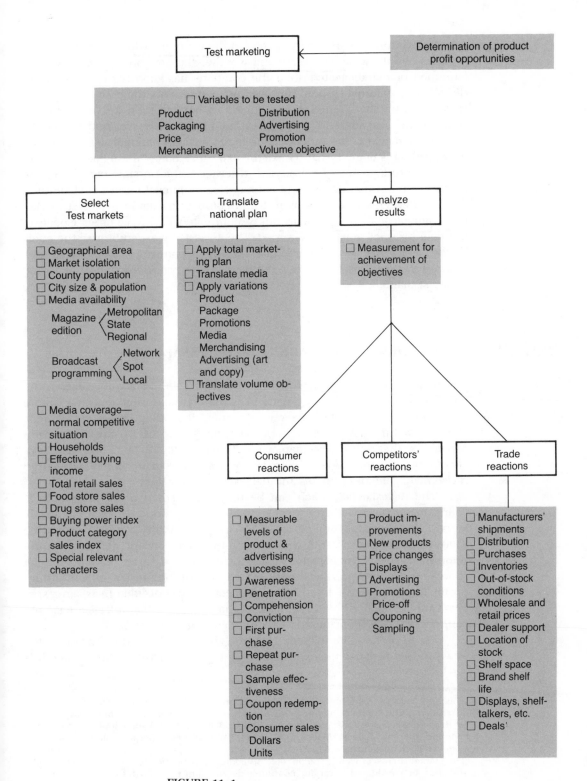

FIGURE 11–1

The relationship and composition of test marketing

Source: Remus Harris, "The Total Marketing System," *Marketing Insights* (30 January 1967): 14.

proceed without a market test. Second, if the technology for the new product requires about the same investment, whether the new product is nationally launched or market tested, then the company has good reason to proceed directly to a national launch. However, when this is not the case, and a much larger investment is required for a national launch, then test marketing is probably the best alternative. The investment risk must be weighed against the loss of profit that could otherwise be made through national sales while the new product is being tested. Third, the extra time competition has to develop a similar product must be assessed in terms of the possible benefits of the market test. Competition will, of course, also monitor the test while simultaneously developing their version. If the competition could bypass a market test, then a test should probably not be conducted. Finally, all other aspects of the new product must be examined. The advertising expenditures, the effort by the sales force, as well as the possible negative impact on the firm's reputation if the new product fails must be carefully considered before deciding on a test market.

DETERMINING WHETHER A TEST MARKET IS NECESSARY

One method of evaluating whether test marketing should be undertaken is the Bayesian approach.[2] This approach uses the expected-value decision rule to determine whether to test market a product, depending on several possible states of nature. The Bayesian approach assigns to each state of nature a probability that reflects the management's willingness to act. Then the possible outcomes of the market test are reviewed in terms of how the results would revise management estimates and decision-making ability.

The Bayesian approach can be used for a company—EAA—which is trying to decide whether to test market a new product or to immediately launch it nationally. The cost of test marketing for this company would be $200,000. From the research done during the idea, concept, and product development stages, management has quantified the three possible market shares for the new product (states of nature) and the present value of future profits (consequences). In addition management has decided to assign a specific value to each of the three possible market shares that it feels are significant enough to be considered.[3] These are given in Table 11–1.

Based on the data in Table 11–1, the company decided to introduce the new product because the expected monetary value (EMV) of introducing it is

---

[2]This approach can be found in most statistics or marketing research books. For a detailed discussion of the approach and an application, see Robert Schlaifer, *Probability and Statistics for Business Decisions* (New York: McGraw-Hill, 1959) 333, 338—9. A good application can also be found in Frank M. Bass, "Marketing Research Expenditures: A Decision Model," *Journal of Business* 36, no. 1 (January 1963): 77–90.

[3]The market shares, as well as the possible outcomes of the market test discussed later could, of course, be considered as varying continuously from 0 to 1. Although this is indeed a more realistic case, the general procedure does not differ from the more simplified case presented here.

TABLE 11–1
Market shares, profits, and probability assignments for a new product

Market Shares (S_i)		Probability $P(S_j)$	Present Value of Future Profits (U_{ij})
Capture 15% of market (S_1)		0.6	$15,000,000
Capture 6%	(S_2)	0.1	2,000,000
Capture 0%	(S_3)	0.3	− 7,000,000

$P(S_j)$ = the probability of occurrence of that state of nature
EMV$_{\text{introducing new}}$ = 0.6($15,000,000) + 0.1($2,000,000)
$\qquad$ product $\quad$ + 0.3 (− $7,000,000)
$\qquad\qquad$ = $7,100,000
EMV$_{\text{not introducing}}$ = 1($0)
$\qquad$ new product = $0

sigificantly greater than the EMV of not introducing it. The calculations are as follows:

$$\text{EMV} [A_i] = \Sigma (U_{ij}) \cdot P(S_j)$$

where $\quad$ EMV $[A_i]$ = expected monetary value of action A_i
$\qquad\qquad U_{ij}$ = the outcome of that action (i.e., the present value of future profits)

With this information, EAA management must decide whether the additional information that would be gained from a market test is worth $200,000. It is often valuable to place parameters on the decision—that is, to determine a cost limit for the market test, beyond which the value of the information obtained would not be worth the costs incurred. This entails determining the greatest increase in expected value that perfect information would provide—that is, the expected profit that can be gained if management selected the best possible course of action for the state of nature that actually occurs. For this particular example, if the company captures either 15 or 6 percent of the market, introducing the new product would earn profits of $15,000,000 or $2,000,000, respectively. However, if the new product were to capture none of the market, then the best course of action for the company would have been not to introduce the new product and receive a payoff of $0. If this scenario occurred again and again, the company would make a $15,000,000 profit 60 percent of the time, a $1,000,000 profit 10 percent of the time, and a $0 profit 30 percent of the time.[4] Therefore,

$$\text{Total Expected Profit} = 0.6(\$15,000,000)$$
$$+ \ 0.1(\$2,000,000) + 0.3(\$0)$$
$$= \$9,200,000$$

---

[4]Although, in theory, we are not using probabilities in terms of relative frequency, it is beneficial to think of the probabilities in this way when determining the expected profit occurring when management selects the best course of action.

Subtracting the EMV of the profits occurring when introducing the new product ($7,100,000) from $9,200,000 indicates the value that the cost of the market test exceeds. In this case this value is $9,200,000 − $7,100,000 = $2,100,000.

In order to determine the value of the $200,000 market test, management must determine the probabilities that each of the outcomes of the market test may have on a given state of nature and its selection. To do this, management must indicate the possible outcomes of the market test and then assign the probabilities of each outcome. The results of these management decisions are given in Table 11–2. As can be seen, management feels that three possible outcomes of the market test are relevant: selling at least 15 percent of the market, R_1; selling from 5 percent to 14 percent of the Market, R_2; and selling less than 5 percent of the market, R_3. Management has also assigned probabilities to the research outcome given the various states of nature occurring. These probabilities are designated as follows:

$$P\left(\frac{R_k}{S_j}\right)$$

where
$R_k = k$th survey result
$S_j = j$th state of nature that actually occurs

Table 11–2 indicates that management has assigned a probability of 0.7 to $P(R_1/S_1)$; in other words, management feels that there is a 70 percent chance that the research results will indicate R_1, given the state of nature S_1, is the true state of nature. A probability of 0.2 is assigned to capturing 5 to 14 percent of the market in the test, given S_1 is the true state in $P(R_2/S_1)$. In addition the probability assigned to capturing less than 5 percent of the test market, given that S_1 is the true state of nature (what actually occurs) $P(R_3/S_1)$, is 0.1.

Given this information, management can now determine (1) what the EMV (profit) of introducing the new product would be, given a certain result of the test market, and (2) the EMV (profit) of not introducing the new product,

TABLE 11–2
Conditional probabilities of outcomes of test market of new product

Outcomes of Test Market (R_k)	States of Nature Capture 15 Percent of Market S_1	(S_j) Capture 6 Percent of Market S_2	Capture 0 Percent of Market S_3
Sell at least 15 percent of market in test (R_1)	.7	.2	.1
Sell 5–14 percent of market in test (R_2)	.2	.7	.1
Sell less than 5 percent of market in test (R_3)	.1	.1	.8
Total	1.0	1.0	1.0

given the same test market result. If less than 5 percent of the market in the market test was sold (R_3), and the expected monetary value (profit) of introducing the new product was greater than the expected value (profit) of not introducing it, then the new product would be introduced. If the reverse was true, given the same market test results, the new product would, of course, not be introduced. This analysis is called posterior analysis, in which the probabilities of S_1, S_2, S_3 (being true probabilities conditional upon R_3 occurring) must be determined. These can be computed using the data in Tables 11–1 and 11–2 and the following formula:

$$P\left(\frac{S_j}{R_k}\right) = \frac{P\left(\frac{R_k}{S_j}\right) \cdot P(S_j)}{P(R_k)}$$

This is Bayes' theorem, restated for the problem at hand: the probability that state of nature j will occur is equal to the conditional probability of survey result k, given state j, times the marginal probability that the survey will show result k.

The computational result for each of the three research results and each state of nature is shown in the first three columns of Table 11–3. Notice that the sum of the column values equals the previously assigned unconditional probabilities of Table 11–1. The sum of the row values (indicated in column 4 of Table 11–2) are the unconditional probabilities of $P(R_k)$. The revised probabilities in the last three columns of Table 11–3 were found by dividing each state of nature by the marginal probability of the survey result, $P(R_k)$. This is the value obtained from use of Bayes' theorem. For example, $P(S_1/R_1) = 0.42/0.47 = 0.89$; $P(S_2/R_2) = 0.07/0.22 = 0.32$; and $P(S_3/R_3) = 0.24/0.31 = 0.78$.

These revised probabilities can now be used to recalculate the EMV (profit) of each of the two courses of action. The EMV (profit) of action 1 (the decision to introduce the new product if the company sells less than 5 percent of the market in the test R_3) is 0.19($15,000,000$) + 0.03($2,000,000$) + 0.78($-\$7,000,000$) = $2,550,000. Under the same circumstances, the decision not to introduce the new product would, of course, result in $0. Therefore, if the test results are R_3, the company should choose A_2 (not to introduce the new product) and make $0. However the opposite results occur if the test results are R_1 or R_2. For example if the company market tests the new product

TABLE 11–3
Joint probabilities and revision of prior probabilities for new product

Survey Result R_k	State of Nature (S_j)			Marginal Probability of Survey $P(R_k)$	Revised Probabilities		
	S_1	S_2	S_3		$P(S_1/R_k)$	$P(S_2/R_k)$	$P(S_3/R_k)$
R_1	.42	.02	.03	.47	.89	.04	.07
R_2	.12	.07	.03	.22	.54	.32	.14
R_3	.06	.01	.24	.31	.19	.03	.78
$P(S_j)$	.60	.10	.30	1.00			

and sells at least 15 percent of the market (R_1), then the expected profit for the decision to introduce the new brand (A_1) is 0.89($15,000,000) + 0.04($2,000,000) + 0.07(−$7,000,000) = $12,940,000. Similarly the expected profit for the decision to introduce the new brand (A_1), given that the company sells between 5 and 14 percent of the market in the test R_2, is 0.54($15,000,000) + 0.32($2,000,000) + 0.14(−$7,000,000) = $7,760,000. This is of course, better than the expected profit of A_2($0).

From this posterior analysis, the final step in determining the unconditional EMV (profit) of the decision to market test the new product can be accomplished. The result can be compared to the EMV (profit) of acting without the information gained from a market test. This final calculation—preposterior analysis—is made by summing each of the products of the conditional expected profit of the optimal act times its corresponding probability of being correct. The probabilities are indicated in Table 11–3. The probability of R_1 resulting in the test $P(R_1)$ is 0.47, $P(R_2)$ is 0.22, and $P(R_3)$ is 0.31. The previously calculated expected profits conditional upon R_1, R_2, and R_3 are, respectively, $12,940,000, $7,760,000, and $0. Therefore the EMV of the decision to market test is 0.47($12,940,000) + 0.22($7,760,000) + 0.31(0) = $6,081,800. This result must be compared with the EMV of the decision not to test, which was previously calculated to be $7,100,000. Because the EMV of the decision not to test is greater than that for the decision to test, the market test information is considered not worth the cost; therefore the market test should not be undertaken. However if the EMV of the decision not to test is *less* than the EMV of the decision to test, then the test would be undertaken, as long as its cost did not exceed this difference.

Determining the need for a market test by calculating the EMV under various degrees of uncertainty is a very valuable method of assessing information needs. Use of the process itself is valuable alone, as it forces management to evaluate the market test design while considering all possible relevant results to the decision at hand. This method also ensures that a market test is conducted only if the information gained would significantly aid the decision to market the product.

THE TEST MARKET

Once the decision has been made to test market the new product, a procedure must be formulated that will avoid distortion of the results. Care must be taken so that the conditions under which the new product is tested closely resemble those that will prevail when the product is launched. Management must know not only the sales level but also the nature of these sales and the levels of distribution. Three very important aspects of any market test procedure are selecting the test market, determining the sample size, and designing the test experiment.

Selection of Test Markets

The selection of the test market is primarily determined by the product. The test markets should approach national norms in such areas as advertising, competition, distribution, and product usage. This means that a detailed, national marketing plan is prepared that, at least, covers media selection, sales effort, and promotional budgets before the product actually enters the test market. In this way the test market can concurrently provide evaluation of alternative marketing strategies that will subsequently be used on a national basis.

The first decision to be made is whether to test in a metropolitan or nonmetropolitan area or in a district or regional market. Once this decision is made, the exact cities or regions must be chosen. Although some unique attribute of a particular city or region may require its inclusion, a test market should generally be selected based on criteria in Figure 11–2 applicable to the product to be tested.[5]

The criteria in Figure 11–2 are not all-inclusive, and they should not all be considered in evaluating an area for testing every product category. The nature of the product will emphasize some factors over others, as well as exclude some factors altogether. For example, for a company test marketing a product that it will eventually sell through company-owned stores, the questions concerning distribution may have little meaning. The checklist is merely a means of selecting an area(s) that can be controlled as much as possible so that an approximation of the national market is accomplished, thus providing a basis for prediction.

Another worksheet that can be used in test market selection is shown in Figure 11–3 (see pages 217 and 218). Here criteria are broken down into four general areas: project ability, control, measurement, and cost.[6] Each of these general areas is further broken down into specific criteria to be used in the test market selection process. For example, the degree of projectability of the test market is determined by the market size, the demographic representation of the market, media availability, category (brand development, geographic dispersion, and sales-distribution representation), and historical test market activity. On the other hand, the amount of control in the test market area is determined by the media isolation, sales distribution spill, and competitive balance. Availability of research services, accuracy, and timing determine the third area—measurement. The final area—cost—is critical, as has been previously discussed. The allocation of a typical test market dollar is indicated in Figure 11–4. The majority of each dollar spent in test marketing is in research (45 cents) and advertising and promotion (30 cents). The research and evaluation of the test market can take on a wide variety of forms, such as telephone surveys, consumer diary panels, store sales, and distribution studies. Other costs in a test market dollar include production (15 cents), package design (6 cents), and distribution (4 cents).

Given budget constraints and criteria, a city is selected that reproduces the typical American city. The sales results and market feedback from this typical city can then be used to project results on a national basis. Fifteen of

---

[5]For a comprehensive discussion of the advantages and disadvantages of testing on a city or regional basis, see "The Nation's Top Test Markets," *Sales Management*, 88, no. 1 (5 January 1962): 55–72 and Sandra Salmans, "New Trials in Test Marketing," *New York Times*, 11 April 1982, 1 F and 21 E.

[6]These areas are discussed in Edward M. Tauber, "Improve Test Market Selection with These Rules of Thumb," *Marketing News* (22 January 1982): 8–9.

City or Region _____

Criteria	Relative advantage			
General	Good	Average	Poor	Does not apply
1. Representative as to population size				
2. Diversified in age, religion, and number and types of families				
3. Typicality in terms of sales potential for the tested product category				
4. Represents industry and employment				
5. Degree of isolation from other areas				

Product oriented	Good	Average	Poor	Does not apply
1. Stability of overall year-round sales				
2. Amount of product category sales				
3. Typicality in terms of sales potential for the tested product category				

Marketing mix	Good	Average	Poor	Does not apply
1. Typicality of wholesale outlets				
2. Typicality of number and type of retail outlets				
3. Representative as to advertising media				
4. Degree of cooperation of available advertising media				

Control	Good	Average	Poor	Does not apply
1. Degree of trade cooperation				
2. Degree of company control over entire test market operation				

FIGURE 11–2
Checklist for selection of a test market

Test Market Selection Worksheet

	Candidate Markets				
	1 St. Louis	2	3	4	5
I. PROJECTABILITY					
1. Market size ___% of U.S. households in ADI	1.29				
2. Demographic representation: Index U.S.					
Age—head of household: Under 35	94				
35-54	100				
55 and over	105				
Disposable income;: 0-$14,999	93				
$15,000 and over	107				
Ethnic composition: Spanish-American	17				
Non-White	100				
Effective buying income/household in dollars	$17,623				
3. Media availability:					
Number of TV stations	6				
% Cable penetration	12				
Number of metro radio stations required for 50% share of adult listeners	5				
Number of daily newpapers in ADI	8				
4. Category/brand development: CDI BDI					
5. Geographic dispersion: Census area:	West North Central				
6. Sales-distribution representation: % ACV Expected (or now for company's other brands)					
7. Historical test market activity: Rank of use as test market during recent period (1977-1979)	24th				
II. CONTROL					
1. Media isolation:					
% spill-in	2%				
% spill-out	5%				
2. Sales distribution spill-out/in:	With TRIM:				
% shipped outside ADI	0%				
% shipped into ADI	0%				
3. Competitive balance: (market share of major competitors indexed to national/regional) 1. 2. 3. 4.					

FIGURE 11–3
A method for test market selection

Test Market Selection Worksheet

	Candidate Markets				
	1 St. Louis	2	3	4	5

III. MEASUREMENT
  1. Availability of research services:
     Warehouse withdrawals
     Audits
     Scanner item movement
     Mail diary panel
     Scanner consumer panel
	SAMI
	Nielsen/custom
	TRIM
	custom
	TRIM/MSA				

2. Accuracy Comments:	Excellent with TRIM scanner audits and MARKETRAX scanner panel				
3. Timing Comments:	With TRIM: Back data immediate, test data weekly				
IV. COST Estimated cost for period of the test	$ _____				

FIGURE 11–3 (Continued)

Source: Edward M. Tauber, "Improve Test Market Selection with These Rules of Thumb," *Marketing News* (22 January 1982): 8.

FIGURE 11–4

Test market costs

Source: Sandra Salmans, "New Trials in Test Marketing," *New York Times*, 11 April, 1982, 1 F.

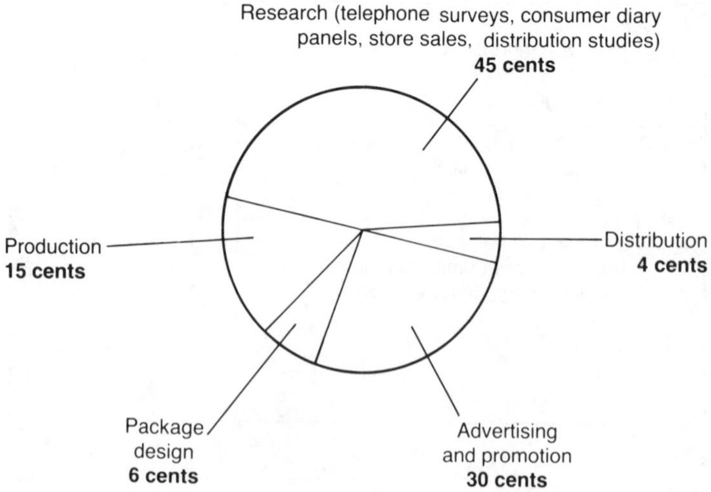

Research (telephone surveys, consumer diary panels, store sales, distribution studies)
45 cents

Production
15 cents

Distribution
4 cents

Package design
6 cents

Advertising and promotion
30 cents

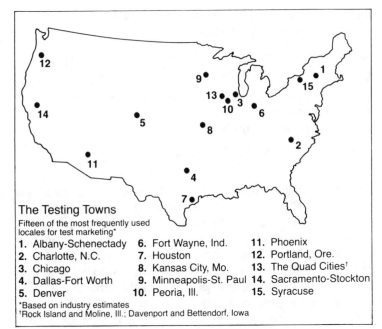

FIGURE 11–5
Most frequently used cities for test marketing
Source: Sandra Salmans, "New Trials in Test Marketing," *New York Times*, 11 April, 1982, 1 F.

The Testing Towns

Fifteen of the most frequently used locales for test marketing*

1. Albany-Schenectady	**6.** Fort Wayne, Ind.	**11.** Phoenix
2. Charlotte, N.C.	**7.** Houston	**12.** Portland, Ore.
3. Chicago	**8.** Kansas City, Mo.	**13.** The Quad Cities†
4. Dallas-Fort Worth	**9.** Minneapolis-St. Paul	**14.** Sacramento-Stockton
5. Denver	**10.** Peoria, Ill.	**15.** Syracuse

*Based on industry estimates
†Rock Island and Moline, Ill.; Davenport and Bettendorf, Iowa

the most frequently used locales for test marketing are indicated in Figure 11–5. Each of these locales, such as Albany-Schenectady, Charlotte, Chicago, Dallas-Fort Worth, Denver, and Fort Wayne, has provided results in past test markets that reflect national norms.

Sample Size Determination

In test marketing the new product, it is very important to make sure that the number of stores used is adequate to represent the "universe." The methodology in sample size determination for store audits is different from that used in consumer studies in evaluating the new product in both the idea and concept states (discussed in Chapter 10). Because the retail stores used in the test market have different properties, a proportionate sampling methodology is employed. One fundamental formula for sample size determination is as follows:[7]

$$n = \frac{\left(\dfrac{ts}{d}\right)^2}{1 + 1/N \left(\dfrac{ts}{d}\right)^2}$$

where n = number of stores in sample

t = the value determined by the confidence coefficient selected (i.e., t = 1.00 for 68 percent confidence coefficient, t = 1.65

---

[7]This formula, as well as its proofs and derivations, can be found in William G. Cochran, *Sampling Techniques*, 2nd ed. (New York: John Wiley & Sons, Inc., 1963), 74–76.

for 90 percent confidence coefficient, and $t = 1.96$ for 95 percent confidence coefficient)

d = the limit of error found by multiplying the stated error preference by the mean sales of a similar product

s = the standard deviation of the universe obtained from previous knowledge of the universe or from an outside independent audit supplier

N = number of stores in the population

We can use this formula to determine the sample size needed for a new product for Kelly's Toys and Games, Inc. The company has been continually evaluating its newest product—a game called *Impeachment*. This game is a political, adult game that benefits customers through a recreational and in- structional medium. It allows consumers to put themselves in the position of a president about to be impeached and, as such, has an historical as well as contemporary flavor. The historical basis is provided in the form of a booklet explaining past impeachment proceedings and the legislative, judicial, and executive processes involved. The contemporary aspect of the game is that it allows each participant a chance to win the game by using spaces on the board and the two sets of cards. A player's skill in determining the right time for key decisions is a major factor in winning the game.

The company has already tested this product in both the idea stage and concept stage and feels that a test market is needed before limited company resources can be committeed to an introductory campaign on a national basis. Management of Kelly's is trying to determine how many stores should be included in the test market. In its marketing of an adult game two years ago, the company found that the universe of stores has a standard deviation of 2.24, with mean product sales of 4.3. The number of stores in the market that would carry the product is 204. The management at Kelly's wants the number of stores to be selected at the 95 percent confidence level and the 10 percent error level.

From the preceding information, the number of stores needed can be calculated as follows:

$$n = \frac{\dfrac{(1.96)^2(2.24)^2}{(0.43)^2}}{1 + 1/204 \left(\dfrac{(1.96)^2(2.24)^2}{(0.43)^2} \right)} = 68$$

where

N = universe size = 204

t = confidence level = 1.96 at 95 percent confidence level

s = standard deviation = 2.24

d = error limit = 0.10(4.3) = 0.43

In order to be 95 percent confident that the number of stores is representative within a ±10 percent error of the mean sales for the new product, 68 stores are needed. Three factors influence the sample size (i.e., the number of stores needed): error level, confidence level, and the standard deviation of the stores.

If everything else remained the same in this scenario, except that the acceptable error level was increased to ±15 percent, then the number of stores needed would be reduced to 36:

$$n = \frac{\dfrac{(1.96)^2(2.24)^2}{(0.65)^2}}{1 + \dfrac{1}{204}\left(\dfrac{(1.96)^2(2.24)^2}{(0.65)^2}\right)}$$

On the other hand, if only the confidence level was changed—that is, reduced from 95 percent to 90 percent—the number of stores needed would be reduced to 54:

$$n = \frac{\dfrac{(1.64)^2(2.24)^2}{(0.65)^2}}{1 + \dfrac{1}{204}\left(\dfrac{(1.64)^2(2.24)^2}{(0.65)^2}\right)} = 54$$

The final factor that influences sample size is the standard deviation. Assuming that the confidence level, error level, and the size of the universe remained at 95 percent, 10 percent, and 204, respectively, but the mean product sales increased to 5.7 and the standard deviation decreased to 2.06, the sample size needed would be reduced to the following:

$$n = \frac{\dfrac{(1.96)^2(2.06)^2}{(0.57)^2}}{1 + \dfrac{1}{204}\left(\dfrac{(1.96)^2(2.06)^2}{(0.57)^2}\right)} = 34 \text{ stores}$$

As indicated in these examples, the number of stores needed in a selected test market depends upon the average product sales, the standard deviation, the confidence level, the error level, and, to a slight extent, the number of stores in the universe.

Design of Test Experiment

Once the company has selected the city and number of stores it will use in the market test, it is important to decide the appropriate testing procedure or experimental design. The classic experimental model is indicated in Table 11–4.

The number of stores is randomly assigned to either the experimental group or the control group of stores, with the stores being similar in terms of the product being tested. Measures of product sales are obtained before the

TABLE 11-4
Experimental model for test marketing new products

Source of variation	Time of Measurement	
	Before	After
Experimental group of stores	A	B
Control group of stores	C	D

experimental variable (i.e., the new product) is introduced and then after it is removed. If the difference in sales between A and B is significantly greater than the difference in sales between C and D, significant sales have been achieved under the marketing factors tested.

When designing a test experiment for some marketing aspect of a new product, each of the possible formal experimental designs should be evaluated in terms of the problem at hand. The four most frequently used experimental designs are as follows: completely randomized, randomized block, Latin square, and factorial.[8] Management involved in test marketing new products should understand the procedure and limitations of each design as well as the situation(s) in which each is most useful.

Table 11-5 indicates that the experimental designs proceed from completely randomized to factorial in terms of the number of variables tested and difficulty of computation. A completely randomized design is used when only one variable, such as different new packages, is going to be tested. A randomized block design provides information on two test variables, such as price and package of the new product. A Latin square design is used to test three variables, such as package, price, and point-of-purchase display.

All these designs are called single-factor experiments and do not evaluate the interaction or joint effect the new product variables may have together. Imagine, for example, a point-of-purchase display for a new cake mix featuring a large picture of a cake. There is a coupon attached by which the consumer can mail three box tops to the company and receive a 50¢ refund. In addition the company is interested in knowing the impact of a price of 69¢/box versus a price of 49¢/box. Not only would the sales promotion display affect sales by itself but the price would also. Then, too, the joint impact or interaction between the point-of-purchase display and price would affect sales. Whenever management wants to measure the interaction between the variables, factorial design must be employed. Table 11-5 indicates a two-variable factorial design and therefore one interaction. This can be expanded to a three-, four-, or five-variable factorial design. Of course, every time an additional variable is added, more interactions occur. For example, with three variables, there are four interactions, and with four variables, ten interactions. Eventually factorial design becomes very unwieldy in terms of determining the significant interaction effects of the variables.

---

[8] The theoretical base, as well as the depth of presentation of these experimental designs, can be found in many books dealing with data analysis. See, for example, Keith K. Cox and Ben M. Enis, *Experimentation for Marketing Decisions* (Scranton, Pa.: International Textbook Co., 1969), and Charles R. Hicks, *Fundamental Concepts in the Design of Experiments* (New York: Holt, Rinehart & Winston General Book, 1964), 1–94.

TABLE 11–5
Types of experimental designs used in test marketing

Type of Experimental Design	General Model	Variables Tested
Completely Randomized	$x_{ij} = u + T_j + E_{ij}$	one
Randomized Block	$x_{ij} = u + B_i + T_j + E_{ij}$	two
Latin Square	$x_{ij} = u + B_i + T_j + \lambda_k + E_{ijk}$	three
Factorial	$x_{ijk} = u + A_i + B_j + AB_{ij} + E_{k(ij)}$	two and interaction

FACTORS AFFECTING TEST MARKETING

When implementing and evaluating the results of test marketing, consideration must be given to a wide variety of factors that could affect the test. One essential factor is the general level of business and economic activity. The overall economic atmosphere can affect consumer attitudes, such as readiness to buy and interest in capital expenditures, and therefore can alter the evaluation of the new product. For example, if business activity is low and if a firm is evaluating the market response to a major new capital item, the evaluation of this new product might be worse than it would be in a period of usual business conditions. Similarity the uncertainties of inflation and unemployment might bias consumers in their evaluation of a new, major durable good.

Other important factors affecting test marketing are the existing market structure and competitive conditions. The number and types of competitive products, the degree of competition, and the effectiveness of competitive marketing strategies can significantly alter test market results. These forces interact to varying extents, influencing the results of a test. Similarly the prevailing consumer preferences and available marketing channels will contribute to the complex interacting market structure.

Competition may attempt to influence the results of a market test by offering special prices or premiums, or increasing promotional efforts during the test period. Whether this interference is committed knowingly or unknowingly, the effects must be taken into account in order to have a reliable new product evaluation. For example a company recently test marketing a new consumer product in two cities disregarded the results in one city because of an extensive promotional campaign launched for a competing product.

Finally, the originality and appeal of the presentation, as well as other marketing variables, such as the package and price, can affect the market test's results. Care must be taken to test only those new product variables desired, making sure that no other factors under the firm's control interfere with the results.

TRIALS IN TEST MARKETING

Several problems arise when conducting a market test of a new product.[9] A major problem is the reliability of the results. Many new products that achieve

---

[9]For a discussion of some problems, see Carl H. Hendrickson, "Some Pitfalls of Sales Test," *Sales Management* 65, no. 9 (10 November 1950): 124–29.

favorable test market results are not commercial successes. Even though test marketing is used extensively, the results are often neither useful nor accurate, especially as a predictive device. One study on the predictive value of test-marketing indicated large margins of error.[10] Care must be taken to ensure greater consistency between the results of a market test and the national market.

Another problem is the cost of test marketing. In addition to making large investments in the procedure itself, the company must also pay the salaries and expenses of employees, any outside agency, and distribution channels. For example, in the case of company employees, time is used in designing, monitoring, and evaluating the new product's test and diverted from other products or programs for the new product. In addition, high per-unit costs are involved in the production of a modest number of units for use in the market test.

A major cause of inaccurate projections and the related costs are the shortcomings in the basic design of the test marketing program. Not only must the design be sound and follow developed plans, but it must also be closely adhered to. Companies commonly make the mistake of using unqualified personnel to conduct the testing and not establishing criteria for adequate evaluation of the results.[11]

The problems that arise in the collection and evaluation of data are so important that these two topics merit individual attention. The sales measurement is frequently a problem in a lengthy, full-scale market test. For example, a firm attempting to measure the acceptability of a new consumer product may have to use a wholesaler, who then distributes to the retailers. Initially "sales" will reflect an inventory build-up by the wholesaler and retailer. Many weeks may elapse after actual retail sales are made to the final consumer before the distributor's reorders reach the producer. With various inventory decision rules being practiced, it may be difficult, if not impossible, to relate reorders to purchases by final users. Even when retail sales are audited in the test market, inventory misallocation problems are common.[12] Another evaluation problem is in determining whether the new product success is a result of its testing plan, mere coincidence, or the strength of the firm. For example a consumer may be influenced to buy a new product because it was introduced by a large corporation with a good distribution system, large sales force, substantial advertising investment, and good image.[13] Finally there is the almost insurmountable problem of evaluating the exact nature of the purchasers of the new product. Were they first or repeat purchasers? Were they formerly loyal purchasers of a competitive product who switched, or were they habitual product switchers? Did the purchasers buy the new product to test it, or did

---

[10]Jack Gold, "Testing Test Market Predictions," *Journal of Marketing Research* 1, no. 3 (August 1964): 8–11.

[11]For a discussion of this problem in test marketing, see Harry C. Groome, "Take the Risks Out of Test Marketing," *Sales Management* 92, no. 8 (17 April 1964): 33, and A. R. Kroeger, "Test Marketing: The Concept and How It Is Changing," *Media Scope* (January 1967): 51–54.

[12]An example of this problem in obtaining reliable information is given in Edwin M. Berby, Victor L. Cole, and Jack Gold, "Testing Test Marketing Predictions," *Journal of Marketing Research* 2, no. 2 (May 1965): 196–200.

[13]For a discussion of this difficulty, see E. B. Weiss, "The Trials and Tribulations of Test Marketing," *Advertising Age*, 16 January 1961, 84.

they see some merit in the product? Accurate measurement of who purchased the new product and why is invaluable to the producer.

An all-encompassing problem in test marketing is time. Time spent in test marketing a new product is not only costly for the firm, but it often gives competition an advantage as well. While a company is carrying out its test marketing, its competitors may be gaining enough time to respond in a manner that will seriously reduce the new product's full-life profitability. As noted in Chapter 1, with product life cycles becoming shorter and shorter, few, if any, companies are so far advanced in technology that a long lead time of a new discovery can be expected. For example Sunbeam's electric skillet was copied so widely that the market was saturated within a few years. Although Lestoil, a liquid household cleaner, had a tremendous sales record of $25 million, the advent of such products as Handy Andy by Lever Brothers, Mr. Clean by Procter & Gamble, and Liquid Ajax by Colgate-Palmolive caused Lestoil's sales to substantially decrease to $16 million in a short period of time. These are only a few examples in which time spent for test marketing meant reduced total profits for a new product because the exclusivity had been reduced.

The time problem in test marketing is indeed a paradox. If the market test period is too short, national distribution strategy may be based on inaccurate, incomplete, or inconclusive data. If a new product is tested for too long a time, the competition may be able to respond with a substitute product.

Competitors can also distort the results of a test market by carrying out a heavy promotional campaign in the test area or by actually purchasing large quantities of the new product. Such strong, competitive activities can affect a test market so that new product results are of little value. The degree to which the competition is allowed to evaluate the new product should be carefully considered before deciding to test market a new product.[14]

ALTERNATIVES AND REMEDIES FOR TEST MARKETING

One solution to some test marketing problems is to use data from consumer panels to measure a test product's staying power. Markov chain analysis can give the test marketers insight into the dynamics of competition and the forces influencing consumer brand-switching. This method can be used to determine a new product's share of the market for the total product category at a future time by tracing its market share through preceding, successive periods of time. The product's share through the successive stages is a function of transitional probabilities: that is, the sum of the probabilities of switching from one brand to another is the square of the number of brands in the product category. For example, a company wants to determine the probability that a customer would purchase its new product at a particular time in the future. The customer did one of the following when making the last purchase of the product: bought the new product, bought the leading competing brand, or bought a nonleading but competing product. From consumer panel data showing each customer's purchases over an extended period of time, the

---

[14]For examples of competitive action distorting test market results, see "Espionage and Sabotage: How They Disrupt Testing," *Printer's Ink* (27 August 1965): 90–95.

company can estimate the conditional probabilities that a customer will switch from one brand to another when given the outcome of a previous trial.[15]

Another alternative to test marketing is to implement research that is between concept testing and full test marketing. A model test market can be employed in which the total marketplace is modeled, but both the time and costs of a full market test are reduced. This step is particularly useful for new products not having conclusive concept test results, new products of small companies who cannot afford a full-scale test market, and new products requiring considerable capital investment to test market. Cadbury Limited has a very unusual model market test. A representative panel of housewives is mailed a sales catalogue describing all the products being tested, the leading brands in the product category, and any promotional efforts occurring. This material is accompanied by an order form. The housewives then receive a weekly visit at home from a salesperson with a mobile shop, and the women make their weekly selections.[16] An alternative to this model of a market test would be to set up a company store and have the representative panel of housewives shop there.

Another development to aid and refine test marketing new products is to simulate market reality with a computer. Of course, the key to success in using test marketing simulation (TMS) is the choice of the proper model. A test marketing simulation can be particularly valuable in assessing a new product's survivability and evaluating strategic marketing plan elements.[17] In one case at Johnson and Johnson, repeated simulations of the Reach toothbrush allowed a comprehensive model to be developed. The evaulation of a new competitive entry by this model indicated that the competitive product was not as strong as expected, and $600,000 was trimmed from the defensive spending budget of Reach. Successful application of computer simulation techniques to test marketing problems has, in part, been due to the problems in the quantification of consumer behavior, as well as to extensive costs. In addition, if any important variables are omitted or if the underlying assumptions are inaccurate, the results will be invalid. One such simulation model is DEMON.[18] The purpose of this system is to ensure a viable national new product introduction by assuming the best possible marketing plan, conducted in the least amount of time, with maximum profit, risk, and payback period efficiently balanced. DEMON has the advantages of reducing risk (and consequently the rate of new product failure), reducing the time required for new product introduction,

---

[15]For an elementary description of the Markov process, see S. Dutka and L. Frankel, *Markov Chain Analysis: A New Tool for Marketing* (New York: Audits and Surveys, 1966); for a more extensive discussion, see Ronald E. Frank, "Brand Choice as a Probability Process," *Journal of Business* 35, no. 1 (January 1962): 43–45; Alfred A. Kuehn, "Consumer Brand Choice as a Learning Process," *Journal of Advertising Research* 2, no. 2 (December 1962): 10–17; and A.S.C. Ehrenberg, "An Appraisal of Markov Brand Switching Models," *Journal of Marketing Research* 1, no. 4 (November 1964): 347–62.

[16]This is described in N. D. Cadbury, "When, Where, and How to Test Market," *Harvard Business Review* 54, no. 3 (May–June 1976): 100.

[17]The use and application of test marketing simulation is discussed in "Test Market Simulation Evaluates New Products Sets Realistic Expectations—If Right Model Used," *Marketing News* (22 January 1982): 10.

[18]David B. Learner, "DEMON New Product Planning: A Case History," *New Directions in Marketing*, (Proceedings of the American Marketing Assocation Conference, Chicago, June 1965), 489–508.

reducing or possibly eliminating test marketing, lowering the costs of new product introductions, and ensuring maximum profits on a new product. This network system has the capacity to model the whole world simultaneously by altering, integrating, and weighting all variables against a set profit and goal and possible sales, thereby obtaining results that could otherwise only be obtained after extensive test marketing. DEMON is only one of several simulation models used in new product decisions.[19]

Another method to solve some of the problems in conventional test marketing is a mixture of test marketing and national introduction. In this "rollout method," a company selects a test area for distribution, such as a sales territory, and markets the new product using the same marketing mix variables as will be used for national distribution. When sales are substantial enough in the selected market, the company "rolls out" to the next sales territory and eventually to national distribution. The method, of course, requires extensive pretest market activity in the idea, concept, and product development stages previously discussed.

SUMMARY

The practice of test marketing new products is increasing, even though it requires large expenditures. Although there are many factors and problems to be considered, test marketing a new product helps ensure success in national distribution.

One critical aspect of test marketing is selecting the appropriate market for testing the new product. Specific criteria should be established and used in this selection process. Test results should reflect these factors and be projectable onto the national market. Both the sample-size determination and the design of the test experiment will greatly influence whether the success of the new product is accurately predicted.

A basic design exists for every market test situation. In spite of this, problems affect the test marketing process. Factors, such as the existing market structure, competition, and presentation appeal, can distort the results so that they cannot be reliably used to forecast the national sales of the new product. In certain cases procedures other than the market test are more useful in predicting product success.

SELECTED READINGS

ENDICOTT, CRAIG. "This Survey Seeks to Find the Surveyed." *Advertising Age*, 9 February 1981, S22.
Relates the experience of the author who recently tried the door-to-door interviewing he had always sent others to do. An amusing yet enlightening article.

---

[19]Two other approaches are given in Marshall Freimer and Leonard S. Simon, "The Evaluation of New Product Alternatives," *Management Science* 13, no. 6 (February 1967): B279–92, and Henry J. Claycamp and Lucien E. Liddy, "Prediction of New Product Performance: An Analytical Approach," *Journal of Marketing Research* 6, no. 6 (November 1969): 414–20. Various aspects of these and other models were discussed in Chapter 3.

FORD, DAVID and RYAN, CHRIS. "Taking Technology to Market." *Harvard Business Review* 59, no. 2 (March–April 1981): 117.
Suggests that management must market its firm's technology more aggressively throughout the life cycle of that technology. This encompasses all phases of technology development, ensuring that customer demand is completely and accurately met.

GALGINAITIS, CAROL. "What's Beneath a Test Market?" *Advertising Age*, 9 February 1981, S2.
Reviews a new product manager's uses for a test market. A key element is the cooperation of a minimum number of stores to limit competitive interference.

HOWARD, NILES. "A New Way to View Consumers." *Dun's Review*, 117, no. 8 (August 1981): 42.
Reviews SRI International's psychological system. Categorizes individuals by values and life styles, and focuses on the customer's state of mind, instead of age, income, and sex.

JOBBERS, D. "Test Marketing—An Evaluation of More Recent Micro-Marketing Methods." *Management Decisions* 17 (1979): 441–449.
Reviews several methods for evaluating test markets. Planning is considered key to the test market process, and competitive reaction measures are evaluated in detail.

LENTINI, CECILIA. "Test Marketing: Tomorrow, and Tomorrow . . . Whither the Method?" *Advertising Age*, 22 February 1982, M9.
Describes how increased emphasis on very early pre-screening has led to an explosion of market research services. Yet, budgets are remaining the same while companies get better new product introductions because of research.

SALMANS, SANDRA. "New Trials in Test Marketing." *New York Times*, 11 April 1982, 11.
Explains that as the economy has declined, competition has increased, and competitive reaction to test markets has been especially strong. The "right" test market cities for test marketing are reviewed in light of the changing economy and the competitive environment.

SANDS, SAUL. "How Much Should You Spend on a Marketing Pre-Test? A Shortcut Approach." *Interfaces* (August 1981): 62–66.
Pretesting of proposed marketing actions is a common type of marketing research activity. A key question is whether the information collected in the pretest is worth the cost. Described as a quick and easy way of deciding whether to buy additional information in the case of a two-action, two state-of-nature decision problem. Simplicity makes it useful for widespread use.

SHERMAN, STRATFORD P. "A Supermarket at Your Sofa." *Fortune* (21 April 1980): 121.
Examples of existing electronic shopping systems and their success indicate to the author that consumers may some day never leave their home to shop. This might provide test marketers with new options to enhance their surveys' accuracy and coverage.

TAUBER, EDWARD M. "Improve Test Market Selection with These Rules of Thumb." *Marketing News* (22 January 1982): 8.
Controlled store testing is often used for marketing mix changes in established products. Lists several heuristic rules that should serve as guidelines for marketing managers who are overseeing test market decisions.

Test Market Simulation Evaluates New Products; Sets Realistic Expectations—If Right Model Used." *Marketing News* (22 January 1982): 10.

Market models can be used to simulate test markets if certain factors are properly set and model assumptions clearly outlined. The authors' market research for ArmourDial Co. suggests that issues like price, positioning, packaging, formulation, source of volume, sampling effectiveness, and shelf location are capable of being simulated.

Testing Shows Marketers Where to Put the "Big Bucks." Round Table Discussion, *Sales and Marketing Management* (16 March 1981): 50.
Reviews discussion by experts, and indicates that test markets are often helpful in allocating marketing dollars to the right target markets.

WELLER, DON. "Testing the New Product." *International Trade Forum* (January–March 1981): 16.
Determines that international test markets add an extra dimension to the potential new product test market success or failure. Some suggestions are offered for avoiding international product mishaps.

YOVOVICH, B. G. "Competition Jumps the Gun." *Advertising Age*, 9 February 1981, S20.
Reviews how and why competition can interfere with your firm's test markets. Explains that elements of strategy sometimes can be implemented to avoid such confrontation.

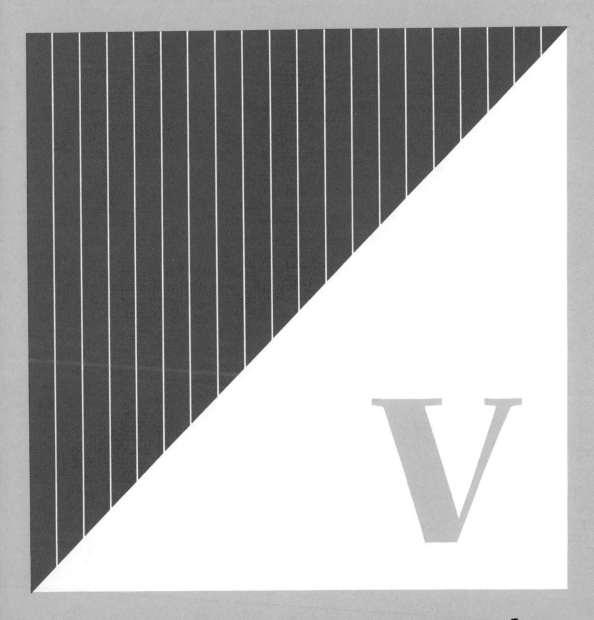

V

The New Product Acceptance and Demand Determination

Market Analysis and Market Segmentation for New Products

By any measure, American Express Company (Amex) has been a "cash machine." With its Travelers Cheques, money orders, charge cards, insurance companies, investment house, and overseas commercial banks, it has persistently crossed the traditional barriers that existed between financial institutions. In each case it carved its markets carefully, selecting niches where competition was unspecialized and restrictions on multiservices were nonexistent. Without a U.S. bank, for example, it was able to sell its Travelers Cheques nationwide, using other banks as its marketing partners. And although its insurance holdings keep it from owning a bank in the United States, it has banks around the world.

It promoted its specialized products with shrewd marketing tactics and high service standards that have brought record earnings each year for the past three decades of the company's 127-year history.

But today the financial institutions themselves are knocking down the barriers that existed between them. Spurred by new electronic technology, bank credit cards are tapping investment accounts at brokerage houses and credit union accounts, as well. Some savings banks and mutual funds are offering cut-rate life insurance. On the drawing board are plans to use the Visa card for loans against cash values in insurance policies. Looming over all is an electronic funds transfer system (EFTS) that represents the promise—or threat—of national branch banking. And along with the new technology, U.S. financial institutions are becoming more aggressive, offering variations of Amex's products to Amex's own markets.

So far, Amex has managed to outpace the competition by enhancing its products and sharpening the focus on its markets. But it becomes increasingly expensive to do so. To keep its edge, Amex has added to its business mix an international merchant bank that, though relatively small, opens a whole new market.

As for other companies that have carved out specialized markets only to find themselves challenged by equally aggressive, savvy competitors, the best response is to look for additional products for its affluent market or to find other businesses that fit its specialized mold.[1]

McDonald's Corp.'s success owes a lot to the fast-food chain's powerful appeal to children. And that appeal, in turn, has a lot to do with Ronald McDonald, the clown that represents the chain in TV ads and in appearances at McDonald's units. Now Burger King Corporation, the nation's second largest hamburger seller, is setting out "to make a frontal assault on Ronald McDonald," according to E. Christian Schoenleb, Burger King's group vice-president for marketing. Schoenleb aims to use a two-pronged marketing attack to boost sales by 20 percent in Burger King's 2,000 units. To win over the kids, this Pillsbury Co. subsidiary has created the "King," who will make his debut in the chain's national television ads. Schoenleb maintains that the King, who performs magic tricks, will rival Ronald McDonald in popularity. Later, 50 actors will be outfitted in King costumes, ready for personal appearances at

---

[1]"American Express: Why Everybody Wants a Piece of Its Business," *Business Week* (December 1977): 56.

Burger King outlets across the country. To back up the King, the chain will launch a premium campaign, giving away different prizes each week—whistles, iron-on patches, and other trinkets—to children who cajole their parents into taking them to Burger King outlets. All told, says Schoenleb, Burger King will spend more than one-third of its $45 million ad budget to promote the chain as the place for children to eat.[2]

As the preceding scenarios indicate, a company must identify the group of customers that will have a high probability of purchasing the product. The company should direct the new product and all the associated marketing activities towards this group. If this is not done, the probability of success for the new product decreases significantly. This chapter discusses the undifferentiated approach versus the market segmentation approach in target-market selection. Following a discussion of the various segmentation variables, the chapter concludes by presenting some new segmentation strategies for various markets.

UNDIFFERENTIATED APPROACH

Some companies define the target market for their new product as the total market for this product category. In other words, the company assesses the needs of all customers in the target as being very similar, eliminating the need to subdivide the market according to any product or market attribute. Although this undifferentiated approach is rarely used, it may have application for a new product that consumers would think of as being homogeneous.[3] New products such as staple food items like sugar or salt would be possibilities for the undifferentiated strategy. This general total market approach is illustrated in Figure 12–1. In most instances a market segmentation approach is a much better strategy.

FIGURE 12–1
Undifferentiated market approach for new products

OVERVIEW OF MARKET SEGMENTATION

Every company must determine what customer group needs its product, whether it be a consumer, a firm, a group of individuals, or a governmental unit. A balance must be achieved between the number of markets to which a firm appeals and the new product ideas forthcoming. Since there are many sources for new product ideas (discussed in Chapters 8 and 9), and since a firm's resources are generally limited, new product ideas must be carefully analyzed (see Chapters 10 and 11). An excellent input for this analysis is

---

[2]"McDonald's Corporation's Success," *Business Week* (28 November 1977): 72.
[3]See, for example, Wendall R. Smith, "Product Differentiation and Market Segmentation as Alternative Marketing Strategies," *Journal of Marketing* 21, no. 1 (July 1956): 3–8, and Alan R. Roberts, "Applying the Strategy of Market Segmentation," *Business Horizons* (Fall 1961): 65–72.

delineating and analyzing possible target markets and then selecting the best one.

Benefits of New Product Segmentation

The concept of market segmentation for new products has been almost universally adopted. Previously firms developed new products and marketed them without considering the consumer who would be buying the product. Firms concentrated on the product and all its attributes, rather than on the market characteristics. Preference or need variations within the markets were usually ignored, since it was felt that everyone would want the new product.

As competition became more intense, as consumers became more educated, and as communications systems improved, this nonsegmentation approach no longer assured market success for new products. This was evidenced in part by the high failure rate of new products (discussed in Chapter 1). As a result, some firms began to look to the market, define its needs, and design their new products and all their attributes based on this analysis. Figure 12–2 depicts this market-oriented concept. The actual market can be evaluated on a variety of relatively definitive bases. In other words, customer groups could be distinguished. Market segmentation is the process of dividing a total market into smaller groups composed of entities (people, firms, governmental units) with similar needs and characteristics. Through market segmentation, various groups of potential buyers of the new product, each with its own set of buying desires or requirements, are identified.

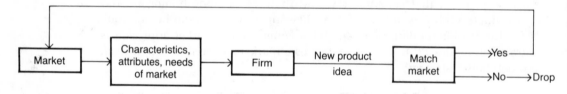

FIGURE 12–2
Market-oriented approach for new product

Market segmentation is a management tool that enables the total market to be divided into viable customer groups that can be served by specific market programs. A segment is viable when the size is just large enough to support revenues in excess of the additional marketing costs incurred in reaching that segment. Through this the company can more easily spot and compare marketing opportunities, adjust products and marketing appeals to distinct target groups, and develop marketing programs and budgets with a better idea of the response patterns of the target market. Three basic strategies can result from segmentation analysis: undifferentiated marketing, differentiated marketing, and concentrated marketing.

With undifferentiated marketing, the firm views the total market as homogeneous and aims the marketing effort at the whole potential market. With differentiated marketing, the firm offers modified products and targets different segments of the total market. With concentrated marketing, the firm competes for a large share of a limited or distinct segment, directing products and marketing efforts towards the response characteristic of that segment.

By identifying the market and its characteristics for new products in this manner, a firm benefits in several ways. First, the needs of the market are identified and therefore can be matched with the proposed new product; the market needs can be examined in terms of the degree to which currently available products are filling these needs. Needs of market segments that presently have low levels of satisfaction can be evaluated in terms of the proposed new product. Second, if a certain market is clearly identified, then specific characteristics for the new product can be developed and these desired attributes can be emphasized in the company's promotional strategy. Finally, the entire marketing program can be designed and funded to satisfy the different parts of the market.

Conditions for Product Segmentation

It may appear that the golden rule for new product success is *No Matter What Else You Do, Be Sure to Segment*. With this rule in mind, significant expenditures should be devoted to determining the target group of customers. While it is always better to segment than not to segment, market segmentation for a new product should be carried out according to the factors, benefits, and costs involved. For example there is no benefit in determining that the only market for a new dog treat (a snack food given a dog) are owners of mongrel dogs who sit up and bark. That is, there is no reason to break a market down further than the benefits derived warrant. The goal of market segmentation is to achieve maximum segmentation information per segmentation base. This goal for a new product is achieved when three conditions are met.

The first condition is size. The market segments identified must be large enough to be worth considering. The smallest segment that should be considered is one that, if saturated, would result in meeting the sales and/or profit goals established for the new product. Because effective segmentation requires considerable time and costs, it would not be worthwhile to isolate a market segment that is not large enough to justify the expense.

The second condition is the firm's capabilities. The firm must be capable of successfully marketing the new product to the market segment. If the chosen segment cannot be effectively serviced by the firm, then it is not worth the cost of identification. For example one small electronics firm in Boston determined that its best market segment for a particular new product was in Los Angeles. Yet the company had no distributors, service outlets, or any familiarity with this region. This particular market segment was totally inaccessible to the firm, given its budget constraints.

The final condition is quantifiability. The segmentation criteria for the final market segment must be based on market information that is either already available or that can be obtained through primary research. There is no value in delineating a market segment based on unquantifiable characteristics. Appropriate segmentation criteria for a gas grill would be ease of cleaning and ease of starting. The status that a consumer would attach to owning the grill, however, would be a criterion that could not easily be measured and would, therefore, be inappropriate.

TABLE 12–1

Major new product segmentation (variables and corresponding breakdowns)

Variable	Segmentation Breakdown for New Products
Demographic	
Age	5 and under; 6–10; 11–17; 18–24; 25–34; 35–49; 50–65; over 65
Education level	Some high school or less; high-school graduate; some college; college graduate; advanced college-degree work
Family life cycle	Single; married with no children; married with children under six; married with children over six living at home; married with adult children living out of home; widowed
Family size	1; 2; 3; 4; 5; more than 5
Income	Under $10,000; $10,000–$14,999; $15,000–$19,999; $20,000–$29,999; $30,000–$49,999; $50,000 and over
Nationality	American; British; French; German; Italian; Spanish; other
Occupation	Executive or managerial; professional or technical; teacher or professor; salesperson; government or military; secretarial, clerical, or office worker; craftsman, mechanic, factory worker; homemaker; student; retired
Race	Caucasian; Black; Asian
Religion	Catholic; Protestant; Jewish; other
Residence	Rent; own condominium; own home
Sex	Male; female
Social class	Lower-lower; upper-lower; lower-middle; middle-middle; upper-middle; lower-upper; upper-upper
Geographic	
City (SMSA) size	Under 19,999; 20,000–99,999; 100,000–249,999; 250,000–499,999; 500,000–999,999; 1,000,000–3,999,999; 4,000,000 and over
Climate	Northern; Southern
County size	A; B; C; D
Market density	Urban; suburban; rural
Region of country	New England; Atlantic; Central; Mountain; Pacific
Terrain	Mountains; plain
Psychological	
Life style	Swinger; non-swinger; other
Motives	Emotional; economic
Personality attributes	Extrovert; introvert; dependent; independent; other

TABLE 12–1
(Continued)

Variable	Segmentation Breakdown for New Products
Product-related	
Benefits desired	Durability; dependability; economy
Brand loyalty	Strong; minimal; none
Controllable marketing elements	Advertising; sales promotion; price; service
End-use	Consumption; capital
Volume of consumption	Heavy; medium; light; none

BASES FOR NEW PRODUCT SEGMENTATION

A large number of variables can be used in effective segmentation for most new products. Most of these, along with their categories for consumer products, are listed in Table 12–1. As indicated, the variables can be grouped into four major classes: demographic, geographic, psychological, and product-related.[4] Of course not all variables are appropriate for every market. Relevancy by market type will be discussed later in this chapter.

Groups of people, firms, or governmental units can be assigned to these segmentation classes so that the total market is segmented. As illustrated in Figure 12–3, the firm can then develop a new product and its marketing mix to appeal specifically to a particular segment or segments.

FIGURE 12–3
Market segmentation approach for new products

Selection of Variables

A segmentation variable and its components are the basis for the division of a market into segments. New product segmentation variables (indicated in Table 12–1) are characteristics useful for segmentation.

In deciding which segmentation variables should be used for a particular market, several factors should be considered. In evaluating these factors, one should keep in mind that the choice of the appropriate segmentation variable

---

[4]There are other possible classification schemes available. For example, one good alternative classification scheme can be found in Philip Kotler, *Marketing Management: Analysis, Planning, and Control*, 4th ed. (Englewood Cliffs, N.J.: Prentice-Hall Inc. 1980) 194–212.

is a critical step in sound market segmentation for the new product.[5] Its importance becomes even more evident when one considers the need for good market segmentation for new product success. The factors previously discussed in relation to conditions for product segmentation are similarly important in variable selection. The variables should be selected for market size. It would not be beneficial to use segmentation variables that indicate a market of insufficient size to be profitable. In other words, a sufficient market size should be delineated through any segmentation used. It should be selected in terms of the firm's ability to successfully market the new product if the segment identified so warrants. If it is not within the firm's capabilities and resources to market to the segment indicated, then the segmentation variable should not be used. The segmentation variable selected should also lead to quantifiable results. In other words, information should be available so that sales can be realistically estimated for each of the market segments defined. Finally, underlying each of the factors previously discussed, the segmentation variable has to be selected for the needs, desires, and product use of the total market of consumers, industrial concerns, or governmental units; that is, the total market and its characteristics should be the prime factor in selecting a particular segmentation variable.

Segments are formed by the aggregation of different variables—such as demographics or socioeconomics—shared by customers that can be expected to respond similarly to any marketing mix. The most crucial step in segmenting markets is defining appropriate segmentation variables. As explained below, there are four widely used methods available for defining these variables. When using these methods in determining the appropriate segmentation variable, the competitive environment and market structure must be considered.

A priori   This method defines the segment variables before any research is done. The variables are chosen according to management experience or the competitive environment.

Post hoc/clustering   This method involves conducting marketing research to test the choice of variables. While at least one variable must be chosen a priori, others are derived through a clustering process using customer profile data. Common variables used in this method are attitudes, needs, lifestyle, personality characteristics or benefits sought.

Flexible segmentation   This method is especially good for new products. A preference ranking for the proposed product is performed by a sample of

---

[5]The concept of selection of the best segmentation variable has been evaluated in many ways. It is, of course, a variant with at least the product, market, and usage rate. For example, see Robert D. Hisrich and Michael P. Peters, "Selecting the Superior Segmentation Correlate," *Journal of Marketing* 38, no. 3 (July 1974): 60–63; H. Lee Mathews and John W. Slocum, Jr., "Social Class and Commercial Bank Credit Card Usage," *Journal of Marketing* 33, no. 1 (January 1969): 71–79; John W. Slocum, Jr. and H. Lee Mathews, "Social Class and Income as Indicators of Consumer Credit Behavior," *Journal of Marketing* 34, no. 2 (April 1970): 69–74; James H. Myers, Roger R. Stanton, and Arne F. Hauf, "Correlates of Buyer Behavior: Social Class versus Income," *Journal of Marketing* 35, no. 4 (October 1971): 8–15; James H. Myers and John F. Mount, "More on Social Class versus Income as Correlates of Buying Behavior," *Journal of Marketing* 37, no. 2 (April 1973): 71–73.

consumers. Current brands are then ranked using the same attributes. Finally demographic, psychographic, or other characteristics are correlated with the information from these two rankings.

Componential segmentation This method predicts which personality types will be most responsive to specific product and marketing features. Respondents are chosen on an a priori basis, and their evaluations are obtained on various comparative scales.

Types of Variables

A company that wishes to use a segmentation approach to successfully market a new product can select from a wide variety of segmentation variables. Demographic segmentation variables are the broadest and most widely used set of characteristics to segment markets. For the consumer market, they include age, family size, education level, family life style, income, nationality, occupation, race, sex, and social class (see Table 12–1). For the industrial market, commonly used demographic variables are the number of employees, volume of company sales, amount of profit, and product line. Wide use of this group of variables also occurs in the government market, including additional variables, such as the type of agency, degree of autonomy in decision-making, and size of budget.

Demographic variables are very often used in all markets because generally they are closely related to the needs as well as the purchasing behavior of the market; that is, they generally correlate very closely to sales of many product categories. As we would expect, sales of major durable consumer goods, such as high-quality furniture, are highly correlated with income. Similarly sales of any of the automobile manufacturers are highly correlated with the purchase of their raw materials, such as steel. Another reason for the widespread use of demographic variables is that most are easier to quantify than other types of variables. In fact a significant amount of published data exist for many variables (see, for example, both the *Census of the Population* and the *Census of Business*).[6]

For example DuPont used demographic segmentation when analyzing the market for a new plastic resin—a resin the company believed to be the most rugged plastic developed to date. Because of this characteristic and an initial price of $1.45/pound for orders of 40,000 pounds or more, DuPont felt that the new resin would be used in products such as helmets, agricultural equipment, sporting goods, high-impact gears, toys, and the housings of power tools.[7] The company segmented its market for their resin on the basis of the type of product line of companies in the industrial market.

Geographic variables are the next most widely used set of characteristics to segment markets. Besides being easily quantifiable and having data available, geographic segmentation can provide potential customer categories on

[6]U.S. Department of Commerce, Bureau of the Census, *Census of the Population* (Washington, GPO, 1982), and U.S. Department of Commerce, Bureau of the Census, *Census of Business* (Washington, GPO, 1978).

[7]DuPont Unveils Plastic It Terms Toughest Ever, *Wall Street Journal*, 17 December 1975, Eastern edition 17.

the basis of sales force or manufacturer's representatives' locations. In other words, because the sales force must know what the target market is, the market will have to be divided on the basis of existing sales territory divisions. In addition, through geographic variables, a firm can choose a market in which it enjoys a comparative advantage. A beautician will distinguish between his or her "local" clientele and more distant customers. Until recently, Coors Brewery served primarily Denver and the surrounding area. The Boston Symphony Orchestra primarily serves the Boston area when playing at Symphony Hall but all of New England when playing at Tanglewood during the summer.

The various breakdowns of this variable, as well as the others discussed, are summarized in Table 12–1. As indicated, geographic variables, such as region of the country, market density, and city size, are used in the consumer market. Markets for new products are often divided into regions of the country according to sales territories so that target customers can be introduced to the new product. In addition one or more geographic factors may cause one region to differ from another. A company marketing a product to the consumer market could divide the United States into five regions: New England, Atlantic, Central, Mountain, and Pacific. An even more precise breakdown could also be done. A firm operating in a smaller-than-national market could use more specific geographic segmentation variables, such as counties or zip code areas.

Market density indicates the number of potential customers for the new product per unit of land area, such as a square mile. Although market density is related to city size, it is not necessarily a proportional relationship. For example, in two different geographic markets of about equal size and population, the market density for a new baby's toy will be much higher in one of the areas if that area contains a significantly greater number of babies. Division of the consumer market by density may help the company decide on the advertising, distribution, and sales activities necessary for the new product to reach low-density markets as contrasted with high-density markets.

City size (standard metropolitan statistical area—SMSA) can also greatly influence the amount of exposure per advertising dollar. In addition, a firm can evaluate cities for inclusion in test market selection (discussed in Chapter 6) by sorting the cities into categories according to population, such as under 19,999; 20,000–99,999; 100,000–249,999; 250,000–499,999; 500,000–999,999; 1,000,000–3,999,999; and 4,000,000 and over. The size of the city may influence the need for a new product. For example the demand for a new security device may increase as the city size increases.

The final geographic segmentation variable commonly used in the consumer market is climate. Climate often has a significant impact on customer behavior and needs for new products. Examples of such new products include air conditioning and solar heating equipment, a new type of ski, and new building materials.

For the industrial and government markets, fewer geographic variables are used. Region of the country is commonly used for the industrial market. This is due in part to sales territory divisions, as well as to the geographic clustering that tends to occur for some types of businesses. For example a large part of the textile industry is located on the east coast. The center of the automobile industry is Detroit, Michigan.

The government market, on the other hand, can be divided on a regional basis or on a federal, state, or local basis, depending on the uses of the new product being marketed. This results in different buying practices because the federal, state, and local governments each have individual procurement procedures.

Although there are many different psychological segmentation variables that could be used, perhaps the three most common in the consumer market are life style, motives, and personality traits. Although a psychological variable can be used alone to segment a market, it is usually combined with the other types of segmentation variables, such as demographic and geographic ones. It should be kept in mind that when psychological variables are used in market segmentation, primary research is almost always required because little, if any, published data exist.

When using life style, the firm is evaluating market possibilities for its new product on the basis of the market group's attitudes toward consumption, work, and play. This can result in delineating such groups as "swingers," who seek the latest in new products of various product lines that reflect a high status in society; and "straights," who primarily seek new products that accomplish what they are supposed to. In spite of the fact that life style segmentation usually requires a significant expenditure on research, firms are using it increasingly to define markets for their new products.[8] Volkswagen, for example, introduced what it termed *life-styled automobiles.* For the conservationist there is a Volkswagen that is economical, safe, and ecologically sound. In addition there is a car that handles well and is sporty for the car buff.

Aside from life style, motives are another commonly used psychological variable. Since motives are what move an individual toward a goal, they will influence in varying degrees whether a person will buy the new product. The motives can vary from emotional ones (prestige, belonging, love) to economic ones (product dependability, economy, and convenience). For example, if convenience is the primary motivating factor in whether a new product is purchased, then the number of stores that carry the new product becomes extremely important. Although motives are an excellent basis for psychological segmentation, in that they explain the forces motivating a person's new product purchase, their accurate quantification is, at best, very difficult, and, in some cases, virtually impossible.

The final commonly used psychological segmentation variables are embodied in the general term *personality traits.* Many different personality traits can be used to define potential markets for new products, including extroversion, introversion, degree of dependency, aggressiveness, ambitiousness, or competitiveness. Personality traits are very useful in segmenting the market for a new product when the new product is similar to competing products already on the market; however they are not useful in every case. They seemed to have little effect, for example, on a buyer's choosing a Ford versus a Chevrolet automobile. In the 1950s Fords and Chevrolets were promoted as having different "personalities." Ford buyers were considered independent, masculine, and self-confident; and Chevrolet buyers were considered conservative,

---

[8]See Mark Hanan, *Life-Styled Marketing* (New York: American Management Association, 1972).

thrifty, and middle-of-the-road. In an attempt to measure whether this was true, owners of the two car types were given an Edwards' personal preference test. The Ford owners did not differ to a significant extent from the Chevrolet owners on any attribute except dominance. Therefore it was concluded that "the distribution of scores for all needs overlap to such an extent that [personality] discrimination is impossible."[9] However critics have disputed this finding on the basis of the small localized sample and the statistical methodology employed. Indeed, even though it appears reasonable that personality characteristics would affect buyers' product choice, numerous research efforts since the Evans study have yielded very little supporting evidence.[10]

The usefulness of these psychological variables in the industrial and government markets has had virtually no statistical evidence. However, it is possible that the degree of industrial leadership of the industrial market and the degree of ability to change for the government market could prove to be useful psychological segmentation variables for a new product for each of these markets.

Although psychological variables can provide a very effective mechanism for delineating market segments for the new product, their function as segmentation variables will probably continue to have limited use. These variables are very difficult to measure accurately. Sorely needed, for example, is a better set of tests to measure personality. In addition the relationship between a given psychological variable and actual purchase of a new product is often obscure and untested. Still, the theory behind the promotion of many products, such as Smirnoff vodka and Schweppes basic water, is an appeal to special personality types. Only time and statistical evidence will determine the real applicability of this group of segmentation variables for determining the new product's market.

The final class of segmentation variables for new products is product-related variables. This class of segmentation variables divides the new product market on the basis of the product's attributes and the consumer's relationship to the product. Three main criteria are used: benefits, volume of use, and controllable marketing elements.

One form of product-related segmentation is the use of various benefits the consumers expect from the particular product category. Although the needs of the consumers underlie any segmentation variable used, consumers' needs in respect to product characteristics are the basis of benefit segmenta-

---

[9]The whole series of articles dealing with this study include Franklin B. Evans, "Psychological and Objective Factors in the Prediction of Brand Choice: Ford vs. Chevrolet," *Journal of Business* 32, no. 4 (October 1959): 349–369; Gary A. Steiner, "Notes on Franklin B. Evans' 'Psychological and Objective Factors in the Prediction of Brand Choice,'" *Journal of Business* 34, no. 1 (January 1961): 57–60; Charles Wineck, "The Relationship Among Personality Needs, Objective Factors, and Brand Choice: A Reexamination," *Journal of Business* 34, no. 1 (January 1961): 61–66; Franklin B. Evans, "You Still Can't Tell a Ford Owner from a Chevrolet Owner," *Journal of Business* 34, no. 1 (January 1961): 67–73; Alfred A. Kuehn, "Demonstration of a Relationship Between Psychological Factors and Brand Choice," *Journal of Business* 36, no. 2 (April 1963): 237–41; and Franklin B. Evans and Harry V. Roberts, "Fords, Chevrolets and the Problem of Discrimination," *Journal of Business* 36, no. 2 (April 1963): 242–49.
[10]See Ralph Westfall, "Psychological Factors in Predicting Product Choice," *Journal of Marketing* 26, no. 2 (April 1962): 34–40; *Are There Consumer Types?* (New York: Advertising Research Foundation, 1964) 5–16; and W. T. Tucker and John J. Painter, "Personality and Product Use," *Journal of Applied Psychology* 45, no. 5 (October 1961): 325–29.

tion. In other words, consumers are identified according to the importance of the different benefits they may be seeking from the product.[11] For example one group of automobile buyers may be seeking a product that is economical to maintain, easy to operate, and inexpensive to purchase and repair.

There may be various benefits for the consumer market. Some of these are durability, dependability, convenience, and status from ownership. Although some criteria, such as dependability and durability, are also useful in segmenting the industrial and government markets for new products, other bases take on more importance in these markets. For example the buyer of a new product in the industrial market may be looking for reliability in the seller and support services, efficiency in operation, and/or enhancement of the firm's earnings. For the buyer in the government market, reliability of the seller and the provision of support services, as well as product dependability, are very significant bases of segmentation.

Choosing the benefits to be used as the basis for market segmentation for the new product can be very difficult. This is particularly true in the consumer market. The benefits actually sought by the customers must be identifiable, and these benefits must be the real reasons underlying new product purchase. Although home gardeners may indicate that they buy a particular new brand of fertilizer for ecological reasons, the real reason may be economy. In addition the segment of the market determined through benefit segmentation must be quantifiable as well as accessible to the firm. For example, although a certain segment of the market may want an economical, dependable, no-frills watch, this market may not be accessible to a quality watch manufacturer because of its image and distribution system.

The second type of product-related variable is volume segmentation. In this type of segmentation the market is first divided on a use and nonuse basis. The use segment is further classified by the amount of product consumed; that is, the users of the new product can be divided according to heavy, medium, or light usage or some other more-detailed consumption basis. This segmentation basis is applicable to all three markets.

Once determined, all breakdowns according to volume of use should be evaluated in terms of a viable potential market for the new product. For example nonusers comprise two types of people: those who do not use any product at all in this product category and those who use one. Even those people who do not use any product at all in the product category should be dropped as a possible market segment only after careful deliberation. Similarly, just because a group potentially has the largest consumption rate for a new product, this segment may not be the most profitable to reach because of the required heavy expenditures, brand loyalty, or the high degree of competition. Regardless of the segment finally chosen as the target for the new product, volume of use is a very useful criterion for new product market segmentation.

The final type of product-related segmentation variable commonly used for new products is controllable marketing elements. The firm divides the market into groups that are responsive to different marketing elements within

---

[11]See, for example, Daniel Yankelovich, "New Criteria for Market Segmentation," *Harvard Business Review* 42, no. 2 (March–April 1964): 83–90, and Russell J. Haley, "Benefit Segmentation: A Decision-Oriented Research Tool," *Journal of Marketing* 32, no. 3 (July 1968): 30–35.

TABLE 12–2
Market segmentation for new products by type of market

Segmentation Criteria	Basis for Type of Market		
	Consumer	Industrial	Government
Demographic	Age, family size, education level, family life cycle, income, nationality, occupation, race, religion, residence, sex, social class	Number of employees, size of sales, size of profit, type of product line	Type of agency, size of budget, amount of autonomy
Geographic	Region of country, city size, market density, climate	Region of country	Federal, state, local
Psychological	Personality traits, motives, life style	Degree of industrial leadership	Degree of forward-thinking
Benefits	Durability, dependability, economy, esteem enhancement, status from ownership, handiness	Dependability, reliability of seller and support service, efficiency in operation or use, enhancement of firm's earnings, durability	Dependability, reliability of seller and support services
Volume of use	Heavy, medium, light	Heavy, medium, light	Heavy, medium, light
Controllable marketing elements	Sales promotion, price, advertising, guarantee, warranty, retail-store purchased service, product attributes, reputation of seller	Price, service, warranty, reputation of seller	Price, reputation of seller

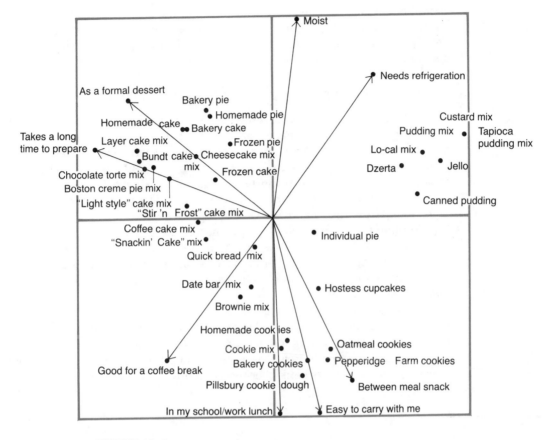

FIGURE 12–4

Example of CATALYST—a mapping and measurement procedure

Source: "CATALYST Measurement, Mapping Method Identifies Competition Defines Markets," *Marketing News* (14 May 1982): Section 1, p. 3.

the firm's control, such as price, advertising, sales promotion, warranty, guarantee, or service. As indicated in Table 12–2, the most important marketing element will, of course, vary by type of market—consumer, industrial, or government. But regardless of the market, once the best market segment is determined on the basis of controllable marketing elements, these elements should receive the most emphasis in the new product introductory marketing plan.

Market segmentation and analysis can also be accomplished through more complex multivariate techniques that were discussed in terms of strategic planning in Chapter 3. While there are many techniques available, they all have similar treatments and outputs. One new technique for identifying competitive structures in broadly defined consumer markets is CATALYST (CATegory anALYST).[12] Figure 12–4 indicates dessert categories as determined by CATALYST. Arrows represent key usage occasions and product attributes. Each product's proximity to each other product, as well as to the arrows, indicates the degree of substitutability. This identifies the product category

---

[12]For a discussion of this technique, see "CATALYST Measurement, Mapping Method Identifies Competition, Defines Markets," *Marketing News* (14 May 1982): 3.

from the consumer's viewpoint: cookies are at the bottom; pies and cakes in the upper-left quadrant; and custards, gelatins, and puddings in the upper-right quadrant. A new product's position on the map will indicate its perceived market niche and opportunity, as well as its competition.

Market-Gridding in New Product Market Selection

A useful technique for evaluating the potential target markets for a new product is the market grid. A market grid concept depicts the total market in a two- or three-dimensional manner, based upon selected, relevant market characteristics. An example of a market grid is shown in Figure 12–5. Notice that only potential customers in the target market are shown in the lined area. Customers who are not part of the potential market for the new product but who are a part of the total market are in the "all others" category. It should be kept in mind that the usefulness of the market grid as an evaluative technique is solely dependent upon the segmentation variables selected and their respective classifications. This is the most important part of successful market segmentation for the new product and should, therefore, be subject to a great deal of painstaking analysis.

Using a market grid for selection of the target market(s) for a new product can be well illustrated in the banking market. The specific grids are shown in Figures 12–6 and 12–7. When looking at the grids for this market, one should keep in mind that we are evaluating potential customers for a new banking concept, not the various bank offerings presently on the market.

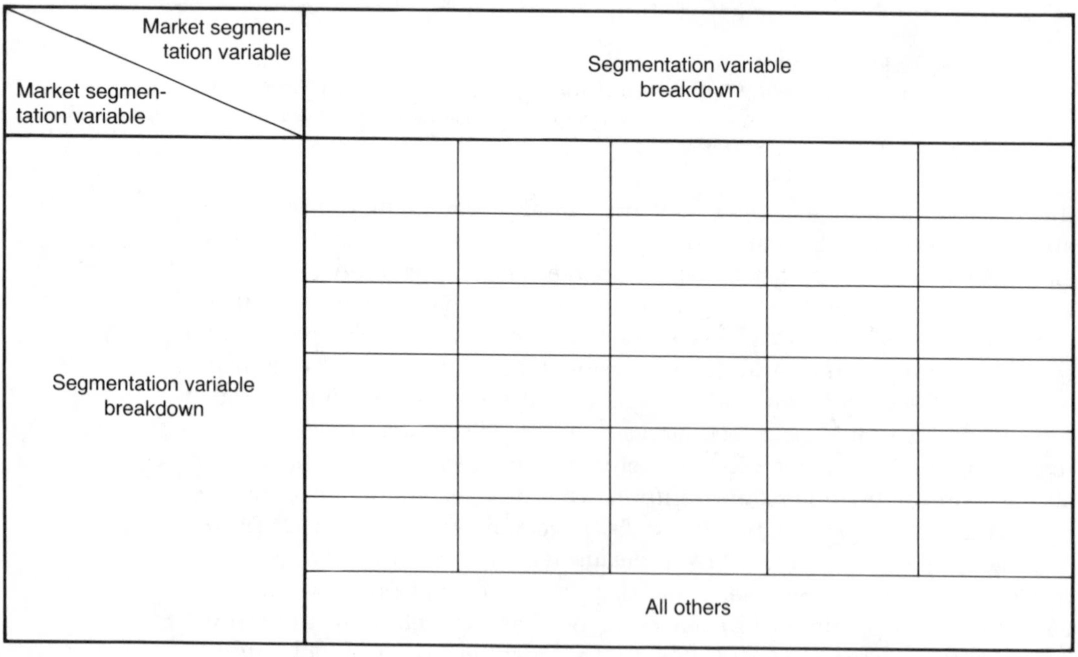

FIGURE 12–5
Market grids for new product market selection

Type of customer / Region of country	Urban	Suburban	Rural
Consumer market			
Industrial market			
Government market			
All others			

FIGURE 12–6
An example of a general market grid for the services of a new commercial bank

Type of business / Size of business	Small	Medium	Large
Manufacturing			
Retail trade			
Service industries			
Transportation			
Utilities			
Wholesale trade			
Construction			
Insurance and real estate			
All others			

FIGURE 12–7
Market grid example for the services of a new commercial bank for the industrial market

Type business / Bank service	Insurance and real estate	Construction	Wholesale trade	Utilities	Transportation	Service industries	Retail trade	Manufacturing
Commercial loans								
Checking accounts								
Savings accounts								
Safe-deposit box								
Branch banking								
After-hours depository								
Bank-by-mail								
Lock box plan								
Account reconcilement plan								
Freight payment plan								
Payroll accounting plan								
Commercial mortgages								

FIGURE 12–8

Market grid example for the service of a bank for the industrial market

Executive welcome service							
Industrial development services							
Corporate trust services							
Custodian services for securities							
Credit inquiries							
Wire transfer of funds							
Foreign and domestic collections							
Payroll and excise tax depository							
Trust services							
Commercial real estate mangement							
Pension and profit sharing plans							
All others							

FIGURE 12–8 (Continued)

To ensure that every possible customer is analyzed, customer characteristics are used first. As shown in Figure 12–5, the type of customers (a demographic segmentation variable) is analyzed by market density (a geographic segmentation variable). Assuming that the industrial market is the prime target for the bank, further analysis of this market is needed.

Two different approaches to the analysis of this market are shown in Figures 12–7 and 12–8. Figure 12–7 shows an analysis of the industrial market for a commercial bank by two demographic segmentation variables—size of business and type of business. Note that the size of business is differentiated on a very general basis: small, medium, and large. It can be made much more specific by using categories such as the number of units sold, number of people employed, or dollar amount of profit. In addition, not all possible types of businesses in the industrial market are shown. Only those considered to have enough potential to be in the target market are included. Other businesses without this potential would be included in the "all others" category.

Figure 12–8 illustrates another approach to the analysis of the industrial market as the potential target market for the commercial bank. This example uses a demographic segmentation variable (type of business) and a controllable marketing element segment variable (services that could be offered by the bank). This grid is an attempt to determine the amount of use each type of business would have for each of the potential services of the bank. The resultant analysis would show the prime candidates in the business market for all of the bank's services, as well as indicate the potential usage rate of each service. Note that none of the sections of the selected segmentation variables is all inclusive—hence, the need for the "all other" category.

SUMMARY

The importance of market segmentation for success in marketing new products cannot be overemphasized. This is particularly true as competition becomes heavier and more new products inundate the market.

The benefits of new product segmentation are many, particularly when the proper conditions for product segmentation are met. The markets for the new product can be segmented according to demographic, geographic, psychological, and/or product-related variables and the use of market grids. To be ultimately useful in the successful marketing of the new product, the segments must be of sufficient size, related to the firm's capabilities, and quantifiable. Indeed the importance of segmentation for new products has never been greater.

SELECTED READINGS

BOOTE, ALFRED S. "Market Segmentation by Personal Values and Salient Product Attributes." *Journal of Advertising Research* 21, no. 1 (February 1981): 29–35. Describes an application of an approach to market segmentation that relies on personal values as the key variable in the underlying prediction model. Ties values to three other psychographic variables and recommends mix changes based on the findings.

COSMAS, S. C., and SHETH, J. N. "Identification of Opinion Leaders Across Cultures: An Assessment for Use in the Diffusion of Innovations and Ideas." *Journal of International Business Studies* 11, no. 1 (Spring–Summer 1980): 66–73.
Presents Delphi-type interviews conducted with five cultural groups of students. Reveals that there is a set of common dimensions by which opinion leaders are evaluated, but that different cultures assign different weights to the dimensions.

CURRIN, I. S. "Using Segmentation Approaches for Better Prediction and Understanding from Consumer Mode Choice Models." *Journal of Marketing Research* 18, no. 3 (August 1981): 301–9.
Investigates the diagnostic and predictive efficacy of market segmentation and the relative power of two segmentation schemes using a market share probabilistic choice model (LOGIT) as dependent variables. Research concludes that the models provide fairly accurate estimates of market share and that using the segmentation concept affords diagnostic and predictive advantages.

JOHNSON, HAL G., and FLODHAMMER, ABE. "Some Factors in Industrial Market Segmentation." *Industrial Marketing Management* 9, no. 3 (July 1980): 201–5.
Examines the complex buyer-seller relationships of industrial marketing systems, proposes a model for segmenting industrial markets, and summarizes implications for organization buying behavior.

KESTENBAUM, R. D. "List Segmentation: Profit vs. Privacy." *Direct Marketing* 44 (August 1981) 24.
One of the problems with market segmentation techniques is the extent to which personal privacy is jeopardized. The author reviews the tradeoffs between privacy invasion and profit accumulation.

LAURENT, C. R. "Image and Segmentation in Bank Marketing." *Bankers Magazine* (July–August 1979): 32–37.
Discusses how perceptual mapping of the qualitative attributes of banks shows that private depositors do indeed have distinct images of competing banks, that these images can vary substantially, and that segmentation based on these perceived differences is crucial to marketing success.

STEWART, DAVID and HOOD, NEIL. "A Methodology for the Evaluation of Segmentation Policy: An Empirical Application in the Car Industry." *Journal of the Market Research Society* 23, no. 3 (July 1981): 137–49.
Reviews car-industry segmentation practices and contends that the methodology has general-application potential.

WATANABE, MARK. "A Profile Grows to New Heights." *Advertising Age*, 6 April 1981, S1.
Ad Age Report on the emergence of Spanish-speaking people as a market segment. Includes reports by agency executives, broadcast executives, print media experts, tailoring products to Spanish-speaking audiences, and a market profile.

WINTER, FREDERICK W. "A Cost-Benefit Approach to Market Segmentation." *Journal of Marketing* 43, no. 4 (Fall 1979): 103–11.
Argues that it is important to include cost-benefit issues in determining the appropriate level of aggregation in market segmentation plans. Offers an alternative view of normative segmentation and suggests modifications to standard segmentation practice that are necessary for decision-making.

WINTER, FREDERICK W. "March Target Markets to Media Audiences." *Journal of Advertising Research* 20, no. 1 (February 1980): 61–66.

Asserts that the most valid means of matching segments to media involves simultaneous measurement and direct matching. Simple cross-classification or calculations of cross products are most valid.

YOUNG, S. "Fine-Tuning Your Sell for '80's Me-Generation." *Marketing Times* 10, no. 1 (January 1980): 41–44.

The Me-generation has attracted increased market attention. Suggests several fine-tuning techniques to manage this large and expanding market to enhance your firm's sales.

ZUFRYDEN, F. S. "ZIPMAP—A Zero-One Integer Programming Model for Market Segmentation and Product Positioning." *Journal of the Operational Research Society* 30, no. 1 (January 1979): 63–70.

Proposes a model to help search for and develop new product brand ideas. The model, based in individual consumer behavior constructs, determines a position for a new brand in the perceptual space of product attributes, maximizing company sales.

13

Diffusion and Adoption
of New Products

A s discussed in Chapter 1, one of the major reasons for product failure is inadequate marketing strategy and planning, particularly with regard to product decisions, distribution, and promotion. Since one of their primary roles is to integrate new products into the cultural composition of the society, marketers should have a more comprehensive understanding of the process of new product adoption. This understanding of new product adoption can be enhanced in part through research that focuses on those features that influence the diffusion of new products into the social system.

HISTORICAL PERSPECTIVE OF DIFFUSION RESEARCH

The process by which an innovation develops from its inception to its actual use by purchasers is referred to as diffusion. Diffusion research studies have been carried out for about 50 years in many different disciplines. Interest began with rural sociologists attempting to understand how new farm technology could be more rapidly diffused in the farm community. This study was necessary because of technological changes occurring in our society in the early 1940s. At that time the average American farmer had to help feed a growing number of people, necessitating large increases in the average crop yield through new technological development. Innovations were designed to improve the yield, but there was a considerable time lag between their introduction and their general acceptance. The sociologists sought techniques and strategies by which to hasten this diffusion.[1]

Although it took the initiative in diffusion research, rural sociology is not the only discipline that has contributed to understanding the diffusion process. Research studies have focused on innovations in anthropology, sociology, education, medicine, economics, and industrial and consumer goods marketing.[2] Marketing's interest in diffusion stems from an attempt to increase the probability of success of a new product through an increased understanding of the factors governing the diffusion process. However, until recently, research in the marketing area has been minimal. Its importance as a means of improving the new product development process is only now being fully appreciated.

THE DIFFUSION PROCESS

The diffusion process is often confused with the adoption process. The adoption process is the development of the consumers' awareness of a product—

---

[1] For a comprehensive presentation of the diffusion processes see Everett M. Rogers, *Diffusion of Innovations* (New York: The Free Press, 1962), and Thomas S. Robertson, *Innovative Behavior and Communication* (New York: Holt, Rinehart & Winston General Book, 1971).
[2] A listing of all published diffusion research in all disciplines is maintained at the Documents Center at Michigan State University. These publications are listed in Everett M. Rogers, *Bibliography on the Diffusion of Innovation* (East Lansing, Michigan: Department of Communication, Michigan State University, 1971).

ranging from the point when they first become aware of a new product's existence to the point when they actually use or adopt the product on a regular basis. Hence the diffusion process is the aggregate of all individual adoptions over time.

Elements of Diffusion

Four crucial elements are identified in the diffusion process: (1) the innovation, (2) the communication from one individual to another, (3) the relevant social system of which these individuals are a part, and (4) the time dimensions in the process.

The innovation   An innovation is an idea or a product that the buyer perceives as being new. As discussed in Chapter 1, the basis of defining a product as "new" seems to depend on the disrupting effects it has on established behavioral patterns. It appears that more meaningful theory development could occur if research studies in marketing were classified according to the relative newness of the product. Then marketers could refer to research findings categorically related to the new product they are introducing. These findings would be useful to their new product introduction, since similar results may be expected with products having similar degrees of newness.

The communication   The essence of the diffusion process involves human interaction where one person transmits information regarding an innovation to another person. This communication, which can occur through formal or informal channels, is necessary in order for diffusion to take place.

The social system   A social system is defined as a population of individuals who are functionally differentiated and engaged in collective problem-solving behavior. Thus the social system for a new product would be all those individuals or firms in a specified area who can use the product. This specified area may represent a test market where a new product is being tested for possible large-scale market introduction. (See Chapter 5 for a discussion of test marketing new products.) The information on the diffusion of a new product in the test market could be quite useful in the design of a marketing plan by the firm attempting to achieve large geographical or national distribution. When all individuals or units in a specified area have adopted the innovation under study, the diffusion process is complete.

Some confusion exists as to what is meant by the adoption unit in the social system. There actually exists a continuum of adoption units ranging from individual choice to a group decision. At one extreme we find innovations that are adopted by an individual; although this individual may be influenced by others in the social system, he or she basically makes the final choice. At the other extreme are innovations that are adopted by a group. Many community decisions would fall into this latter class.

The time dimension   Not all people adopt an idea or new product at the same time. Thus time in the diffusion process refers to how long after introduction an individual or unit decides to adopt an innovation. For example Figure 13–1 shows that housewife A heard a commercial for the new Silkience

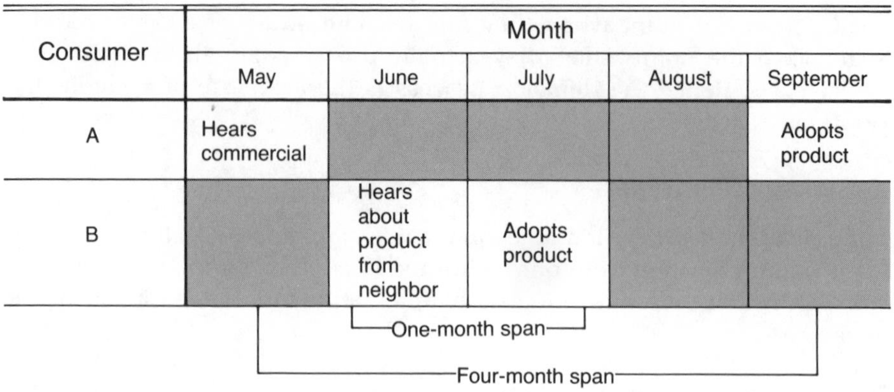

Consumer	Month				
	May	June	July	August	September
A	Hears commercial				Adopts product
B		Hears about product from neighbor	Adopts product		

One-month span

Four-month span

FIGURE 13–1
Examples of time spans incurred before actual adoption

shampoo but did not actually begin using this product on a regular basis until September. Housewife B heard of Silkience from her neighbor in June, but she began to purchase this product on a regular basis in July. It can be seen that the time needed to adopt the new product by each housewife differs significantly. Marketers would be quite interested in knowing not only why it took housewife A so long to adopt the innovation, but any individual and environmental variables distinguishing A from B.

One attempt to establish adopter categories was developed by rural sociologists classifying buyers according to when they adopt an innovation. These categories are discussed in the following paragraphs.

Adopter Categories

Figure 13–2 depicts the diffusion curve partitioned into categories of adopters over time. The distribution of adoption of any innovation over time has been found to resemble a normal distribution. This is supported by Rogers, who found that in eight different studies adopter distributions all approached normality.[3] The categories established by the rural sociologists assume this normality condition since each category is represented as an area that is a specified number of standard deviations from the mean. Thus the area under the curve is used to determine the percentage of people or units who have adopted in that time span. The innovators represent the first 2.5 percent to adopt, the early adopters represent the next 13.5 percent and so on. Although this categorization is arbitrary, it has been meaningful to rural sociologists in studying the adoption of agricultural innovations. However much of the marketing diffusion research deviates from the percentages and labels that rural sociologists established to identify each adopter category. For example marketing researchers have been mainly concerned with distinctions, such as those between innovators and noninnovators or between early adopters and later adopters. In each case the categories were identified using different percentages. The use of any percentage is strictly arbitrary and usually is chosen

---

[3]Rogers, *Diffusion of Innovations*, 158.

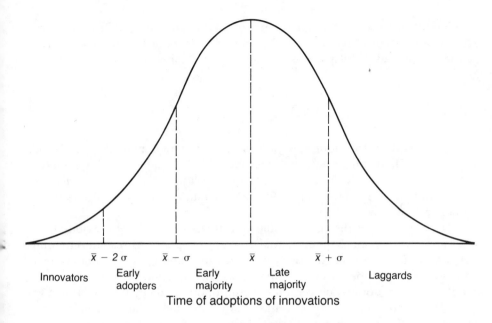

$\bar{x} - 2\sigma$ $\bar{x} - \sigma$ $\bar{x}$ $\bar{x} + \sigma$

Innovators | Early adopters | Early majority | Late majority | Laggards

Time of adoptions of innovations

FIGURE 13–2
Market segments identified by time of adoption of new product
Source: Everett Rogers, *Diffusion of Innovations*, The Free Press, 1962, 162.

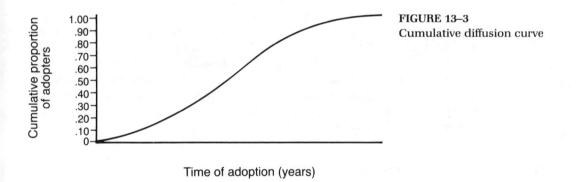

FIGURE 13–3
Cumulative diffusion curve

for its convenience to the researcher. This issue is discussed further at the end of the chapter.

As shown in Figure 13–3, the diffusion process can also be depicted as a cumulative curve resembling an S-shape pattern very similar to the familiar product life cycle discussed in Chapter 1. However the cumulative curve is based on the proportion of consumers in any given potential market who adopt the innovation, whereas the vertical axis for the product life cycle represents absolute dollars of sales.[4] The identification of the potential market (depicted as 1.00 in Fig. 13–3) would depend on what the firm felt it could achieve given certain market conditions. One way of not only determining this

---

[4]For a more detailed discussion of the product life cycle and cumulative diffusion curve, see Robertson, *Innovative Behavior*, 30.

percentage but also of studying the nature of the cumulative diffusion pattern would be to use test markets and then make projections based on an analysis of the sample data. This process is described in Chapter 6.

THE ADOPTION PROCESS

As stated earlier, the adoption process is concerned with the individual, whereas the diffusion process is concerned with aggregate behavior. Researchers studying the individual have discerned certain behavioral tendencies or mental patterns that have enhanced the understanding of why and how consumers adopt new products.

However before discussing some of these mental and/or behavioral tendencies, we want to focus on the actual meaning of adoption. In some situations *adoption* refers to the single purchase of the innovation, but in others it refers to a series of purchases of the innovation by the same person. The key to identifying adoption depends on the commitment made by the consumer. For example when a consumer purchases any large-ticket item he or she has actually made a commitment to that product for an extended period of time. However the single purchase of a new household cleaner may not represent any actual commitment at all, since the consumer could switch brands without ever repurchasing the new product in question. Thus for a product having a short life-span, it it necessary to identify some commitment on the part of the consumer by determining the number of purchases of the new product he or she makes. Of concern is what number of repeat purchases represents adoption. Some researchers have distinguished adopters on the basis of usage rate of any new product. For example, Frank and Massy categorized buyers of Folger's coffee according to the proportion of Folger's coffee purchased to the total pounds of coffee purchased in a given period of time.[5] The definition of adoption for frequently purchased products varies according to the parameters established by the researcher.

Stages in The Adoption Process

From the time when an individual first hears of an innovation to the time when adoption occurs has been recognized as consisting of five sequential stages:

1. *Awareness stage* The individual is exposed to the innovation but lacks complete information about it, or the individual is aware of the innovation but is not yet motivated to seek further information.
2. *Interest stage* The individual becomes interested in the new idea and seeks additional information about it. The innovation is favored in a general way, but is not yet judged in terms of its utility to a specific situation.

---

[5]Ronald E. Frank and William F. Massy, "Innovation and Brand Choice: The Folger's Invasion," *Proceedings of the American Marketing Association* (December 1963): 102.

3. *Evaluation stage* The individual mentally applies the innovation to his or her present and anticipated future situation and then decides whether to try it.

4. *Trial stage* The individual uses the innovation on a small scale in order to determine its utility in his or her own situation.

5. *Adoption stage* The individual decides to continue the full use of the innovation.

Note that any innovations may be rejected at any stage of the adoption process. For example a woman may hear about Silkience but, since she uses a dandruff shampoo, rejects the innovation at the awareness stage.

Rejection of an innovation can also occur after an individual has adopted it. This behavior is referred to as a *discontinuance*. Discontinuances in marketing are quite common, especially since consumers often purchase everyday items without any special loyalty to one brand. Thus, in marketing, the importance of defining adoption as a function of the commitment made by the consumer becomes more critical.

A consumer can also move through several of these stages simultaneously, as in the case when he or she buys something impulsively or receives a free sample of a new product. Such instances do not destroy the argument for the existence of adoption stages, but merely compresses them into a shorter time span. The strict sequencing of these stages has been criticized. The beginning stage, for example, may be the perception of a problem that seems to concur with the interest stage discussed earlier, making it possible for the problem or interest to precede awareness.

Increased understanding of the sequencing of adoption stages and the types of communication most effective in each stage will enable the firm marketing a new product to develop a promotional campaign consistent with the behavior of the consumer. Initially more mass media should be employed to relay product awareness and information on product utility. As time passes, promotion should include a more personal appeal, since the consumer generally becomes more dependent on others for evaluating, trying, and adopting the new product. The strategy employed, however, must give some consideration to the nature of the new product. Since the characteristics of the new product will affect the adoption process, it may be necessary to alleviate any negative characteristics of the product that may slow its rate of adoption in the marketplace.

Effects of Consumer Risk in Adopter Categories

The movement through stages in the adoption process seems to depend on two important variables: need and risk. A consumer is much more likely to go through the adopter categories for a necessary product than for an unnecessary product. Risk on the other hand plays an important role in all stages of the adoption process.

Consumer risk is subjective and varies from one individual to another because of the process of perception. One person may perceive a new product as having a high risk and another person may perceive it as having a low risk. These individual perceptions will vary from one extreme to another, depending on the product category. Risk is perceived relative to the ability of the product in question to satisfy certain individual consumer requirements. Thus a product is risky not just because of safety considerations but because of financial, social, and psychological implications. That is, the product may not fit the image a consumer wishes to convey to others, may not be acceptable in the consumer's peer group, or may be too costly given that an acceptable alternative already exists.

Management must consider consumer risk in the introduction of any new product. Media planning, demonstrations, comparisons, testimonials, free samples, coupons, trial sizes, and guarantees can all be used to minimize risk. As risk in the purchase of a new product is reduced, more commitment or adoptions are likely to occur.

Characteristics Of the Innovation

In marketing a new product special emphasis must be given to particular product attributes and the brand image created by the manufacturer. The perceived attributes or characteristics of the innovation may regulate the rate at which diffusion occurs. These characteristics are relative advantage, compatibility, complexity, divisibility, and communicability.

Relative advantage   This is the degree to which an innovation is superior to one it supercedes or competes with in the marketplace. A detergent that did not contain phosphates, for example, represented a perceived relative advantage over the old product it replaced or any competitive product that still maintained its traditional formula. However the addition of green crystals to a detergent was less likely to be perceived by the consumer as having any relative advantage. The strategy used by most firms is to seek product differentiation that results in a relative advantage over any existing products. A new wonder drug that could cure a sore throat within hours would certainly have an advantage over any other available product and would probably increase the rate at which this product would be diffused. It should be emphasized, however, that the relative advantage of an innovation depends extensively on the perception of the members of a given social system.

Compatibility   This refers to the degree to which a new product is consistent with existing values and experiences of the consumers. A new product whose perceived image is not consistent with cultural norms will diffuse less rapidly than one that is consistent. Also, new products that are similar to other products that have been failures will negatively affect the rate of diffusion.

Instant coffee, for example, experienced some initial barriers in achieving its predicted success. The product was promoted as a time-saver for housewives. But the housewife during this period often associated the use of time-saving products with laziness and thus tended to reject the instant coffee. Research revealed this attitude and advertising was corrected to be more compatible with existing norms and values.[6] When television was introduced,

---

[6]Mason Haire, "Projective Techniques in Marketing Research," *Journal of Marketing* 24 (April 1960): 649–52.

it was found to be more consistent with the values of lower classes and hence diffused more quickly at that level.[7]

It appears that the marketer should emphasize that the new product is consistent or compatible with existing norms or values so that the probability of its success and rapid diffusion may be enhanced.

Complexity   This term refers to the degree of difficulty in understanding or using an innovation. If the consumer is unable to evaluate (evaluation stage) a new product's utility, diffusion in a given market will be delayed. Products that require new knowledge also appear to take longer to diffuse. For example Graham concluded that one reason that canasta and television diffused at different rates was the difference in complexity of the two ideas.[8] Canasta generally required some personal explanation, whereas television usage was relatively simple and required little explanation.

Divisibility   This is the degree to which a new product may be tried on a limited basis. The ability of the consumer to purchase small, trial-size packages of new products at a low cost may encourage the rate of adoption. The amount of risk associated with such a small-scale purchase is minimal and allows consumers to examine an alternative to their present brand. Other products that are sold with the added feature of a ten-day, free home trial attempt to accomplish the same effect as the small trial size. Many other products, however, cannot be sold on a small scale or with the ten-day, free home trial. Industrial products, such as computers and machines, are difficult to move and thus are more permanent purchase decisions.

Marketers in recent years have offered more trial sizes for their new products. Free samples given by manufacturers have also shown substantial increases. One example of this increase occurred in 1967 when Colgate-Palmolive mailed free sample boxes of Axion to 50 million of the nation's 60 million households. Gillette has recently chosen to extend its sampling of products (e.g., Dry Idea) that are in the third year of commercial marketing. Indications are that this approach costs less than the same dollar amount of advertising needed to get the consumer to the trial stage in the adoption process. Marketers must be concerned with making it as easy as possible for the consumer to try a new product with little risk. Thus, as long as a product category is growing, sampling may be used to enhance the diffusion process.

Communicability   This refers to the degree to which information regarding a new product may be easily communicated to other people in the marketplace. Products that are visible to others, such as clothing and automobiles, and products that can be easily demonstrated, such as televisions, stereos, and other small appliances, generally fall into the high-communicability category. Generally products that are very complex will be most difficult to communicate to others. The toy industry is a good example of a situation where product communicability is critical. Even though a new game has been thoroughly

---

[7]Saxon Graham, "Class and Conservativism in the Adoption of Innovations," *Human Relations* 9 (February 1956): 91–100.
[8]Saxon Graham, "Class and Conservatism in the Adoption of Innovations," *Human Relations* 9 (February 1956): 91–100.

tested for understanding, interest, and willingness to buy, consumers will not usually purchase a new game or toy unless they perceive it as being simple to use and understand. Consumers do not like to read complicated instructions and, unless the benefits of a new game can be clearly communicated, the product's adoption rate will be severely affected.

All of the above factors have been found to have a significant effect on the diffusion process.[9] The relative influence of each would of course vary depending on the product and industry. Any one of these factors could destroy a new product, even though the product satisfied the other four factors. Thus it is important for the firm to recognize the factors inherent in the purchase decision and to focus its marketing efforts accordingly to ensure the success of their new product introduction. The firm also must identify the relative weight or importance of each factor in the purchase decision process. For example a new product may have a relative advantage over competitive products but be too complex to convey these benefits to the marketplace or incompatible with existing cultural norms. Alternatively a new product may be consistent with cultural norms, simple to understand, and divisible but offer no new advantage over other competitive products.

Management should carefully evaluate the relevance of each of these five factors and the relative importance of each to the consumer decision-making process. This important information can then be used in planning and implementing marketing strategy. If the product is complex, promotional strategy, selective selling techniques, labeling, and so on can be incorporated in the marketing program to accommodate this negative attribute. Divisibility may not always be possible because of the nature of the product. As an alternative, however, the firm may offer a guarantee, free maintenance, home trial, or some other service to alleviate the possible concern of the consumer. Promotional strategies, such as demonstrations, can also improve the communicability problem. Nevertheless the new product must be carefully analyzed to identify any of the preceding attributes so that marketing strategies to remedy these consumer perceptions can be employed.

CORRELATES OF INNOVATIVENESS: EMPIRICAL FINDINGS

The findings pertinent to the development of a diffusion research tradition in marketing are discussed in the following paragraphs. These contributions should improve new product introductions—in particular by helping to develop a more effective marketing strategy. As you may recall, a poor marketing strategy has proved in the past to be one of the major reasons for new product failures.

Rural Sociology

Most of the pertinent research in the diffusion area is found in the rural sociology literature. A general summary of these findings involving farm inno-

---

[9]Everett Rogers and David J. Starfield, "Adoption and Diffusion of New Products: Emerging Generalizations and Hypotheses," In *Applications of the Sciences in Marketing Management*, ed. Frank M. Bass et al. (New York: John Wiley & Sons, Inc. 1968), 227–50.

vations indicates that earlier adopters, in contrast to later adopters, are generally younger, have a higher social status, have a more favorable financial position, have more specialized farm operations, have a different type of mental ability are more cosmopolitan, and have more opinion leadership.

A study by Rogers, specifically concerned with personality traits, found that certain traits were statistically significant in relation to innovativeness.[10] Of seven personality variables studied in relation to adoption, rigidity, change orientation (general attitude toward new technological practices), innovative proneness (desire to seek out changes), and adoption self-ratings were found to be significantly related. Although change orientation, innovative proneness, and adoption self-ratings were positively related to adoption, rigidity was negatively related to adoption.

Economics

Most of the research concerned with industrial product adoption has been published by economists. However the results are quite relevant to marketers who want to increase their understanding of the diffusion of new industrial products.

In one study Mansfield found that the size of the firm and the profitability of an innovation to the firm explained about 50 percent of the variation in time of adoption.[11] Brozen, on the other hand, suggested that a positive relationship between size of the firm and innovativeness may not exist, and therefore the largest firms in an industry may not be the most innovative.[12] Carter and Williams found that the firm's growth rate was positively related to the "technical progressiveness" of the firm. The higher the growth rate, the more likely the firm was to have adopted innovative management practices and policies.[13] These studies focused strictly on environmental variables and not on the individual characteristics of the buyer(s).

Marketing

In the marketing diffusion literature, studies have focused on both consumer and industrial products. Although the majority of these studies were concerned with consumer goods, more attention is presently being given to industrial products. Empirical findings in both areas relevant to marketing will therefore be discussed separately.

Consumer Goods

While the existence of stages has been researched extensively by rural sociologists, marketers have recognized that variations in diffusion theory do occur when crossing discipline boundaries. Thus some research has focused on validating the existence of stages in the consumer goods adoption process.

---

[10]Everett M. Rogers, "Personality Correlates of the Adoption of Technological Practices," *Rural Sociology* 22 (September 1957): 267–68.

[11]Edwin Mansfield, "The Speed of Response of Firms to New Techniques," *Quarterly Journal of Economics* 77 (May 1963): 290–311.

[12]Yale Brozen, "Invention, Innovation, and Imitation," *American Economic Review* 41 (May 1951): 239–57.

[13]C. F. Carter and B. R. Williams, "The Characteristics of Technically Aggressive Firms," *Journal of Industrial Economics* 7 (March 1959): 87–154.

Robertson found that stages did exist and that innovators and noninnovators responded differently to information in each of these stages.[14] In their study of a new drug called Gammanym, Coleman, Katz, and Manzel found that stages did exist and that, contrary to rural sociology findings, personal sources of information were more important at the awareness stage. In the first six months of the product's existence, most doctors sought advice from their colleagues when first becoming aware of the new drug. These networks seemed to disappear after six months when the doctors were in the later stages of the adoption process. During these later stages, the doctors sought information from outside sources, such as medical journals and detail representatives before actually adopting the product.[15]

The existence of stages provides some important insight for management in the introduction of a new product. It indicates that certain types of communication will be more effective at different stages in the adoption process. Thus, as in the new drug product where personal sources were more influential in creating awareness, management may need testimonials from early innovators to appeal to other potential users. In other instances a more informative appeal using mass media may be necessary to achieve awareness and interest in the new product.

Most of the research in marketing has focused on the correlates relevant to the adoption process, with the intent of profiling consumers based on their time of adoption (i.e., whether they were innovators or early adopters). These studies have found that there were significant differences in innovators, early adopters, later adopters, early majority, and laggards. Typically major differences have been found in demographic variables, such as age, occupation, education, income, spouse's employment status, family size, and ethnic group; shopping variables, such as stores shopped in, amount of product purchased, number of shopping trips, and influence of advertising; and personal and social characteristics, such as venturesomeness, social mobility, cosmopolitanism, social integration, social class, and perceived risk.[16]

Each of the adopter categories provides some meaningful basis for market segmentations in the consumer goods category. As long as there are indications that one adopter category is significantly different from another, management can develop marketing strategy to effectively reach each segment. Timing is a critical factor in the successful marketing of new products and, if profiles of adopter categories are meaningful, marketing managers will find the information very useful in developing their marketing plans to coincide with the diffusion process.

---

[14]Thomas S. Robertson, "Purchase Sequence Responses: Innovators vs. Non Innovators," *Journal of Advertising Research* 8 (March 1968): 47–52.

[15]James Coleman, Elihu Katz, and Herbert Menzel, "The Diffusion of an Innovation Among Physicians," *Sociometry* 20 (December 1957): 253–70.

[16]Demographic variables are cited in William E. Bell, "Consumer Innovators: A Unique Market for Newness," *Proceedings of the American Marketing Association* (December 1963): 85–95. Shopping variables are cited in Frank and Massy, "Folger's Invasion," 105. Personal and social characteristics are cited in Thomas S. Robertson and James N. Kennedy, "Prediction of Consumer Innovators: Application of Multiple Discriminant Analysis," *Journal of Marketing Research* 5 (February 1968): 64–69; Donald T. Popielarz, "An Exploration of Perceived Risk and Willingness to Try New Products," *Journal of Marketing Research* 4 (November 1967): 368–72; and Lyman E. Ostlund, "The Interaction of Self-Confidence Variables in the Context of the Innovative Behavior," *Proceedings of the American Marketing Association* (Fall 1971): 351–357.

Industrial Goods

As previously mentioned, little research has focused on the diffusion process in the purchasing of new industrial products. Only two published studies have looked at the existence of stages in the industrial product process. A pilot study by Ozanne and Churchill found some evidence to validate the five-stage model. In a reversal of previous results, however, these researchers found that personal sources of information were more prevalent in the later stages.[17]

A second study by Peters and Venkatesan also supports the existence of the stages. However, in this study involving a small computer, personal sources of information were important in all stages of the adoption process.[18] Therefore it appears that the relevant sources of information at each stage may be dependent on the nature of the innovation. An innovation in which personal sources of information are more important at the later stages will probably require trade and media advertising, brochures, and perhaps trade shows as a means of building awareness and interest. These methods of communication would precede the personal selling effort. For other innovations, the personal selling effort will occur simultaneously, as support from media promotion may be necessary to obtain actual adoption by the consumer.

Most of the research in this area is concerned with identifying the significant variables in the industrial product adoption process. A study conducted by Robinson, Faris, and Wind indicated that determinants of an industrial buyer's behavior include psychological and behavioral characteristics, as well as organizational and environmental variables. Perceived risk and attitude toward the supplier were apparently relevant to the new industrial product adoption.[19] Cardozo found in a preliminary study that variables, such as the buyer's perceived risk and self-confidence, and environmental factors, such as profit margin, competitive advantage, and type of firm, were related to adoption behavior.[20] Webster also supports the importance of behavioral variables in the industrial product adoption process.[21]

Peters and Venkatesan found that the adoption of a small computer was related to 1) individual behavioral variables, including perceived risk, specific self-confidence, attitude toward supplier, and open-mindedness; 2) demographic variables, including the buyer's education, knowledge of computers, and number of prior jobs held; and 3) environmental variables, including the size of the firm, growth rate, industry type, and prior data processing (EDP) equipment for service used.[22] Doubts of validity and reliability, however, still surround much of this data. Future studies that duplicate this research will give us more conclusive evidence regarding new industrial product adoption.

---

[17]Urban B. Ozanne and Gilbert A. Churchill, "Adoption Research: Information Sources in the Industrial Purchasing Decision," *Proceedings of the American Marketing Association* (August 1968): 352–60.

[18]Michael P. Peters and M. Venkatesan, "Exploration of Variables Inherent in Adopting an Industrial Product," *Journal of Marketing Research* (August 1973): 312–15.

[19]Patrick J. Robinson, Charles W. Faris, and Yoram Wind, *Industrial Buying and Creative Marketing* (Boston: Allyn & Bacon, Inc., 1967).

[20]Richard N. Cardozo, "Segmenting the Industrial Market," *Proceedings of the American Marketing Association* (August 1968): 433–40.

[21]Frederic E. Webster, Jr., "New Product Adoption in Industrial Markets: A Framework for Analysis," *Journal of Marketing* 33 (July 1969): 35–45.

[22]Michael P. Peters and M. Venkatesan, "Characteristics of Adopters and Non-Adopters," *Journal of Marketing Research* (August 1973): 7.

Such evidence may then be utilized in the development of a marketing plan for the new industrial product.

A major problem with the research to this point is the ability of the researcher to identify a practical and workable definition of the innovator, early adopter, and so on. When does the next adoption make an early adopter rather than an innovator? An appropriate percentage is an obvious difficulty in diffusion research. In the rural sociology research, a normal curve with breaks at the standard deviations was used. Hence 2.5 percent was set as the breaking point between the innovator and early adopter. In the marketing literature a 10 percent penetration of the potential market was often used to identify the innovator.

Whatever the percent used, clear distinctions can be made if researchers consider some of the following points:

1. Diffusion of an imitation product will be beyond the innovator category, and marketing should be carried out in accordance with the characteristics of later adopters.
2. In some product categories the diffusion rate for a new product may vary significantly, depending on such consumer variables as social class, ethnic group, or age group. An individual may be an innovator for his or her social class but not for the national market.
3. Researchers must work with aggregates in profiling consumers since there may be no distinction between the last innovator and the first early adopter.

It seems apparent that in today's market it would be more meaningful to define the adopter categories using the product life cycle concept as a basis. If a product is expected to have a certain defined life span, then it would be possible to divide this time span to represent the adopter categories (e.g., buyer in just six months is an innovator).

The significance of diffusion research depends considerably on how the market is segmented. Firms should first consider the alternative market segments before attempting to profile adopter categories. This information will provide an important basis for the marketing plan and future marketing strategy decisions. In particular, manufacturers of certain consumer goods and industrial products are concerned with the heavy users of that product. A recent study indicates that heavy users are most likely to be early triers (innovators or early adopters) of a product.[23] Because of the large volume that the heavy-user market represents, research should be conducted to profile this segment. Test marketing should also be built around these heavy users, testing such activities as media planning, awareness tracking, market share performance, and other marketing mix variables.

The test market discussed in Chapter 11 should also provide management with valuable data on the diffusion process for the new product. Assuming that the test market is representative and is retained over a number of repurchase cycles, the correlates related to adopter categories—that is, innovators, early adopters, early majority, late majority, and laggards—can be pro-

---

[23]James W. Taylor, "A Striking Characteristic of Innovators," *Journal of Marketing Research* 14 (February 1977): 104–7.

jected to the national market. This analysis will give the firm an opportunity to prepare for each adopter category over time and to modify its marketing strategy to enhance acceptance.

In addition to the adopter categories, various tests to determine the most effective means of communication to enhance movement through the stages of awareness to adoption may be conducted. In particular, this promotion should consider the problems of the new product's relative advantage, compatibility, complexity, disability, and communicability.

SUMMARY

This chapter has synthesized a substantial portion of the literature available on the diffusion of innovation. Many of the important findings relevant to each research tradition were discussed.

The chapter looks at both the diffusion process, which involves aggregate behavior, and the adoption process, which involves individual behavior. The diffusion process includes four crucial elements: the innovation, communication, the social system, and time. Adopter categories are identified in the diffusion process and are depicted as fitting a normal distribution. Categories are labeled and identified differently in different disciplines. Most significant are the factors or correlates related to each adopter category. These correlates generally relate individual behavioral variables and/or environmental variables where appropriate.

Characteristics of the product also appear to affect the rate of diffusion. Factors such as the relative advantage, compatibility, complexity, divisibility, and communicability may determine the ease with which consumers accept a new product.

The adoption process is believed to follow a five-stage sequential process, beginning with the consumer's actual awareness of a product's existence and ending with his or her adoption or commitment to that product. Information sources relevant to each stage appear to differ in each discipline and hence may be a function of the nature of the innovation in question.

The foremost reason for studying the diffusion and adoption process is to increase the level of understanding of how, when, and why new products are accepted or rejected. It is believed that a more comprehensive understanding of this process will enhance the success of future new product introductions.

SELECTED READINGS

BROWN, LAWRENCE A. *Innovation Diffusion: A New Perspective.* New York: Methuen Inc., 1981.
Examines innovations and the processes by which they spread in the marketplace. Reviews various traditions of diffusion research and presents a framework. Specific innovations, such as Friendly Ice Cream Shops, cable television, the bank credit card, Planned Parenthood affiliates, and four agricultural innovations are illustrated.

DONNELLY, JAMES H., JR., and IVANCEVICH, JOHN M. "A Methodology for Identifying Innovator Characteristics of New Brand Purchasers." *Journal of Marketing Research* 11 (August 1974): 331–34.

By using a longitudinal research design, overcomes weaknesses in traditional diffusion research that compares early triers with later purchasers. Results indicate that initial buyers of both new brand and similar established brands differ from later buyers.

PETERS, MICHAEL P., and VENKATESAN, V. "Exploration of Variables Inherent in Adopting An Industrial Product." *Journal of Marketing Research* 10 (August 1973): 312–15.

One of the few diffusion research studies that considers industrial products. Examines profiles of users and non-users of a small computer system. Results indicate that industrial buying is also an appropriate area for studying the diffusion process.

ROBERTSON, THOMAS S. "Determinants of Innovative Behavior" *American Marketing Association Proceedings*, Edited by Reed Moyer, 1967.

One of the first marketing applications of diffusion research. Studies the characteristics of adopters of the tonel tone telephone. Compares and analyzes profiles of the adopters and non-adopters. Makes an important contribution to marketing studies on the diffusion of new products.

ROBERTSON, THOMAS S. *Innovative Behavior and Communication.* New York: Holt, Rinehart & Winston General Book, 1971.

Synthesizes the relevant research and applications of diffusion in marketing.

ROGERS, EVERETT M. "New Product Adoption and Diffusion." *Journal of Consumer Research* (March 1976): 290–301.

Summarizes research on the diffusion of innovations that contributes to understanding new product adoption. Also discusses how the background of diffusion research affects its contributions and shortcomings. Future research priorities on the diffusion of innovations are identified.

ROGERS, EVERETT M., and SHOEMAKER, F. FLOYD. *Communication of Innovations: A Cross-Cultural Approach.* 2nd ed. New York: The Free Press, 1971.

Revision of the original *Diffusion of Innovations* published in 1962. Not only updates and integrates the literature but provides much more conceptual clarity regarding the application of diffusion research. Separately presents the various research traditions and provides needed research, implications, and a bibliography.

SHOEMAKER, ROBERT W., and SHOAF, R. ROBERT. "Behavioral Changes in the Trial of New Products." *Journal of Consumer Research* 2 (September 1975): 104–9.

Based on a longitudinal analysis of 1,480 individual new brand buyers, analyzes the theory of perceived risk. Examines the theory of risk reduction in the context of five new consumer products.

TAYLOR, JAMES W. "A Striking Characteristic of Innovators." *Journal of Marketing Research* 14 (February 1977): 104–7.

Study of eleven new consumer packaged goods products that have been introduced into test markets over a three-year period. Total household purchases of the product class are related to the time of trial of the new product in order to determine if heavy users are early triers of new products in that category.

Forecasting New
Product Sales

For a company with healthy profits, low debt, and a long-lived product line, Milton Bradley Co. is in a surprisingly serious predicament. The Springfield, Massachusetts toymaker made a strategic blunder by not getting into the electronic video game market early and is now struggling to catch up. This market is the hottest toy category to date, but changing technology makes leadership in the field a moving target.

Four years ago, Milton Bradley developed a television-connected game, but company executives failed to market it because they thought such items constituted an overpriced fad. Last year alone, however, United States sales of video game systems and software cartridges tripled, to $1.2 billion, according to the trade association Toy Manufacturers of America, Inc. In 1982 these sales are expected to go to $2 billion. Sales of all other toys and games rose only 6 percent, to $4.9 billion, in 1981 and will probably show modest growth this year. Now Milton Bradley Chairman James J. Shea, Jr. is trying to come up with a new-generation video system. He is also seeking the company's first consumer products acquisition outside toys, in case the come-from-behind effort does not accomplish enough.

Milton Bradley placed its bet on portable electronic games rather than video consoles—a decision that seemed smart at first. From 1977 to 1980, company revenues grew at a compound rate of 29 percent per year, fueled by demand for such offerings as Simon, an unsophisticated but successful table-top game. But the phenomenal rise of consoles, including Mattel Inc.'s Intellivision and Warner Communications Inc.'s Atari, has siphoned off huge chunks of parents' toy-buying dollars. Last year portables were blasted like so many alien targets zipping across the television screen, and Milton Bradley's electronic-toy revenues plunged 33 percent. Gains made by the company's old standbys, board games such as Candy Land and Battleship, helped overall profits stay relatively healthy for a recessionary year. Still earnings dropped 37 percent, to $20 million, on sales down 9 percent, to $381.4 million.

The growth in video games might be great enough to allow one more competitor to slip in profitably, and no one doubts Milton Bradley's ability to produce electronic hardware or software. But success is by no means assured. And while the company waits, the current leaders in video games are finding that their products appeal more and more to players younger than 14—Milton Bradley's prime market for its traditional toys. The company's hand-held electronic games, meanwhile, still sit on many store shelves, threatening the company's image among retailers. "We haven't ordered any hand-held games yet this year, other than Coleco's Pac-Man," says Luke Kirby, vice-president of Caldor, Inc., a discount chain. "Once burned is enough."[1]

It began quietly in San Francisco on January 5, 1975. Corcetti Liquors sold a bottle of Christian Brothers brandy for $7.20—a price $2.38 below the legal minimum. Richard Corcetti was promptly hauled before California's regulators. On appeal competitors were stunned by a ruling that he had committed no crime and that, in fact, fair trade laws setting minimum liquor prices

---

[1]"Milton Bradley: Playing Catch-Up in the Video-Game Market," *Business Week* (24 May 1982): 110–14.

were illegal. The decisions stood—all the way to California's Supreme Court in 1978.

The protection that allowed this cumbersome and costly distribution system is going the way of the hip flask and the half-pint. This change not only brings dramatic price reductions for consumers, but also brings retailers out from behind the counter and into the world of competition. New York, for example, is the nation's second-largest liquor market, and a year ago a state appeals court outlawed minimum pricing on wine. Now a bottle of Bolla, which costs a store manager $3.13 and used to go for $4.69, often sells as a loss leader for as little as $3.13—the same as the wholesale price. That kind of discounting puts considerable pressure on the state's 4,240 retail outlets, 700 of which are on the wholesalers' list of high-credit risks—up by 300 stores from last year.

What's shaping up in the liquor business today is a classic case of survival of the fittest—just as with the airline, trucking, and savings and loan industries. "Those that can adapt, will. Those that can't, perish," said an executive of a California-based liquor company. "That's the way the free enterprise system is supposed to work."[2]

These scanarios indicate the importance that forecasting plays in the product planning and development process. During the development stages, each decision regarding the new product is based on a forecast or estimate, regardless of the forecasting method used. In planning the launching of the new product, several sales forecasts are made, each giving an estimate of probable sales for a certain level of marketing activity and prevailing uncontrollable environment.

This chapter discusses the many favorable and unfavorable factors affecting new product sales forecasting and their relationship to the actual sales forecast. Several specific forecasting methods are presented. The chapter concludes with a discussion of the many problems in accurately forecasting new product sales.

FACTORS AFFECTING NEW PRODUCT FORECASTING

Two sets of variables are involved in making any new product forecast: the uncontrollable variables and the controllable elements in the new product's marketing plan. Each of these sets is composed of favorable and unfavorable factors that need to be identified for a specific new product forecasting situation.

Favorable Factors

Favorable factors of course, are those that will positively affect the new product's sales. By carefully analyzing the new product and market situation, management can identify factors that would be beneficial to sales. These factors are highly dependent on the type of product market as well as on whether the

---

[2]Thomas O'Donnell with Janet Bamford, "Jack Daniels Meets Adam Smith," *Forbes* (23 November 1981): 163–69.

primary focus is a consumer, industrial, or government market. Certain economic and business activity indicators will also help indicate the sales of new products to varying degrees. These indicators include increased demand for goods; low unemployment rate; increasing consumer, industrial, and government purchases; reasonable balance of trade; and inventories in line with sales.

Unfavorable Factors

On the other side, there are factors that will probably negatively affect the sales of the new product. These elements should be even more carefully analyzed than favorable ones since often they increase the probability of a new product failure. Again the effect of these items will vary, depending on the product and market characteristics of the new offering. The following indicators generally have a negative effect on sales: high interest rates, rising prices and threat of inflation, decline in construction (particularly home building), decline in automotive sales, labor discontent and strikes, restrictive monetary policy, and decline in the stock market.

Judgment must be used to assess the degree of impact each factor will have on the sales of the new product. Those factors that have a significant effect on the sales should be carefully analyzed in terms of present and future trends. This analysis is particularly helpful in forecasting new product sales because it requires a thorough investigation of the current market situation and an evaluation of the future of important components.

SERIES INDICATORS

One factor influencing the sales of any new product, regardless of its market, is the general economic situation. It is always important to have at least a broad gauge of general business activity. For successful launching, it is equally important to have an accurate prediction of future business conditions. This prediction requires having a complete understanding of the mechanisms of the economy and their interrelationships. This evaluation becomes even more critical when international sales are involved.

Each company establishes favorite indices of those general economic conditions most relevant to its product category. These indices should be examined for any indicators that may affect the sales of the new product. Although indicators are product-specific, there are three general categories of indicators, as indicated in Table 14–1. Naturally the group of indices that lead general business activity is usually examined most closely. The movement of a leading index (upward or downward) precedes that of the sales of the new product. Unfortunately this movement does not consistently mean an upswing or downturn in new product sales. It is particularly difficult to interpret the meaning of leading indicators when there is some inconsistency or contradiction among several indicators, as there often is.

Simultaneous (coincident) and lagging indicators have less importance in the decision to launch a new product. Coincident indicators are those that are in harmony with the peaks and troughs in the new product's sales. Al-

Leading Indicators

Average work week of production workers

Value of manufacturer's new orders (durable goods industries)

Construction contracts awarded for commercial and industrial buildings

Contracts and orders for plant and equipment

Newly approved capital appropriations

Net change in the business population (new businesses incorporated and failures)

Corporate profits after taxes

Index of stock prices

Change in business inventories

Value of manufacturers' new orders

Simultaneous (Coincident) Indicators

Unemployment rate

Index of help-wanted advertising in newspaper

Index of industrial production

Gross national product

Personal income

Sales of retail stores

Index of wholesale prices

Lagging Indicators

Business expenditures on new plants and equipment

Book-value of manufacturer's inventories

Consumer installment debt

Index of labor cost per unit of output

TABLE 14–1
Principal business indicators for predicting new product sales

though these will be useful in establishing marketing strategies in various stages of the product life cycle after the new product has been on the market, they have little use in the initial go, no-go decision. For a number of important economic indicators, peaks and troughs follow turning points in general business activity. Even when their activity lags behind the general level of business activity, in some cases, they can have some predictive value for new products. For example a firm that must borrow funds to launch the new product should be aware that interest rates usually go up some time after a coincident series peak has been reached. If a trough occurs, then interest rates will often reach their lowest level a few weeks later, which would be the ideal time to secure financing.

COMPETITIVE EVALUATION

One important determinant of the new product's sales that warrants very careful consideration is competing companies, in general, and their products, in particular. As discussed in Chapter 1, misunderstanding competition is often the cause for new product failure. To help reduce the possibility of misunderstanding, it is often desirable to analyze each competitor individually. Specifically the firm should decide whether the new product can be competitive on the basis of various controllable marketing elements. This analysis is very helpful in ensuring that the new product does not have "quality features" that the market does not want and therefore is unwilling to pay for. Analysis of competition, if done in all stages of the development process, is particularly beneficial in orienting research and development toward the market, as was discussed in Chapter 2.

Competitive evaluation needs to be done on both an overall-company as well as a specific-product basis.

The overall competitive company evaluation should be based on several considerations, as indicated in Table 14–2. These include engineering/production, sales/distribution, management, and general marketing. The capabilities of the company in each of these areas needs to be realistically appraised in terms of the capabilities of the competition. The results of this evaluation will not only help in determining a realistic sales forecast but will also help the company make any needed modifications.

Management's assessment of major competitors is very important, especially in determining the competitive reaction to one of the most vulnerable areas of the new product: price. (See Chapter 16 for a discussion of pricing.)

After the overall company evaluation, specific competitive products need to be similarly evaluated. A useful approach is indicated in Table 14–3, where each competitive product is compared with the company's offering in terms of its design/performance features, overall product features, promotion features, and price features. Care must be taken to evaluate all products that fill even a slightly similar market need. It is particularly important to evaluate the design/performance features of competitive products in terms of quality, utility, convenience, versatility, reliability, durability, serviceability, sensory features, safety, and uniqueness. This part of the competitive product analysis will reveal the potential market position of the new product.

The market position will of course be directly reflected in the new product's sales.

MARKET SIZE DETERMINATION

Throughout the entire new product development process, from conception to market introduction, a primary concern is the size of the market. This size must be determined on both a present and projected basis.

The problem of measuring market size for the new product varies with the degree of newness of the product. It is easier to determine the market potential for a continuous innovation, particularly if the innovation is very similar to products the company is currently marketing. Market size deter-

TABLE 14–2
Competitive company evaluation

Consideration	Capabilities of Major Competitors		Company Capabilities	Results
	Competitor	Competitor		
Engineering/Production				
Research and development				
Applicable experience				
Engineering capabilities				
Production capabilities				
Sales/Distribution				
Applicable experience				
Sales capabilities				
Distribution channels				
Management				
Applicable experience				
Management capabilities and resources				
General Marketing Advertising				
Sales promotion				
Overall image				
Present market position				
International capabilities				

TABLE 14–3
Competitive product evaluation

Consideration	Features of Competitive Product		Features of Company's Product	Results
	Product A	Product B		
Design/Performance Features				
Quality				
Utility				
Convenience				
Versatility				
Reliability				
Durability				
Serviceability				
Sensory features				
Safety				
Uniqueness				
Other Product Features				
Package				
Service				
Brand recognition				
Guarantee/Warranty				
Assortment				
Promotion Features				
Advertising expenditures				
Type of advertising				
Publicity				
Sales promotion				
Sales force				
Price Features				
Cost				
Pricing and policies				
Terms and conditions				
Actual price				

mination becomes easiest when the new product is a line extension. For discontinuous innovations, market determination is more difficult, since not only does the company lack experience, but, of course, no published data exists.

Companies often overestimate the sales of the new product because they underestimate competitive reaction, or are too optimistic about the new product's marketing plan, or consumers overstated their opinion and degree of buying interest when queried about the new product.

The overall market size must also be evaluated for its growth potential. When the new product is in a growth market, the probability of success greatly increases. However measuring long-term growth potential poses even more difficulties than determining the initial market size. Again the degree of difficulty increases with the newness of the product; that is, the more discontinuous and revolutionary the new product is, the more difficult it is to establish the growth potential. Even when the overall growth pattern for a product class can be estimated, it still may be difficult to estimate the growth for a single product within that class. This is especially true in rapidly changing markets. Because of these difficulties, the first one-year forecast often occurs late in the development of the new product.

In determining market size for the new product, it is often valuable to use a form such as the one indicated in Table 14-4. This form is useful in forecasting not only sales volume, but profit and market share as well. The form can be filled out by sales personnel and product managers. Their estimates can be pooled either by straight averaging, or by a weighted averaging method if certain estimates are considered more valid. It is important to note that the price per unit is requested for each year that a sales volume estimate is made. This forces the estimator to more carefully and realistically determine the sales that will be achieved.

Another useful form is shown in Table 14-5. The rating scale provides a succinct means for sales people, product managers, or members of the new product committee to evaluate the potential sales and profits of the new product. Each of the individuals would rate every category from 0 to 100 percent. Each of the ratings (column 2) is then multiplied by the respective weight (column 3) (preassigned by a management committee) to obtain its weighted rating (column 4). When all the weighted ratings are summed, an overall product rating is obtained for each product being evaluated. The sum of weights (column 3) should be 100 percent.

The total weighted rating can be used to estimate the new product's sales volume. An example of sales volume determination for several new products is shown in Table 14-6. The rating of each product (A) is obtained from Table 14-5. Those total 400. The preference ratio (B), the number that indicates which product is best, is found by dividing each product rating by the total of all product ratings—that is, by 400. For example the preference ratio for product 104 is $\frac{80}{400} = 0.20$. The total sales expected for the next period (C) is a very difficult but essential estimate to make. While it can be in either dollar or unit figures, it is better to use unit figures whenever possible, as it eliminates the import of any price changes. It is often obtained from data on similar new products or from the sales of other products in the same product class. In this case total new product sales were estimated to be $10,000.

TABLE 14-4
New product sales and profit determination

Product Experience and Forecast

Categories Determining Value of Product	Unit of Measurement	Test Market and Sales	Forecast Years 1	Forecast Years 2	Forecast Years 3
Price per unit	$				
Sales volume	No. of units				
Sales volume	$				
Variable margin per unit	$				
Variable margin per year	$				
Net profit per unit	$				
Net profit per year	$				
Market share	%				
Sales as percentage of firm's total volume	%				

Source: Adapted from Norbert Lloyd Enrick, *Market and Sales Forecasting*, 2nd ed. (Huntington, N.Y.: R. E. Krieger Pub. Co., Inc.), 128.

TABLE 14–5
Rating potential sales and profit for a new product

| Future Expectations for the Product | Rating R_1 (%) | | | | | | | | | | | | Weight W (%) | Weighted rating $W \times R_1$ (%) |
| | Low | | | | | | | | | | High | | | |
	0	10	20	30	40	50	60	70	80	90	100			
Market potential that can be realized														
Required amount of promotional expense														
Profit per unit														
Contribution to sales of other products														
Other contributions to firm's overall program														
Total														

Source: Norbert Lloyd Enrick, *Market and Sales Forecasting*, 2nd ed. (Huntington, N.Y.: R. E. Krieger Pub. Co., Inc.), 193.

TABLE 14-6
New product sales and profit determination—conversion of rating to volume

Procedure	Product Number							Total
	101	102	103	104	105	106	107	
A. Rating of product (%)	40	80	50	80	60	20	70	400
B. Preference ratio = $\dfrac{\text{Product rating (\%)}}{\text{Total of ratings (\%)}}$	0.10	0.20	1.125	0.20	0.15	0.05	0.175	1.0
C. Total sales expected for next period ($)								10,000
D. Product sales expected $ = B \times C$	1,000	2,000	1,250	2,000	1,500	500	1,750	
E. High expectation $ = 1.20 \times D$	1,200	2,400	1,500	2,400	1,800	600	2,100	12,000

Source: Norbert Lloyd Enrick, *Market and Sales Forecasting*, 2d ed. (Huntington, N.Y.: R. E. Krieger Pub. Co., Inc.), 192.

This total sales expected ($10,000) can now be allocated to each of the new products under consideration by multiplying it by the respective preference ratios. For example, for product 104, the total product sales is $2,000 ($10,000 × 0.2). Usually three sales forecasts are calculated: the most optimistic, the most likely (calculated above), and the most pessimistic. In the case at hand, from examination of past new product sales, management felt that the most optimistic sales could exceed the most likely forecast by as much as 20 percent (E). Therefore multiplying each new product's sales by 1.2 yields the most optimistic sales for each new product under consideration. For product 104, this figure is 2,400 (2,000 × 1.2). A pessimistic expectation sales figure is also calculated so that profit determinations can be made for low, most likely, and high sales.

SPECIFIC FORECASTING METHODS

Since no single forecasting method gives accurate forecasting of new product sales in all situations, it is better to use several methods simultaneously. Not only are better forecasts obtained this way, but each forecast acts as a check on the results of another forecast.

Probably the most important determinant in deciding on which forecasting methods to use is the amount of data available. This amount is usually dependent on whether some type of market test was undertaken, as was discussed in Chapter 11. A corporate survey identified criteria used in selecting a forecasting method, including cost, the characteristics of the specific product situation, the user's technical ability, and the characteristics of the methods considered.[3] The user's technical ability indicates that for a company to use a particular forecasting method, it must have an understanding of the method. This was also reflected in the fourth criterion: the characteristics of the methods considered. These delineate the preference of the forecaster, as well as the forecasting skills and background.

Among the methods most commonly used in new product forecasting are the jury of executive opinion, regression analysis, and econometric models. These methods are used to varying degrees depending on the specific product/market situation. The acceptance and use of various techniques in general forecasting situations is indicated in Table 14–7. The companies participating in the survey were asked the number of different forecasting methods they used. Not only did over half of the responding companies account for seven of the eight techniques mentioned, but the "survey also determined that once a company tries a specific method, it is likely to continue to use it."[4] While the results may not be exactly the same if the companies were asked specifically about new product forecasting, there is little reason to believe that there would be a great deviation. The major difference for new product forecasting is in the amount of data available.

---

[3]These and other results are found in Steven C. Wheelwright and Darral G. Clarke, "Corporate Forecasting: Promise and Reality," *Harvard Business Review* 54, no. 6 (November–December 1976): 40–60.
[4]Ibid., 41.

TABLE 14–7

Acceptance and use of alternative forecasting methods

Method	Use of the Method by Those Familiar With It	Ongoing Use of the Method by Those Who Have Tried It	Those Unfamiliar with the Method
Jury of executive opinion	82%	89%	6%
Regression analysis	76	91	8
Time series smoothing	75	84	13
Sales force composite	74	82	10
Index numbers	67	85	33
Econometric models	65	88	12
Customer expectations	57	78	15
Box–Jenkins	40	71	39

Source: Steven C. Wheelwright and Darral G. White, "Corporate Forecasting: Promise and Reality," *Harvard Business Review* (November–December 1976): 41.

Jury of Executive Opinion

The jury of executive opinion—one of the oldest and simplest techniques known—asks top executives of the firm about future sales. By averaging these views, a broad-base forecast is obtained, which is usually a more accurate forecast than that obtained by using only a single estimate. In the jury of executive opinion method, estimates of sales are generally obtained from executives of functional areas, such as marketing, finance, production, and purchasing. The more factual information these executives have at their disposal, the more accurate their estimate of sales will be.

The advantages and disadvantages of using the jury of executive opinion, as well as three other forecasting methods, are indicated in Tables 14–8 and 14–9. The advantage most often cited for the jury of executive opinion method is that it is quick and easy. Perhaps the major disadvantage is that it relies on the opinions of executives who may be only slightly associated with the new product and therefore could only guess its sales. Some new product managers feel that this method of forecasting the sales of the new product should be used only in the total absence of internal and external data—that is, when no other method is available.

Sales Force Composite

The sales force composite method compiles the sales force estimates of future sales. The salespeople sometimes make these estimates alone, using a specially designed form. Other times, the sales estimates for the new product are made in consultation with the sales manager. The latter approach is generally preferred for new products because the salespeople may have limited familiarity with product concept. The sales manager can provide any needed information and evaluate the forecast with the salesperson. Any necessary modifications can then be made jointly.

The results of the salespeople are totaled for the district or region. The district (or regional) managers then evaluate this estimate by comparing it

TABLE 14–8
Advantages of techniques for forecasting new products

Buyer's Expectation Method	Delphi Method	Jury of Executive Opinion	Sales Force Composite Method
1. Forecast is based on opinions of individual whose buying activities will determine the actual sales results achieved.	1. Forecast is based on opinion of experts in the field.	1. Forecast is based on different specializations, judgments, experience, and viewpoints.	1. Forecast is based on specialized knowledge of individuals closest to the market where the sales occur.
2. Forecasting process reveals general attitudes and feelings about product from potential users.	2. Technique is a very effective method for forecasting sales of new high-technology products.	2. Method is easily and quickly implemented.	2. Forecast is based on opinions of individuals who are directly responsible for the actual sales results achieved.
3. Technique is a very effective method for determining demand for a new industrial product.		3. Forecast is based on opinions of individuals who can make and implement decisions that will affect the actual sales results achieved.	3. Forecast is easily broken down on a customer, product, or territory basis.

TABLE 14–9
Problems with techniques for forecasting new products

Buyer's Expectation Method	Delphi Method	Jury of Executive Opinion	Sales Force Composite Method
1. Forecast is based on expected future actions and events that are subject to change.	1. Forecast requires a great deal of time and some money to implement.	1. Accuracy of forecast is dependent on the quality of the opinions solicited.	1. Forecast is based on estimates of sales person, who is often unaware of broad economic patterns and future company plans.
2. Forecast is based on opinions of buyers who are at various states of knowledge and interest.	2. Accuracy is dependent upon participants' knowledge and level of initial and continued participation.	2. Forecast is based on primary research and original data collection with the associated time and cost.	2. Sales people are evaluated primarily on sales not on subsidiary activities, such as forecasting.
3. Forecast is based on primary research and data collection with the associated time and costs.	3. Forecast requires an able, experienced coordinator.	3. Forecast is often difficult to break down by product or geographic delineations.	3. Forecast is based on primary research and original data collection with its associated time and costs.
			4. Procedures are needed to ensure accuracy of the forecast.
			5. When forecast is used in establishing quotas, sales estimates are understated.

with the accuracy of past estimates of new product sales. Their own experience also enters into their evaluation before the district's estimate is forwarded to the home office, where a total sales estimate is compiled.

While the forecast for the new product is made up of a composite of individual estimates by the salespeople, this does not mean that individual forecasting errors will be negated. A common bias usually operative in the sales force results in low forecasts. Believing the estimate they give for the new product's sales will be the basis of their sales quota when the product is introduced, members of the sales force tend to underestimate the sales potential of the new product. This often occurs even when the sales force is assured that the forecast will not be used in establishing quotas.

Understatement of the sales potential of the new product can cause the firm to lose potential profits by not introducing the new product or by having insufficient production capacity to meet the demand. This tendency for lower forecasts can be corrected by establishing an index of pessimism for each salesperson. The index is derived by comparing all previous estimates of new product sales with the actual sales for each salesperson. The difference, when divided by the estimate, becomes the salesperson's index. By averaging all the indices from previous new product sales estimates, an index of pessimism is established. Multiplying this index by the understated estimate of the new product can yield a proportionately increased estimate, which more accurately reflects the probable sales. In one company the estimates for about 90 percent of the sales people were within ±10 percent of actual sales when indices of pessimism were applied.

The sales force composite method allows the new product's sales forecast to be estimated by knowledgeable people who are closest to the market. However it still has two problems in producing accurate forecasts. First the structure of the market is an important factor in the ability of the sales force to accurately forecast sales. If the salespeople are selling to relatively few accounts, or if relatively few accounts comprise the predominant proportion of their business, then the new product forecasts are usually more accurate than when the salespeople are selling to many small, nondominating customers (see Tables 14–8 and 14–9). In the former situation the salespeople usually have intimate knowledge of key accounts and can better estimate their purchasing behavior with the new product. The second problem with the sales force composite method is that salespeople's evaluation, compensation, and promotion are based on sales, not on their forecasting accuracy. Therefore they usually spend as little time as possible in making the forecast. This problem can be somewhat alleviated if directions from upper-level management indicate the importance of new product forecasts. By establishing a bonus system, the accuracy of new product forecasts by the sales force can increase dramatically.

Buyer's Expectation

Many companies planning the introduction of a new industrial product often ask present, past, and potential customers about their possible purchases if the new product were introduced. This approach can be best implemented when there are relatively few potential customers. These potential customers can be queried about their purchasing plans by mail, telephone, or in person for more accurate and detailed estimates. A major problem occurs if potential buyers really like the new product and overstate their buying intention, however. When this happens, there may be an oversupply of the new product. One technique for discounting artificial demand is called an *index of optimism*. This index is based on the past purchases of new products and is similar to the index of pessimism discussed in the sales force composite technique. However this index is much more limited in its use because buyers change so frequently that there is seldom a buying history for establishing the index.

The buyer's expectation method can also be used for consumer products in conjunction with consumer panels. By employing a consumer panel within a test market, not only can sales of the new product within that area be predicted, but the market test results can be projected to the entire area of final distribution.[5] Under the buyer's expectation method, data on individual purchases of the new product are collected from the panel members in the test areas established. Socioeconomic characteristics and general buying behavior patterns, as well as abnormalities in reporting information, are noted. Growth in the purchase of the product class due to the introduction of the new product into the test market, as well as the repeat purchase rate, can be determined. By extrapolating and then multiplying the growth rate by the repeat purchase rate, a prediction of the market share for the new product in the long run can be obtained. To obtain the total market forecast for the new product, the panel forecast is projected to the test market and, in turn, the test market to the whole market area.[6] Consumer panels can provide a very accurate forecast of new product sales when carefully monitored.

Delphi

The Delphi method is also used to forecast sales for new product concepts as well as to predict future trends. This method employs a panel of experts who answer elaborate questionnaires on the new product's sales potential. Their responses are summarized and used as a basis for establishing the next set of elaborate questions. This process is repeated until a reliable estimate of the sales potential of the new product is obtained. This method, which gained wide attention in 1964 when it forecasted a manned lunar landing by 1970, is particularly useful for determining sales of high-technology products. The advantages and disadvantages of the method are indicated in Tables 14–8 and 14–9.

Correlation and Regression Analyses

Where data exist from test market results, correlation and regression analyses are probably the most widely used procedures to forecast new product sales. These procedures establish the mathematical relationship between sales of

---

[5]This projection and its accuracy are dependent on the degree the test market reflects the area of final distribution. This aspect of test market selection is discussed in Chapter 11.

[6]A useful procedure for estimating new product sales through consumer panels can be found in David H. Ahl, "New Product Forecasting Using Consumer Panels," *Journal of Marketing Research* 7, no. 2 (May 1970): 160–67.

the new product and at least one other variable. Both methods have the underlying assumption that sales as well as the values of the other variable(s) are chosen at random. Correlation analysis is often used to detect relationships between sales and any other variable(s) in the market test data.

After one or more associations are found, regression analysis can be used to predict the value of the dependent variable—the sales of the new product. Regression analysis assumes that the independent variables (those related to the sales of the new product) are predetermined and selected with foresight and judgment. On the other hand, the Y value (new product sales) associated with these independent variables is selected at random—that is, it is distributed around the independent variables.

To illustrate the use of regression analysis, let us assume that we are trying to determine the potential sales of Lifecard. This new product is a plastic card that has a tiny microfilm insert containing the bearer's entire medical history, which can be used as a reference by doctors or hospital staff. There is space on the card to record information such as health insurance, eyeglass prescription, allergies, and the names and phone numbers of people to contact in case of an emergency. The card is being market tested for $9.95.[7] The market test results, as well as the number of heart attacks in the test market area, are indicated in Table 14–10.

Since management feels that the sales of Lifecard is a function of the number of heart attacks occurring, regression analysis will be used to find the relationship, if any, between the number of heart attacks and sales of the product. While there are many types of regression equations to show the application of regression analysis in forecasting new product sales, for ease in calculations, we will assume that a linear regression is the one that best fits the data. This linear regression equation is as follows:[8]

$$Y_c = a + bx$$

The values of a and b are calculated by solving the following two equations:

$$b = \frac{n\Sigma xy - x\Sigma Y}{n\Sigma x^2 - (\Sigma x)^2}$$

$$a = \bar{Y} - b\bar{x}$$

where

n = number of observations in sample

x = independent variable (number of heart attacks)

Y = dependent variable (sales of Lifecard)

a = equation's intercept of Y axis when $x = 0$

---

[7]This product was described in "A Wave of New Products for Work, Play, Travel," *U.S. News and World Report* (29 November 1976): 73–76.

[8]A discussion of regression analysis can be found in N.R. Draper and H. Smith, *Applied Regression Analysis* (New York: John Wiley & Sons Inc., 1966), 1–216, and Charles W. Gross and Robin T. Peterson, *Business Forecasting* (Boston: Houghton Mifflin Co., 1976), 80–124. The formulas presented here are adapted from the latter text.

$$b = \text{slope of equation}$$

$$\bar{x} = \text{mean of } x(\bar{x} = x/n)$$

$$\bar{Y} = \text{mean of } Y(\bar{Y} = Y/n)$$

Using the data in Table 14–10, the regression equation for Lifecard can be determined as follows:

$$b = \frac{n\Sigma xY - \Sigma x\Sigma Y}{n\Sigma x^2 - (\Sigma x)^2} = \frac{12(439) - (72)(71)}{12(445) - (71)^2}$$

$$= 0.521$$

$$\bar{x} = \frac{\Sigma x}{n} = \frac{71.0}{12} = 5.92$$

$$\bar{Y} = \frac{\Sigma Y}{n} = \frac{72.0}{12} = 6.00$$

$$a = \bar{Y} - b\bar{x} = 6.00 - (0.521)5.92 = 2.92$$

$$Y_c = 2.92 + 0.521(x)$$

This sample regression equation indicates that the forecasted sales for the new product is a function of the number of heart attacks occurring. For example if 900 heart attacks were to occur, then Lifecard sales would be $6,609:

$$Y_c = 2.92 + 0.521(9.0) = \$6,609$$

Although the regression equation for predicting Lifecard sales has been established, it is still necessary to determine the standard error of the regression, the significance of b, and the confidence level for the prediction of sales. The standard error of the regression ($\sigma y \cdot x$) is dependent upon the strength of

TABLE 14–10
Test market results of Lifecard

Time in Test Market (biweekly results)	Sales (in thousands of dollars) Y	Number of Heart Attacks (in hundred of units) X	XY	Y²	X²
1st	7.0	6.0	42.0	49.0	36.0
2nd	6.0	7.0	42.0	36.0	49.0
3rd	5.0	5.0	25.0	25.0	25.0
4th	4.0	4.0	16.0	16.0	16.0
5th	5.0	4.0	20.0	25.0	16.0
6th	6.0	5.0	30.0	36.0	25.0
7th	5.0	6.0	30.0	25.0	36.0
8th	6.0	4.0	24.0	36.0	16.0
9th	7.0	8.0	56.0	49.0	64.0
10th	8.0	7.0	56.0	64.0	49.0
11th	6.0	7.0	42.0	36.0	49.0
12th	7.0	8.0	56.0	49.0	64.0
Total	72.0	71.0	439.0	446.0	445.0

the association between the dependent and independent variables. When no association exists at all between the sales of the new product and the independent variable, then the standard error of the regression is the same as the standard deviation of Y. On the other hand, when a strong association exists, there is a small standard error of the regression, with perfect association producing a standard error of the regression equal to zero. In this latter case, the regression equation explains everything about the dependent variable—sales.

The standard error of the regression is calculated by using the following equation:

$$\sigma y \cdot x = \frac{\Sigma Y^2 - a\Sigma Y - b\Sigma xY}{n - 2}$$

$$= \frac{446 - 2.92(72) - 0.521(439)}{12 - 2}$$

$$= 0.843$$

In addition to the standard error of the regression, it must be determined whether the slope of the regression equation is statistically significant. As previously stated, one assumption in regression analysis is that the Y value is normally distributed about the true regression line; this means that b also has a normal sampling distribution.

To test the significance of b, the t statistic is calculated by using the following formula:[9]

$$t = \frac{b - B}{\sigma_b}$$

where

b = the slope for the regression function

B = the population parameter
(B = 0 when testing the significance of b)

σ_b = unbiased estimate of the standard error of b

$$\sigma_b = \frac{\sigma_{y \cdot x}}{\sqrt{\Sigma x^2 - \frac{(\Sigma x)^2}{n}}}$$

This σ_b value is then compared with the critical value of t at the specified confidence level of $n - 2$ degrees of freedom. The null hypothesis is $H_o(B) = 0$ and the alternative hypothesis is $H (B) \neq 0$. If the calculated t is greater than

---

[9]When n is small, the t distribution is used instead of the standard normal distribution. For larger samples, the standard normal distribution is approached, and a z value is computed. The computational procedure of the two methods is very similar. In forecasting new product sales from market test data, there is generally only a small number of observations, so the t distribution is usually used, as is the case in this example.

the table t, the null hypothesis is rejected and a significant association does exist.

Calculating the significance for Lifecard at the 95 percent confidence level:

$$\sigma_b = \frac{\sigma_{y \cdot x}}{\sqrt{\Sigma x^2 - \frac{(\Sigma x)^2}{n}}} = \frac{0.843}{\sqrt{445 - \frac{(71)^2}{12}}}$$

$$= 0.168$$

$$t = \frac{0.521 - 0}{0.168} = 3.10$$

The table t value is found from using any table of critical values of t using $12 - 2 = 10$ degrees of freedom and the 95% confidence level. The table value is 2.228. Since the calculated value (3.10) is greater than the table t value (2.228), b is significant and is the best estimate of the population.

A final calculation for the confidence interval is needed to use regression analysis in new product sales forecasting. The confidence interval gives the range around the true regression line within which the predicted sales value (Y) will fall. The confidence interval is determined through the following formula:

$$Y_c \pm (t \text{ value at specified confidence level}) \, (\sigma_{y - y_c})$$

where

$$\sigma_{y - y_c} = \sigma_{y \cdot x} \sqrt{1 + \frac{1}{n} + \frac{(x_p - \bar{x})^2}{\Sigma x^2 - \frac{(\sigma x)^2}{n}}}$$

x_p = the value of the independent variable used to predict sales of the new product

The confidence interval for the forecasted value of Lifecard is as follows:

$$\sigma_{y - y_c} = 0.843 \sqrt{\frac{1 + 1/12 + (9.0 - 5.9)^2}{25}}$$

$$\sigma_{y - y_c} = 1.02$$

$$6.61 \pm 1.02(2.228)$$

$$4.34 \leqslant y \leqslant 8.88$$

At the 95 percent confidence level, the company can be sure that when heart attacks are at a 900 level, sales of Lifecard will be between $3,290 and $9,930. Similarly other confidence interval estimates can be established for other values of the independent variable.

Specialized Techniques

Specific techniques are often employed in addition to the above techniques for forecasting the sales of the new product. One of these is Data Development Corporation's Sales Waves.[10] Sales Waves provides a measurement of repeat purchasing and the usage characteristics of the new product by offering the product for sale at a company-established selling price. After being screened to ensure that they have recently used the product category, qualified consumers in five cities

> are exposed to a concept board and asked their intention to buy on a four-point buying scale. Those who respond favorably to the top two boxes are given the product to use for two weeks. After the so-called home placement test concludes, respondents are given the opportunity to purchase the product at a retail price set by the manufacturer. The product is offered in this way for several purchase occasions or "waves"—intervals of approximately three months.[11]

Although this technique provides an estimate of the market for the new product, the information is not obtained by measuring sales when a consumer selects the product over competing products on the shelf. Rather it is based on noncompetitive usage and response to a purchase-price established by the company.

Another method to forecast new product sales is the test market laboratory. Various measures of intention to buy and current purchasing behavior are used to predict sales. Several "laboratory" methods are available, such as Yankelovich Laboratory Test Market, Management Decision Systems' Assessor, and Elrick and Lavidge's Comp.[12] Most of these methods resemble the way in which new product sales are projected. After being exposed to a commercial, consumers are involved in a shopping situation in which they can decide to purchase (or not to purchase) the new product. Measurements of consumer awareness and attitude toward competitive products, the degree of satisfaction with the new product, and the effects of alternative prices, containers, or labels are then obtained. Laboratory tests give an indication of the degree of interest in the new product and its competitive edge by allowing the consumer to select it from among competing products and try it in the home. The limitation in forecasting new product sales is that no indication is given of the adoption and repeat purchase behavior beyond the first purchase.

PROBLEMS IN NEW PRODUCT FORECASTING

Forecasting the sales of established products is often a difficult task, but forecasting new product sales is even more difficult. Unless a test market was

---

[10]For a more in-depth description of this technique, see Alexandre Ivahnenke, "Two Forecast Ways Offer Good Rough Estimates of Market," *Marketing News*, 19 November 1976, 4, and *Marketing News*, 31 December 1976, 6.

[11]"Trends in Population, Convenience Foods, Ads, Government Costs Reported to NFBA," *Marketing News*, 31 December 1976, 6.

[12]These techniques are described in Edward M. Tauber, "Forecasting Sales Prior to Test Market," *Journal of Marketing* 41, no. 1 (January 1977): 80–84.

conducted, there is no sales record or company experience upon which to base the forecast. This is particularly true when the new product is either discontinuous or has little similarity with other products in the company's product line. In some cases, a reliable forecast is impossible. Because of the uncertainty in forecasting the sales of the new product and the high risks involved, early sales are closely monitored, and forecasts are continually revised to conform to actual market conditions. Regardless of the product/market situation, the best estimates are usually obtained when a variety of forecasting techniques is used. This allows for constant modifications until a final sales forecast is obtained for the new product.

SUMMARY

One of the most important yet difficult tasks in new product development is estimating the new product's sales at various stages. In forecasting sales it is important to establish and evaluate the relevant factors. Although they differ for every new product, these factors should be identified and classified according to their degree of favorableness. Then the impact on sales of all factors—favorable and unfavorable—should be assessed, along with the probability that each will occur.

The business climate of the new product must also be analyzed. This can best be accomplished by establishing a relationship, using leading economic indicators, between potential company sales and the sales occurring in the general product category of the new product.

Many specific forecasting methods can be used to estimate the sales of the new product, including the jury of executive opinion, sales force composite, buyer's expectation, Delphi, and correlation and regression analyses. Since sales forecasting is such an uncertain task, it is usually best to use several methods concurrently. This can provide the best possible results for very important decisions about the new product.

SELECTED READINGS

"Another Alternative to the CPI." *New York Times*, 1 November 1981, F20.
    States that many forecasts are based in part on the Consumer Price Index (CPI), yet, mixed reactions to its effectiveness have caused businessmen to use several alternatives. Includes descriptions.

BARNES, JAMES H., JR., and SEYMOUR, DANIEL T. "Experimental Bias: Task, Tools, and Time." *Journal of the Academy of Marketing Science* 8, no. 1 (Winter–Spring 1980): 1–11.
    Suggests that researchers, both as reviewers of past findings and as designers of lab experiments, should be aware of the potential bias resultant from time constraints, lack of appropriate decision-making tools, and irrelevant tasks.

BUTTNER, F. H., and LANFORD, H. W. "Technology Development: Determining What to Forecast." *Industrial Marketing Management* 9, no. 3 (July 1980): 187–99.

Outlines what to forecast by the analysis of current events impacting on motivation, specific actions, and sequel events. Presents a model from which one can move into venture planning via three major paths: exploratory, normative, and iterative.

COPPETT, JOHN K., and STAPLES, WILLIAM A. "Product Profile Analysis. A Tool for Industrial Selling." *Industrial Marketing Management* 9, no. 3 (July 1980): 207–11.
Presents Product Profile Analysis (PPA) as a sales tool to be developed jointly between the sales representative and the prospect. The most important benefits of this technique are the specification of the buyer's criteria, and the attention to several criteria rather than one criterion for decision-making.

GENSCH, DENNIS, and SHAMAN, PAUL. "Models of Competitive Television Ratings." *Journal of Marketing Research* 17, no. 3 (August 1980): 307–15.
Reports the results of a time-series analysis of prime-time network television viewing data. Model did not include any program content information but provided very accurate prediction of total network viewing.

GREEN, P. E., CARROLL, J. E., and GOLDBERG, S. M. "A General Approach to Product Design Optimization via Conjoint Analysis." *Journal of Marketing* 45, no. 3 (Summer 1981): 17–37.
Describes some of the features of Product Optimization and Selected Segment Evaluation (POSSE), a general procedure for optimizing product/service designs in marketing research. The approach uses input data based on conjoint analysis methods. The input of consumer choice simulators is modeled by means of response surface techniques and optimized by different sets of procedures depending upon the nature of the objective function.

HELLER, DONALD. "Market Research a Key to Anticipating Trends." *Advertising Age*, 6 July 1981, 46.
Suggests that market research is key to speeding up the communication process among management, especially in targeting new markets.

HERSHEY, ROBERT. "Commercial Intelligence on a Shoestring." *Harvard Business Review* 58, no. 5 (September–October 1980): 22–30.
Explores ways in which small companies can establish programs for monitoring competitors' activities. Some of the most effective methods are also low in cost. Discusses specific methods and approaches for organizing the commercial intelligence function.

PRESSLEY, MILTON M. "Improving Mail Survey Responses from Industrial Organizations." *Industrial Marketing Management* 9, no. 3 (July 1980): 231–35.
Lists mail survey do's and don'ts for notifying the respondent, the outgoing and return envelopes, the cover letter, the questionnaire, and the follow-up.

PRINSKY, ROBERT D. "New Indexes of Commodity Futures." *Wall Street Journal*, 4 January 1982, Sec. 2, 46.
Explains that although forecasting commodities is one of the riskiest financial ventures, new indexes have been recently developed that give incentives to interested investors by comparing spot prices and the new indexes for investment climate.

SAMLI, A. COSKUN, and MENTZER, JOHN T. "An Industrial Analysis Market Information System." *Industrial Marketing Management* 9, no. 3 (July 1980): 237–45.
The United States maritime industry's share of world cargo movements has remained low over the last decade because of the failure of firms to capitalize on market opportunities. Proposes a six-stage market information system's

model for gathering, processing, and using data for firms to make better marketing decisions.

TAYLOR, THAYER C. "Targeting Sales in a Changing Marketplace." *Sales and Marketing Management* 127, no. 2 (27 July 1981): A6.
To effectively forecast sales, many external factors must be included in the forecast. Author contends that management often does not include such factors, relying more on internal signals. Provides several solutions for incorporating external factors.

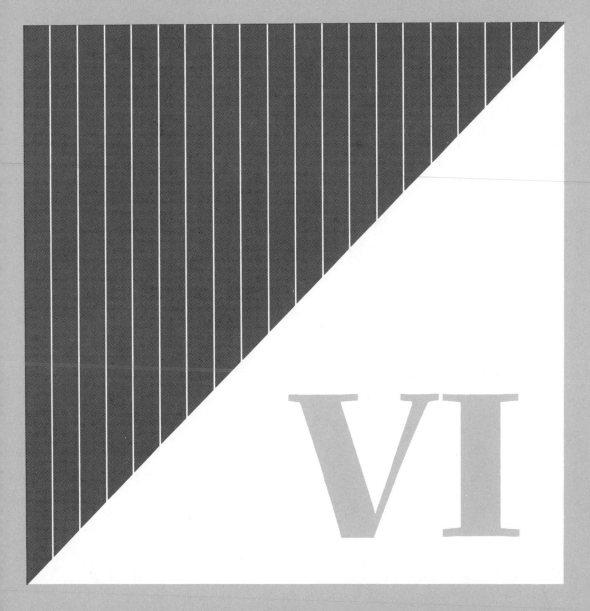

VI

The New Product's Marketing Program

Branding and Packaging
the New Product

Consumers learn to distinguish between alternative goods and services through either actual experience or environmental influences, such as advertising and friends. In this process, the consumer attaches a symbolic meaning to the brand name of the product to help in the recognition and decision-making process. Thus the brand name develops a "personality" of its own that has an effect on whether consumers decide that the product's image is consistent with their needs. In some instances, the brand name alone can imply a meaning—without the aid of advertising or other external information. Brand names such as Mustang, Cling, My Sin, Rabbit, Earth Born, Silkience, Janitor in a Drum, Mr. Muscle, and Rusty Jones project images to consumers without giving much information about the product. If the brand name connotes the right image, it can be a valuable asset in the introduction of a new product.

In many instances, management considers branding a new product to be one of the least important decisions. Most industrial products are given brand names without consideration of their psychological implications. In fact most are either numbers, letters, or meaningless names. For example computer manufacturers have traditionally utilized the corporate name and a model number to identify new computer systems. These numbers may be part of a series, such as the IBM series 370 or Honeywell Level 6.

There are many industrial product brands that have meanings to only certain groups within the industry. These groups represent the potential market segments for the product and would have a great interest in the product. However these technical brand names do have a favorable psychological effect on the marketing effort in that they provide some means of product differentiation.

IMPORTANCE OF BRAND NAME

The most obvious advantage of a brand name is to help the consumer or industrial user identify the most desirable product or service, but it also assists the manufacturer in image building and planning the marketing mix. For example a status or elite-sounding brand name, such as Jontue (a quality perfume), will necessitate a selective distribution channel (e.g., department stores), selective media channel (e.g., women's glamour magazines), a certain advertising appeal (sensual and innocent), a high price, and quality packaging. A brand name is important to both the consumer and manufacturer.

Importance to Manufacturer

Brand names can be used by manufacturers to achieve certain marketing objectives that will support marketing mix strategies.

Aids in Building a Brand Image

New Products are developed, tested, and launched because they satisfy a significant consumer need and will return a profit to the manufacturer. The new product is generally positioned in the market with some distinct characteristics over competitive products. In order to position the new product effectively, the manufacturer must use advertising and promotion to build a brand image that is acceptable to the target market. Thus the brand name becomes the center of promotion and advertising strategy. It can establish acceptance, preference, or loyalty among consumers. Revlon, for example, has been very successful in building a unique brand image for its perfume products. Three of its brands—Charlie, Jontue, and Scoundrel—all have individual brand images that appeal to different market segments. Charlie is designed to appeal to the young, single, carefree, happy woman; Jontue to the young, innocent woman preferring a floral scent; and Scoundrel to the slightly older, mature, sophisticated woman. Successful building of these images through packaging, distribution, pricing, and promotion has given Revlon a significant market share in the perfume market.

Reduces Price Comparisons

With a brand name, the manufacturer can establish a price that is different from the competition. This is largely possible because the brand image will differentiate the product from the competition, allow a price differential, and reduce consumer price comparisons. Most firms prefer to compete on a non-price basis, and branding, to some extent, provides them with the means to do so. Consumers purchasing film for their 35mm cameras, for example, may prefer Kodak film, even though it may be priced higher than other brands. The consumer does not compare film prices and is instead buying a quality, well-known brand.

It has also been shown that branded products have more price stability than nonbranded or obscurely branded products. It is believed that price stability enhances the image of a product. However this conclusion is still open to question.

Facilitates Expansion of Product Line

For the firm expanding its product line, a well-known brand name can be advantageous. It will facilitate consumer acceptance of the new product because of its existing brand reputation. For example if Heinz were to introduce a new product, consumers would generalize that the new product would be similar in quality to other Heinz products, thus enhancing the adoption process. This advantage can easily become a disadvantage if the new product is a failure, however. This issue will be discussed later.

Protects Market

With the ability to achieve identification and a defined image, the manufacturer is assured of some protection of its market. Consumer satisfaction will likely result in repeat purchases, which provides the manufacturer with a possible competitive advantage and some control of the destiny of its product.

Importance to Consumer

Just as the brand name is useful to the manufacturer, it provides important benefits to the consumer. Even in cases where product differentiation is difficult—such as salt—the consumer does develop loyalties and selects unique images of the alternative brands in the category.

Product Identification

The mere existence of a name helps the consumer to determine which alternative consumer product or industrial product is satisfactory. The brand actually perpetuates the consumer choice of the product by making repurchase identification easier. Even "no-name" brands provide the consumer with some means of identification, since he or she recognizes the brand because no name exists and has a certain image of this category of products that helps differentiate them from other brands.

The brand name is particularly important when the competitive products all look alike or cannot be seen. Could you imagine how difficult it would be for a consumer to choose from all of the different brands of salt if there were no brand identification? Hair sprays, spray deodorants, coffee, gasoline, detergents, and so on are all products that are difficult to differentiate because of their similar physical characteristics. However market shares for the various brands in these categories vary over a wide range, because the manufacturer was able to develop brand identification for the benefit of the consumer.

Quality Recognition and Communication

In addition to providing a product with an identity, the brand name helps the consumer to recognize product quality. Brands such as Kodak, IBM, Sears, Sony, and Texas Instruments conjure up an image of quality and reliability. Thus the consumer recognizes a certain quality or reliability when he or she sees these products on the shelf or in a sales presentation.

Identifies New Offerings

Brand names help call the consumers' attention to new products that might be of some benefit to them. This information may enable the consumer to make a purchase decision that satisfies unmet needs.

BRANDING POLICIES AND STRATEGIES

Prior to introducing any product, a branding policy or strategy should be established, especially in view of future new products. The manufacturer may decide to use a family or blanket branding strategy, an individual branding strategy, a mixed family and individual branding strategy, or a combined trade name with an individual brand.

Family or Blanket Branding

In using a family or blanket branding strategy, the manufacturer chooses to apply the same brand name to all products. Firms such as Heinz, Hunt, Campbell, Green Giant, Kodak, and General Electric use this strategy.

This decision is adopted generally when the products are similar in terms of marketing mix strategies; that is they use the same distribution and communication channels and appeal to the same or similar target markets.

The family or blanket branding strategy has some important advantages, provided that the manufacturer is willing to maintain consistency in quality for all items in the line. First of all it facilitates the adoption process and acceptance of new products, since consumers assume that new products have the same quality level as existing ones. Secondly the cost of branding the new product will be less, since brand name research and extensive advertising for brand name awareness and preference will not be necessary. Thirdly the consumer response to the new product will be faster, thereby reducing the introduction stage in the product life cycle where profits are negative.

There are also disadvantages in employing the blanket or family branding strategy. These disadvantages generally center on the ability of the manufacturer to mantain consistency in product quality and the similarity of the products. Lack of consistency in new product quality will result in consumer dissatisfaction, which may carry over to other successful products in the line. The reputation of all products and the manufacturer's image may be negatively affected by one product failure.

When products to be introduced are categorically different (food versus nonfood), a family branding strategy will not be appropriate. For example Swift and Company would not consider the same branding strategy for its hams and bacon as it uses for its fertilizer.

Individual Brands

When a firm produces or sells diverse products, it may not be appropriate to use one family or blanket brand name. Colgate-Palmolive, Procter & Gamble, General Foods Corp., Bristol Meyers, Gillette, Johnson & Johnson, and many others market a wide variety of products that would not be right for family or blanket branding. Tide, a brand name for a detergent, would not be appropriate for a toothpaste or a mouthwash. Most consumers are relatively naive about what specific brands are produced by any one of the above listed firms. This individual branding strategy allows firms to develop different images and positions for each of its products without concern for the negative carryover on the entire product line if any new product should fail.

Individual brands, however, are most costly to introduce because each new product must be heavily advertised to establish brand awareness. Frequently a firm's objective is to achieve a greater degree of market saturation. Specific, individual brands may appeal to one or a few market segments, thus constraining market share objectives. To increase the market share potential, the firm will introduce another product in the same category but appeal to a different market segment to attain broader market coverage and a larger market share. This is a multibrand strategy. For example, Procter & Gamble markets several detergents, such as Tide, Cheer, and Era. Each new detergent introduced by Procter & Gamble claimed somewhat different ingredients and characteristics. Each of these brands when introduced resulted in some reduction of market share of existing brands but, overall, achieved the immediate objective of increasing sales volume.

There are some important reasons for using a multibrand strategy. First, each brand occupies more shelf space, thus leaving less for competitors. Second, most consumers are not loyal to one brand and are likely to switch given the right circumstances, such as a price deal. If the manufacturer did not introduce new brands of the same type of product, it would eventually face a declining share of the market. In order to capture the brand switchers, a multibrand strategy is necessary. Third, each new brand gives the firm an opportunity to appeal to different market segments.

The major problem with a multibrand strategy is the possibility that each new brand in a product category will only obtain a small market share, resulting in reduced profits for the company. For example a firm with three marginally profitable detergents introduces a fourth brand to stimulate its total detergent sales. The new brand attains a small market share, but the reduction in the sales of the other three brands is equivalent to the sales attained by the new brand. The overall result of the new product introduction is no increase in total company sales but increased marketing and development costs. It is possible that total profits may now be negative since the company has spread its resources too thinly, resulting in a weak profit position.[1]

Mixed Branding Strategy

In many large companies, product line expansion occurs in different directions so that new families of products may be planned when a new product is conceived. In this case the firm will use a mixed branding strategy with similar products carrying family brand name and different products with individual brand names. For example Sears, Roebuck & Co. employs a mixed branding strategy by using brand names such as Coldspot, Craftsman, and Kenmore as individual brand names, each with a family of products. Thus each hardware item would be branded as Craftsman, yet all refrigerators and air conditioners are referred to as Coldspot.

This strategy is employed when the product line expands so that new products are quite different from other products, therefore requiring a unique name and marketing strategy.

Trade Name and Individual Brands

The last strategy that may be employed by the firm is to associate the company name with an individual brand for each product. In this case the company name provides some legitimization and the individual name some differentiation for the new product. There are Johnson's Pride and Johnson's Glo-Coat; Post Raisin Bran, Post Cocoa Krispies, and Post Alpha Bits.

This strategy, as family or blanket branding, carries the risk of negative association of all products if the consumer rejects the new item. However, as mentioned earlier, the company may benefit from using its company name by enhancing the adoption process and reducing some of the heavy advertising costs in the introduction stage of the product life cycle.

Whatever branding strategy is employed, the advantages and disadvantages of each of the strategies discussed here should be carefully weighed. Future marketing objectives and target market strategy must also be considered when determining brand strategy.

---

[1] For further discussion on multibrand strategy, see Robert W. Young, "Multibrand Entries," *Plotting Marketing Strategy*, ed. Lee Adler (New York: Simon and Schuster, Inc., 1967), 143–64.

In the packaged food industries, the fact that the choice of the brand name is an inherent part of marketing strategy designed to result in a successful new product launch is well recognized. Research is often conducted to determine the image that a number of alternative brand names may convey to the end-user. The objective of this research is to identify which brand name conveys the desired image before the new product is launched. A negative image conveyed during the product launch cannot be easily rectified and usually results in the product's withdrawal from the market. It may be reintroduced with a new brand name, but the firm would have already incurred damage to its reputation as well as financial losses. To avoid this outcome, it is beneficial to conduct brand name research to identify possible brand name image problems before they occur. In addition to image, the firm may be concerned with such aspects of the brand name as ease of recall, distinctiveness from competition, and relationship to corporate name.

Brand Name Research Techniques

Many different research techniques may be employed to test consumer reactions to alternative brand names. Some of the more widely used techniques are rank order, scaling, and various motivational research techniques.

Rank Order   In this technique, a sample of present or potential users would be asked to rank in their order of preference the alternative brand names for a specified new product. The brand name that receives the highest preference rankings would then be adopted. The major weakness of this technique is that the distance between ranked alternatives—that is, the distance between rank 1 and rank 2, and between rank 2 and rank 3—may not be equal. However the technique is simple to employ and can be completed quickly without great expense.

Scaling   Scaling techniques are often employed to determine end-user reactions to characteristics conveyed by the brand name. Figure 15–1 shows a simplified example of the semantic differential scaling technique. It requires respondents to indicate on a seven-point scale the words that describe the image conveyed by the brand name.[2] If the firm wanted to convey an image of strength, aggressiveness, or masculinity, it would select the brand name that consumers identify with these attributes. Many attributes should be listed in this type of test to determine all possible reactions. The semantic differential is the most popular technique among market researchers.

Other scaling techniques, such as Likert's summated scale, Thurstone's differential scale, and the Q-sort technique are also available but are not discussed here.[3]

---

[2]C. E. Osgood, G. J. Suci, and P. H. Tannenbaum, *The Measurement of Meaning* (Urbana, Ill.: University of Illinois Press, 1957), 76–124.
[3]For a discussion of these scales, see Paul E. Green and Donald S. Tull, *Research for Marketing Decisions*, 3rd ed. (Englewood Cliffs, N.J.: Prentice-Hall, Inc., 1975), 167–208.

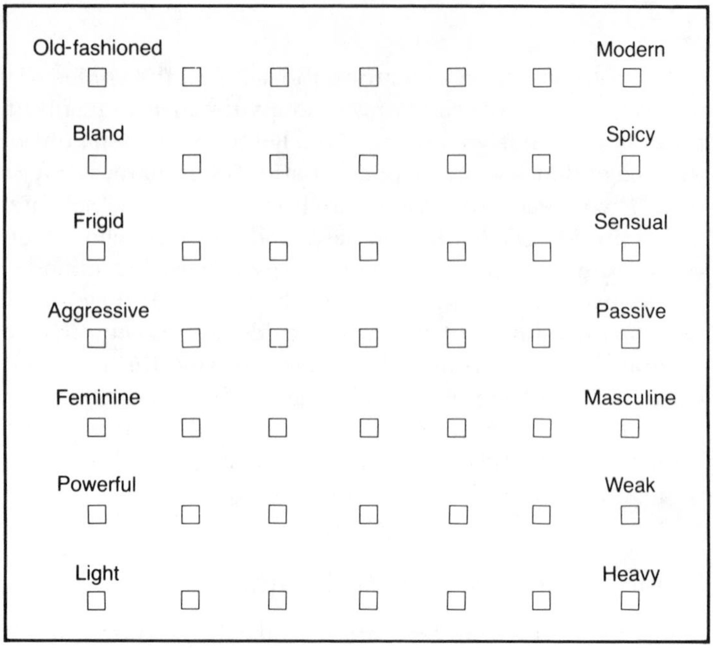

FIGURE 15–1

Scaling example to measure brand name image

Old-fashioned □	□	□	□	□	□	Modern □
Bland □	□	□	□	□	□	Spicy □
Frigid □	□	□	□	□	□	Sensual □
Aggressive □	□	□	□	□	□	Passive □
Feminine □	□	□	□	□	□	Masculine □
Powerful □	□	□	□	□	□	Weak □
Light □	□	□	□	□	□	Heavy □

Motivational Research Techniques The objective of motivational research should not be to select the final brand name but to indicate which one or two words appeal most to the target market. Color, packaging, printing, and other factors used to display the brand name will all affect its performance. However the use of these variables with the brand name in the research makes it difficult for the researcher to separate the actual causes of consumer reaction. Often firms wish to determine underlying psychological suggestions given by the brand name. This can be done through the use of motivational research. Techniques such as word matching, free association, cartoon tests, and narrative projections are used for many different purposes in marketing. For brand name research, the word matching or free association techniques would be most appropriate to determine brand name image.

Free association is the best starting point for any brand name research. The test provides consumers with a list of names, including those being tested, and asks them to identify what is associated with these words. New words can often be derived with this type of research. These are to be used as a brand name that would enhance the new product launch.

In a word matching test, consumers are given a number of words and products, including the ones being tested, and asked to match them. From this test the firm would be able to determine the name or word that best describes the new product. Other products are used in the test to prevent bias by the respondents.

It is also necessary for the firm to determine whether consumers can remember the brand name easily and whether they differentiate the product from other brand names in the industry. The Matex Corporation illustrated this issue when it dropped its Thixo-Tex name for its rustproofing service for automobiles. When Matex advertised its Body By Thixo-Tex system, sales grew at a rapid rate. When the campaign ended, sales declined; consumers had little recall of the name. Research concluded that the name *Thixo-Tex* was not associated with rustproofing, and that a new name with a new theme that would be personalized and remembered was needed. The decision was made to use *Rusty Jones*, with the theme *Hello Rusty Jones, goodbye rusty cars.* Along with a guarantee and an introductory advertising campaign, Matex achieved significant market penetration and eventually the leadership in the rustproofing business.[4]

In general a good brand name should be simple, and easy to remember. It should convey an image to the consumer that is consistent with marketing strategy and the target markets. Satisfaction of these objectives should eliminate some of the underlying causes of new product failure.

THE PACKAGING DECISION

Traditionally the function of packaging was to protect goods. In some industries, this is still considered the primary function. However in many industries, the package has become a promotional tool and image builder for the new product, thereby enhancing its success in the market. This is especially true in cases where the consumer must choose from several alternatives on a shelf where the package then becomes a point of sale display. One of the first marketing decisions to be made is to determine how important the package is in developing a favorable appeal to the consumer.

PACKAGING DESIGN CONSIDERATIONS

The packaging decision must weigh each of four factors in the design phase: a) marketing considerations, b) product protection, c) economic factors, and d) environmental factors. These four considerations are outlined in Table 15–1 and are discussed briefly in the following paragraphs:

Marketing Considerations

The package design must consider the implications of the package on the manufacturer, the retailer (or other middlemen), and the consumer. As shown in Table 15–1, the retailer's major concern is to get the product on the shelf with a minimum of difficulty and to prevent pilferage of high-priced small items. The consumer, on the other hand, is most concerned with such factors as convenience and information provided on the package. The manufacturer must consider the ease of recognition and image conveyed by the package design.

Some of these factors may be contradictory, such as recognition and shelf stacking. For example the manufacturer may decide to use an oddly shaped

---

[4]"What's In a Name? Just Ask Rusty Jones," *Marketing News*, 10 August 1979, 12.

TABLE 15–1
Considerations in design of package

Marketing Consideration	Product Protection	Economic Factors	Environmental Factors
For management	Prevents contamination	Cost of materials	Biodegradable
Recognition	Prevents physical damage	Cost of fabrication	Ability to recycle
Color	Shelf life	Added or reduced manpower	Impact on pollution
Size	Effect of climate	Inventory, shipping, and storage costs	Social pressures
Design	Safety	Refunds and allowances for damaged goods	Legislative pressures
Image	Effect of light	Availability of raw materials	Competition
Attitudes		Equipment needs	Impact on natural resources
For retailer			
Shelf-stacking characteristics			
Ease of price marking			
Ease of case identification			
Quality of packages per case			
Ease of package removal			
Pilferage protection			
For consumer			
Consumer convenience			
Ease of storage			
Handling ease			
Directions for use			
Reuseability			
Disposability			
Information			
Instructions for use			
Expiration data			
Alternative product uses			
Guarantees			
Nutritional Components			

package to attract attention and to differentiate the new product from competition. However the oddly shaped package presents problems to the retailer in terms of shelf space and stacking. Minimization of conflict in these marketing considerations will enhance channel acceptance of the new product.

A common strategy among consumer packaged goods firms is to use a package redesign as an alternative marketing strategy to increase sales and profits. A carefully planned and well-executed program centering around the new package design can cost less than one 60-second television commercial. A case in point was the Columbo Yogurt Co., which realized the need to give its product new visibility on the supermarket shelf. With the proliferation of new yogurt entries, Columbo, the oldest yogurt manufacturer in the United States, felt the need to give its package an uplift. Keeping in perspective the firm's objectives, cost targets, and productive capabilities, a new package was designed that would stand out from the competition on the shelf. The new package significantly enhanced Columbo's shelf impact and helped the firm retain its important market position.[5]

As a marketing tool, packaging offers the marketing executive a flexible and relatively easy alternative to developing significant product modifications and changes. The package of any new product should address the following marketing factors:

1. It should provide visual impact at the point of purchase.
2. It should provide the consumer with an easy identification of the corporate name.
3. It should establish a tie-in relationship with other flavors, sizes, or containers of the same product.
4. It should be simple, not cluttered, and use color to its best advantage.
5. It should be informative to the consumer, indicating clearly what the product is and what is contained in the package.
6. It should complement any advertising campaign to enhance its recognition on the shelf.

Packaging innovations can enhance the market potential of an existing product by increasing exposure and awareness. For example electronic quartz digital wrist watches now are being sold like razors, pens, lighters or lipsticks. Commodore Consumer Products Group, Inc. has been marketing its $7.95 and $9.95 watches on 5 × 8 cards that have been punched for rack or peg-board display. The watches are skin packaged (see-through plastic) and displayed with other mass merchandised products. The firm hopes to increase its market base by selling watches wherever shoppers can be found in large numbers and by backing their product with warranties.[6]

Product Protection

Products subject to spoilage, physical damage, or contamination require laboratory testing prior to designing the final package. Food manufacturers work closely with paper companies, chemical manufacturers, metal processors, and other manufacturers of packaging materials to resolve packaging problems and satisfy the product-protection objectives.

Industrial manufacturers are particularly concerned with physical damage if the product is fragile and expensive. Typical hazards are physical (such as shock, vibration, static electricity, compression), climatic (such as heat, cold, moisture, oxidation, pollution), and biological (such as infestation, bacteria, decay). Because losses due to physical damage of the product are generally the responsibility of the manufacturer, the package must be carefully designed.

Recent innovations in industrial packaging have alleviated some of the problems related to product protection. Air bubble cushioning material that is lightweight and inexpensive is used extensively by industrial firms as a protec-

---

[5]"Packaging Design Seen as Cost-Effective Marketing Strategy," *Marketing News*, 20 February 1981, 1 and 6.
[6]"Low-Priced Digital Wristwatches Mass Marketing Like Ball-Point Pens," *Marketing News*, 7 August 1981, 3.

tive wrap. It can be purchased in large rolls and can be used in bubble bags, void fills, sheets, and anti-static bags, or as a protective wrap.

Another innovation for larger items is the foam-in-place product. This foam product is also light and economical. It is dispersed as a liquid into a carton using a sprayer and immediately begins to expand. A polyethylene film is then placed over the rising foam and the product (e.g., stereo speaker) is placed on top of the poly so that the expanding foam follows the contour of the product. The product is then covered with the film and more liquid foam sprayed on top of it. When the carton is sealed the foam encapsulates the product in a strong, lightweight protective package.

These inexpensive innovations have resulted in significant savings from product damage. Other innovations to protect sensitive products from temperature or moisture have also been developed.

Economic Factors

Because financial loss due to breakage or spoilage can be significant, meeting certain product-protection criteria can minimize costs to the manufacturer. The manufacturer must also consider the actual packaging costs for the new product however. For high-value products, the cost of an expensive package is relatively unimportant, since it is such a small percentage of the value of the product. However for many consumer and industrial products, the cost percentage of the package is significant. This is especially true when there is competition and only a small profit margin.

In the past eight years the cost of producing a glass container rose more than 60 percent. Even more dramatic, the cost of producing a can rose in the same period by over 100 percent. Annual rates of increase for both types of containers are expected to be about 10 percent through the 1980s.

In response to the dramatic increase in the price of containers, some firms have tried to develop flexible packaging, such as the paper bottle. Ocean Spray Cranberries, Inc. recently introduced a 200 ml. ready-to-serve container (flexible packaging) as part of a three-phase program it plans to use to gain acceptance by the consumer for its new packaging concept. Phases two and three will introduce larger-sized packaging, such as that used for its 200 ml. juice drinks.

This type of packaging innovation resulted from using costs of alternative packages. The manufacturer can pass along the cost savings of its innovation to retailers and consumers, which could help to stimulate increases in sales of the product category.

From a strictly economic viewpoint, the manufacturer must consider the costs of package materials, fabrication of the package, manpower, inventory and storage, and equipment needs. To avoid shortages for packaging, management should consider the future availability of the material, especially if the need to change materials may negatively affect the image of the product. The availability of materials is reflected in the cost of these materials. Cost increases in glass and aluminum, and political crises—such as the oil embargo, which caused plastic material prices to rise significantly—forced many manufacturers to look for alternative packaging materials.

Environmental Factors

Without question one of the most significant trends in our society is the consumer's concern with pollution, particularly from discarded packaging. As a result of this trend, many states and local governments have passed legislation to restrict the sale of throw-away containers. These states now prohibit throw-away bottles and require consumers to pay a refundable deposit on all bottled items.

Currently many manufacturers are packaging their products in reusable containers as a means of recycling the package materials. In addition research is continually being carried out to develop new packaging materials that are biodegradable or that minimize pollution problems. Consumable packages have been developed at high cost, but have received poor consumer acceptance. Efforts to develop packaging materials that will ameliorate these environmental problems will continue, particularly in light of the growing shortages of many of the important packaging materials.

PROMOTIONAL ADVANTAGE OF PACKAGING

Packaging offers the manufacturer an important promotional medium, particularly when many of the products in the marketplace are similar and difficult for consumers to differentiate. Packaging may be the dominant force in the consumer's choice when there is little product information available and the size and price are perceived as similar. The importance of the promotional impact of packaging is reflected in the battle by manufacturers to obtain shelf facings in retail stores. Exposure is increased by increasing the number of shelf facings. Thus the package design that attracts the consumer's attention provides an important medium at the point of purchase.

Packaging can work with and extend media advertising by reinforcing, repeating, or specifically targeting advertising messages and images. With the high cost of mass media, the package design has become even more significant as a promotion tool. In fact with the capability of contemporary package graphics and innovative structural design, a package can play an important role in a communication medium.

Hold, a new cough lozenge, has been extremely successful because of unusual and innovative packaging and merchandising. The Beecham Products Co. decided to use stop-sign colors on the package and display cards and to sell the product at the checkout counter, rather than with other lozenges and cough remedies. The objective was to promote the product through packaging and merchandising in a more provocative way. After the initial success of the lozenge, Beecham introduced liquid Hold, taking advantage of the success of its companion product. Thus through a combined effort of its packaging and promotional merchandising, the firm was able to create a uniqueness that separated its product from the competition.

Package Testing

Since packaging complements mass media and also reflects the product's image, it is often necessary to test the package of the new product prior to its test market or introduction. A poorly designed package that does not reflect the desired product's image can do irreversible damage to a new product. It can also waste precious time or significantly raise the costs of a test market.

Pillsbury's Green Giant management feels that the right package is critical to the success of a new product. Using research techniques, such as focus groups and panel testing, the firm tested its package for its new line of frozen Mexican foods. Initially its objective was to develop a package with a line look, focusing on the Fiesta Grande name. The product was to have the Green Giant endorsement but was to be distinct from the Green Giant packaging. The package also was designed so that the product would have strong shelf impact because retailers' freezer space was often limited, especially for an item such as this new product. Thus far the results have been positive and further market expansion is being considered.[7]

SUMMARY

Frequently management overlooks the importance of branding and packaging decisions. Both the brand name and package design contribute to the product's image and must be considered in the marketing mix decision-making process.

The brand name provides the manufacturer with the means to build a company image, reduces price comparisons, and facilitates expansion of the product line. In determining the brand name for the new product, management must decide whether to use a family or blanket branding strategy, individual branding strategy, mixed branding strategy, or combined trade name with the individual brand. Each of these policies should be considered in light of the company's long-range objectives.

Research techniques may be employed to test consumer reactions to various brand names. Some of the more widely used techniques are rank order, scaling, and motivation research.

The packaging decision must consider marketing, product-protection, economic, and environmental factors. Management, in trying to identify effective packaging attributes, should determine whether the package is attention-getting, informative, perceptually pleasing or appealing, and functional. All packaging and brand name decisions are made in conjunction with the firm's marketing objectives.

SELECTED READINGS

"Design Professionals Pick Some Favorites." *Fortune* (7 May 1979): 190–95.
    Features examples of outstanding graphic design in packaging.
DOWNEY, WAYNE. "Geyser Spouts Plastic-Bottle Wine." *Advertising Age*, 15 February 1982, 28.
    Discusses the introduction of a new plastic bottle for wines that is being tested in the San Francisco area by a subsidiary of the Joseph Schlitz Brewing Co.

---

[7]"Ole! for Pillsbury Packaging," *Sales and Marketing Management* 126, no. 4 (16 March 1981): 70–72.

GOLDSMITH, MARK. "Swedish Packing Giant Well Armed for U.S. Entry." *Advertising Age*, 14 June 1982, 4 and 62.
Discusses aseptic packaging, a packaging development being used by Tetra Pak, a Swedish company attempting to expand their penetration of the United States market. This packaging, which does not require refrigeration, is being used for milk and fruit juices and represents an interesting marketing strategy to change consumer habits.

McNEAL, JAMES U., and ZEREN, LINDA M. "Brand Name Selection for Consumer Products." *Business Topics* 29 (Spring 1981): 35–39.
Discusses a survey of consumer manufacturing companies from the Fortune 500 list concerning their brand name selection process. The information compiled from this study includes the criteria used in selecting brand names; stages in the product evolution process in which brand name selection occurs; and the research efforts, in-house departments, and outside firms involved in selecting brand names.

McQUADE, WALTER. "Packages Bear Up Under A Bundle of Regulations." *Fortune* (7 May 1979): 180–89.
Provides a detailed review and interpretation of the federal regulations and lawsuits in the packaging industry. Identifies particular areas of general complaint and offers suggestions on how business may deal with them.

"Introduction to Packaging." *Package Engineering: The 1982 Package Encyclopedia*, Vol 27, 15–39.
This section of the 1982 Packaging Encyclopedia includes discussion of packaging trends, planning and scheduling, organizing the package teams, testing of packaging and package shelf life.

"Sero Likes Scoundrel's Name Too." *Advertising Age*, 8 September 1980, 1.
Discusses how Scoundrel, the name for a new perfume introduced by Revlon, was decided on after reviewing 3,000 possibilities. Revlon had to negotiate for the rights to this name, since Sero shirtmakers had registered it 20 years earlier.

TAUBER, EDWARD M. "Brand Franchise Extension: New Product Benefits From Existing Brand Names." *Business Horizons* 24 (March–April 1981): 36–41.
Explains how some of the most successful companies in the United States have begun to offer new products that capitalize on the firm's well-known trademarks and brand names.

TYLER, WILLIAM D. "Value in Nameonics." *Advertising Age*, 31 May 1982, 34–35.
Argues that the objective of most promotion programs is to make the brand name irresistably memorable. Describes *Nameonics* as a technique to establish a trademark through creative advertising.

UEYAMA, SHU. "The Selling of the Walkman." *Advertising Age*, 22 March 1982, M3 and M37.
Describes the historical sequence of events in the development and marketing of the Sony Walkman. Describes how the product was named and alternative packages were tested.

Pricing the New Product

Ralston-Purina Company turned to a simple attention-getter earlier this year when it wanted to increase its sales of canned cat food: it lowered the wholesale price by about four cents a can.

Until recently this sort of action has been almost unheard of among packaged goods marketers. Their typical sales tactics are cents-off coupons, bigger ad budgets, sweepstakes, refunds, temporary price discounts, or claims of minor product changes (e.g., *new and improved*).

For Ralston-Purina, price is no longer being neglected as a marketing tool. The grocery products division has lowered prices on its Mainstay and Moist & Crunchy dog food.

Such leading consumer products marketers as Procter & Gamble, Kellogg, Coca-Cola, Scott Paper, Mobil, Union Carbide, and Lever Brothers apparently share this view. For various reasons, all have recently cut their prices on certain brands. Some of the companies are fighting private-label and generic products, while others want to reduce temporary discounts, known as *trade deals*, or to revive lagging products.

Procter & Gamble has reduced both list price and trade deals for Folger's coffee, as has Coca-Cola's foods division for Hi-C fruit drinks. Widespread discounting, Coke said in trade publication ads, "led consumers to mistrust the entire pricing structure in the grocery industry."

Because many of the reductions in regular prices are being tied to cutbacks in promotions, shoppers won't necessarily see lower supermarket bills. For example they'll be losing some of the benefits of trade deals, which enable stores to offer discounts on products. Such deals have been accounting for one-third of many companies' marketing budgets.

Whatever their motives, manufacturers that reduce list prices are being praised by retailers. "Consumers are getting fed up with the hocus-pocus of coupons, refunds and price specials," Says Thomas Stemberg, senior vice-president of First National Supermarkets. "They're looking for everyday basic value." James Henson, president of Jewel Food Stores, urges manufacturers to "attack and justify every penny spent in non-ingredient costs."[1]

When Union Carbide recently added a third ply to its Glad trash bags, without raising their price, the company was trying to make consumers think they're getting better value. So are most other packaged goods manufacturers, these days, as they unleash a barrage of temporary price specials, coupons, refunds, and other promotions.

But few marketers are choosing what might seem to be an obvious strategy—cutting their prices permanently. Wouldn't the combination of price and performance be one of the best ways to increase a product's value and mollify price-conscious consumers?

Wilson Harrell thinks so, and he's about to test his belief by introducing spray cleaners he says are as good as Windex, Glass Plus, Fantastik, and Formula 409 but are less than half the price. His technique—removing what he says is unnecessary water.

---

[1]Bill Abrams, "Consumer Goods Firms Turn to Price Cuts to Increase Sales," *Wall Street Journal*, 13 May 1982, 29.

Three-fourths of spray cleaners are just plain water. It adds to the price, takes up space, and is heavy to lug around. Bearing the brand name of 4 + 1, Mr. Harrell's glass cleaner and all-purpose spray are concentrated cleaning liquids; consumers add water (four parts water to one part of the cleaner) and supply their own spray bottle (probably a recycled container such as *Windex* or a similar product).

M&M/Mars found another way to increase value. Faced with a sagging candy market, the company last year increased the weight of several of its bars by an average of 10 percent without changing the price. The candymakers gambled that increased sales volume would more than compensate for thinner margins, and they won: M&M/Mars says sales have increased 50 percent.

Each of these companies delivered value by reducing prices rather than, as Union Carbide did, improving product quality. A third strategy increasingly favored by supermarkets is to offer more generic and private-label goods, which are less expensive than national brands but usually are of lower quality.

But many consumers don't want either reduced or top-of-the-line quality: they want satisfactory quality at an attractive price. "Shoppers don't want quality to go downhill," says Mona Doyle, president of Consumer Network, a Philadelphia market research concern. "But they're also looking for more bang-for-the-buck, what they're getting for their money. Any manufacturer who zeroes in on that is definitely on target."[2]

As these scenarios indicate, price is an important variable in every product marketing decision. The significance of pricing for a new product is indicated in the fact that it tends to hold a major position in corporate affairs (see Chapter 6). The price of the new product is a major factor in determining the revenue and the profits, as well as the volume of sales.

Price is a complex variable because of the many people—manufacturers, wholesalers, retailers, consumers, and public policy makers—who influence and are influenced by it. The relationships among these people as well as the relationship of the price with other elements of the marketing mix—product, distribution, and promotion—must be fully considered before setting the initial price of the new product. In order to answer the basic question of what the price of a new product should be, it is important that carefully designed company objectives and policy be established. This chapter discusses pricing objectives and presents the three fundamental concepts in determining the price for the new product: cost, the consumer, and competition. The chapter concludes with a discussion of the relationship of the initial selling price to other marketing variables that affect the new product, setting a price structure, trade allowances, and consumer demand, and the legal implications of the price.

COMPANY PRICE OBJECTIVES

Before establishing comprehensive pricing policies, a firm must formulate pricing objectives, since the policies are merely the means of achieving the objectives set. Table 16–1 indicates the variety of potential pricing objectives.

---

[2]Bill Abrams, "Spray Cleaner to Test Notion That With Price, Less Is More," *Wall Street Journal*, 21 May 1981, 29.

TABLE 16–1

Possible pricing objectives

1. Maximum long-run profits
2. Maximum short-run profits
3. Growth
4. Stabilize market
5. Desensitize customers to price
6. Maintain price-leadership arrangement
7. Discourange entrants
8. Speed exit of marginal firms
9. Avoid government investigation and control
10. Maintain loyalty of middlemen and get their sales support
11. Avoid demands for "more" from suppliers—labor in practice
12. Enhance image of firm and its offerings
13. Be regarded as "fair" by customers (ultimate)
14. Create interest and excitement about the item
15. Be considered trustworthy and reliable by rivals
16. Help in the sales of weak items in the line
17. Discourage others from cutting prices
18. Make a product "visible"
19. "Spoil market" to obtain high price for sale of business
20. Build traffic

Source: Alfred R. Oxenfeldt, "A Decision-making Structure for Price Decisions," *Journal of Marketing* 37, no. 1 (January 1973): 50.

Although specific pricing objectives will vary from firm to firm—as well as from product to product within a given firm—depending on such factors as market position or competition and demand, they can be classified into three overall groups: profit-oriented, sales-oriented, and other.[3]

Profit-Oriented Objectives

The ultimate objective of any business enterprise firm is, of course, to make a profit, pay dividends to shareholders, and provide a source of funds for investing in research and development and plant and machinery to ensure future growth and continued profitability. Most price models have as a basic tenet that a firm will seek to maximize profits in the short run. This approach is infrequently used, due to the inability to successfully maximize profits; this is caused in part by the uncertainty regarding relevant costs and demand. Another factor causing infrequent use of profit maximization objectives for a new product is that a lower price often allows the new product to gain faster market acceptance and develop a larger, more broadly based market. A profit maximizing objective for a new product may also incite new competing products to be developed more quickly because the high margins would be equally available to competing firms. In fact some firms formally monitor markets for complementary products to find one in which the margins are attractive enough for them to enter. Probably the only case in which short-term profit maximizing concepts are practiced is in situations of competitive bidding

---

[3]For a comprehensive discussion of these and other pricing objectives applied to all product pricing situations, see Robert A. Lynn, *Price Policies and Marketing Management* (Homewood, Ill.: Richard D. Irwin, 1967), 97–114, and Kent B. Monroe, *Pricing: Making Profitable Decisions* (New York: McGraw-Hill Book Company, 1979), 141–222.

FIGURE 16–1
Target rate of return pricing

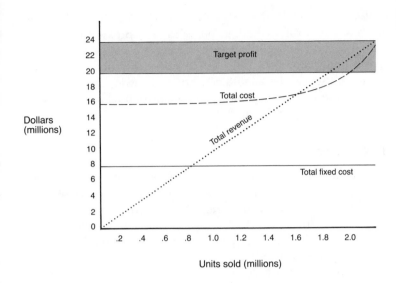

(discussed later in this chapter). A price is set in this case to achieve a specified rate of return based on the total cost of the new product. In order to set this price, the total costs at various levels of output are estimated, along with the most likely level of production during the introductory state. For example let us assume that the K.T.H. company is considering introducing a new industrial adhesive. The company expects to use 80 percent of a total plant capacity of 2 million during the introductory stage; that is it estimates it can sell 1.6 million units. The total cost of producing this amount will be $20 million. Now the management of K.T.H. must determine the associated total revenue curve by first specifying a target rate of return. Due to the research and development costs and the uniqueness of the product, company management wants a 20 percent profit over cost. In dollar terms, K.T.H. management wants to achieve net profits of $4 million during the introductory stage. This cost estimate of $24 million (cost plus profit) is the point on the total revenue curve at a volume of 80 percent of capacity. Another point on the total revenue curve will be zero at a volume of zero percent of capacity, as indicated in Figure 16–1.

The slope of the total revenue curve indicates the price necessary to achieve the target rate of return. For the company in the example above, the slope is 2.5. Thus by selling 1.6 million units of the industrial adhesive, it will achieve its target rate of return of 20 percent or $4 million at a price of $2.50 per unit.

Another profit-oriented objective is a minimum profit objective. This objective is not commonly used by firms since if a new product cannot be sold at the price necessary to achieve the minimum profit desired, it should not be introduced in the first place.

Sales-Oriented Objectives

Sales-oriented objectives often guide the pricing of a new product. Although the adoption of such objectives in no way implies a complete abandonment of profit, these objectives can lead to a price that is below what produces maximum profits. Usually a sales-oriented objective is expressed in terms of attaining a specified market share. For example a target market share may be 10 percent. Since a high volume of sales for the new product results from consumer acceptance, this can be facilitated with a low price. Sales-oriented objectives can often achieve the best profits—especially in the long run—since the orientation is the market, not the firm.

Other Price Objectives

Other objectives, such as meeting competition, conforming with product line, company image, and early cash recovery, can also be used to establish the price for the new product. Many firms adopt a somewhat passive pricing objective—that is, having the price of the new product meet the price of competing product(s) already being marketed. This is a particularly common pricing objective when the new product is not very new to the consumer.[4] In other words if the product does not display distinct characteristics or will not dramatically affect the life style of the consumer, then, to stimulate sales, it may have to be priced close to products presently being marketed. This can lead to a practice called *price following*. Studies have found such firms as National Steel, Republic Steel, Standard Oil of New Jersey, Gulf, and Goodyear to be price followers.[5] When using a meeting competition objective, a firm will rely on other aspects of the marketing mix, such as advertising and sales people's commissions to differentiate the new product.

Instead of meeting competition, a new product can be priced using a product line objective. This means the new product is priced to fit into the product line and/or promote other products in the product line. The new product may not yield a profit itself but, rather, will enhance sales of other products of the firm.

Closely related to the product line objective is the image objective. A company's reputation or image is intangible, but it can still have a great deal of importance and value. Several methods for establishing and dealing with the image of the new product were discussed in Chapter 15. For example, if the new product is identified with the company, then it may have to be priced in line with the prevailing company image to avoid any negative impact on the company's image as well as sales of the company's other products.

The price of a new product must be consistent with the image of the product itself. A major cause for poor sales is that the new product is priced too low relative to other products in that class.

Many firms price a new product so that rapid recovery of cash will result. This objective is more widely adopted in times of high interest rates and poor economic conditions and uncertainty. For the most part firms adopting an early cash recovery objective are either short on capital or regard the future

---

[4]The various aspects of newness and a classification system for products of varying degrees of newness were discussed in Chapter 1.

[5]See R. F. Lanzillotti, "Pricing Objectives in Large Companies," *American Economic Review* 48, no. 5 (December 1958): 924–27, and U.S. Congress, Senate Subcommittee on Antitrust and Monopoly, *Administered Prices*, 88th Cong., 1st sess., 1963, 18138, 18139, 18178.

demand for the product and/or the economy to be too uncertain to slowly cultivate the market.

POLICY FORMULATION

The specific objectives in pricing new products must be formulated into general corporate policies. Of particular importance in formulating a company's pricing strategy and objectives is the long-term financial plan. In defining this financial plan, several key factors should be taken into consideration. First the new product must be priced to yield an appropriate return on invested capital. This makes it necessary to delineate the capital requirements of each new product. Once the capital requirements have been determined, the new product can then be priced to yield a specified return on investment.

Another factor in formulating policy for pricing new products is the expected revenue. The revenue should be on both a total basis and a percentage-of-profit basis. These ratios will indicate the level of flexibility to respond to a price maneuver by competitive products as well as rising costs.

The level and pattern of the cash flow from the new product's sales should also be determined. This is a particularly important factor to consider in policy formulation, since firms often find themselves with insufficient cash reserves to fund promotional efforts to stimulate the new product's sales.

A final area to consider in formulating new product pricing policy is the effect the new product will have on the company's present worth. Since the objective of any pricing strategy is to maximize the positive effect it will have on corporate earnings, this factor should be the cornerstone of all new product pricing.

Within this general policy framework, two specific pricing policies can be implemented for the new product: skimming or penetration pricing policy.[6] Under a skimming pricing policy, a company wants to take advantage of the fact that some consumers will pay a premium to be among the first to enjoy the new product. A company seeks to gain a larger profit from the higher price paid by buyers for whom the new product has a high present value. The price of the new product can then be reduced to appeal to more price-elastic segments of the market.

Several conditions favor implementing a skimming pricing policy for the new product. The most important condition is that there is a sufficiently large number of buyers whose demand is relatively fixed. This is frequently the case for a novel product for which there are no comparable products on the market or for high-technology state-of-the-art products. With no comparable products, consumers are unable to make value comparisons and therefore tend to be relatively more sensitive to price. The main issue to be determined is the size of this market segment. This is particularly important in implementing a skimming pricing policy, since the company plans to forego sales to all those who are price sensitive by setting a high initial price.

---

[6]For a detailed discussion of these two concepts, see Joel Dean, "Pricing Policies for New Products," *Harvard Business Review* 54, no. 6 (November–December 1976): 141–53.

Another condition under which a skimming pricing policy is used is when no economies of scale and savings result from producing and distributing a larger volume of the product; in this case the increased volume achieved from a lower price is not enough to offset the lower revenue received. Therefore a high price achieves better return.

In some cases skimming is the only feasible pricing policy. This is true when a firm is not financially able to produce a large volume of the new product. Since high cash outlays are needed in the introductory stages for production, distribution, and promotion, a firm often can finance the production and distribution of only a limited amount of the product, rather than the amount that would be demanded at a lower price.

Perhaps the greatest concern with a skimming pricing policy in the long run is competitive reaction. When there is a high price resulting in high profits, the development of rival products is stimulated. The probability that this will occur under a skimming pricing policy decreases when the new product has strong patent protection or high development costs.

On the opposite end of the continuum from this market skimming policy is the market penetration policy. Under a market penetration policy a firm will set a relatively low initial price for the new product in order to achieve relatively fast sales penetration of the market. This policy stimulates growth of the market for the new product and allows a large market share to be obtained. A penetration pricing policy is most advantageous and is sometimes necessary when the market for the new product appears to be very price sensitive; this occurs, for example, when a price-responsive market that brings in a large number of buyers at a lower price is forecasted.

Another condition favoring penetration pricing occurs when economies of scale are possible. If a product has a large market—and therefore increased output—the unit cost of both production and distribution decreases. This larger output allows distribution of overhead costs, as well as advertising expenses, over more units, thus reducing the cost per unit.

Finally a low initial price on the new product based on a penetration pricing policy would discourage actual, as well as potential, competition. When high potential competition exists, a penetration policy can be beneficially employed to raise entry barriers for prospective competitors. Firms tend not to enter low-margin markets. This protective barrier is especially important when the new product does not have a great deal of technological advantage and can therefore be easily duplicated.

COST CONSIDERATIONS

Cost is perhaps the most important of the fundamental concepts used in pricing a new product. Cost is the floor below which the price of the new product cannot fall, at least in the long run, in order for the firm to survive.[7]

---

[7]Although this is so in a single-product firm, for multiproduct firms, a new product could be priced below cost in order to stimulate sales of other products in the product line. Yet, even here, care must be taken to ensure that at least the total corporate costs of doing business are covered.

Although it is possible to overemphasize the importance of cost in new product pricing and fail to consider other essential elements, it is necessary for management to know intimately the composition and behavior of product costs. Accurate cost estimates can aid the design of a most profitable new product, with the total cost of the new product being computed before any significant investment in research and development takes place. The price of the new product based on cost should serve as a go or no-go guideline along the product's path toward the market. While cost indicates whether the new product can be marketed profitably at alternative prices, it does not indicate the acceptance level by buyers—whether these be intermediaries or the final consumer.

Classification of Costs

The new product's costs can be classified into categories, such as direct and indirect, controllable and uncontrollable, manufacturing and nonmanufacturing, and, most broadly, fixed and variable. Of particular importance are fixed and variable costs, as it is important to determine the relationships between cost and quantity of output. Fixed costs are those that do not vary, regardless of output. Items such as property taxes, interest, some elements of depreciation, and exempt salaries are ordinarily fixed.

Variable costs are those related to output. Materials and labor are the most typical and insignificant variable costs of the new product. However some costs are variable in one new product situation but not in another. A cost is variable, not because of what the item is, but because the cost changes with output. Although economies of scale initially occur, reducing these costs, variable costs often rise as the capacity of the plant is approached.

Cost Estimating

The cost of the new product in most cases is not an exact amount. It depends on concurrent manufacture of other products, as well as the economies of joint distribution and advertising. For example if a manufacturer locates a new branch outlet near an existing branch, both outlets can jointly share the cost of newspaper advertising serving the two adjacent towns.

New product costs are generally estimated by analyzing costs at various levels of sales of similar products. When a large number of observations exist at different volume levels, a cost function can be fitted to these data after adjustments are made for any variations. This is often done through regression analysis, discussed in terms of forecasting in Chapter 14.

The most difficult problem in cost estimation is allocating those costs that cannot be traced to a particular product. Regardless of the allocation procedure, such nontraceable costs must be distributed among all products so that no one product, particularly a new one, bears more than its fair share of the cost burden. If it is uncertain how much of the general nontraceable costs should be allocated to the new product, the amount allocated can be based proportionately to a traceable cost. For example the nontraceable costs of the plant can be allocated in the same proportion as the traceable direct labor costs.

One way to forecast costs for the new product is to use a cost compression curve.[8] By relating manufacturing cost per unit of value added to the cumulative quantity produced, a curve results that can be used to determine advantages or economies of production. Its use is particularly valuable in pricing high-technology industrial products.

CONSUMER CONSIDERATIONS

While cost serves as the floor for any new product pricing, the ultimate concern is the response of the market to the new product's price. The demand for the new product is the amount that buyers will purchase at each of several possible prices. Some of the estimating techniques discussed in Chapter 14 are also used to predict the demand at various prices. From this demand analysis, the price elasticity of the new product can be calculated and a price determination model formulated.

Price Elasticity

Price elasticity describes the relationship between a change in price and the change in quantity demanded. It indicates the degree of sensitivity that the consumer has to the price of the new product. If the coefficient for the new product is less than 1, demand is relatively elastic. The formula for this determination follows:

$$\text{Coefficient of elasticity} = \frac{\%\ \text{change in quantity}}{\%\ \text{change in price}} = \frac{\dfrac{Q_2 - Q_1}{Q_2 + Q_1}}{\dfrac{P_2 - P_1}{P_2 + P_1}}$$

When demand for the new product is relatively elastic, total revenue will fall as the price increases. Conversely, if the demand is inelastic, total revenue rises if the price is increased.

It is usually very difficult and therefore expensive to get a measure of the elasticity of a new product, but several factors can be used in estimating this factor. First, if the new product is not truly unique, and therefore a wide variety of acceptable substitute products already exist on the market, demand tends to be more elastic. If the need that the new product is designed to satisfy is easily filled, demand is inelastic.

Inelasticity of demand occurs when the new product is needed urgently. In this case, price has no effect on the purchase decision. For example suppose it began to rain during a professional football game. A fan who had two pocket raincoats only needed one, so the other was available to meet another fan's need. The price of this second raincoat is indeed elastic, since the second fan

---

[8]For a more thorough discussion see, for example, Dean, "Pricing Policies." Some of the techniques discussed in Chapter 11 can also be used to establish this demand curve.

needs it to stay dry and there are no other raincoats available (the vendor had sold all his products).

A final determinant of elasticity is the durability of the new product. Since the purchase of durable goods can easily be postponed, this factor usually makes the demand for the new durable-good product more elastic.

Regardless of whether it is elastic or inelastic, the demand for a new product must be analyzed and related to production and distribution costs. One way of analyzing these relationships is by using break-even analysis.

Break-Even Analysis

Break-even analysis is a clear and simple method to determine the effect of change in prices and volume on profits. As indicated in Figure 16–2, the break-even point is the level of sales revenue at which profits are zero. Management can use this graph and the related formulas to evaluate the relationship between fixed cost, variable cost, selling price, and profit. The basic formula for break-even analysis is as follows:

$$P(x) = FC + VC(x)$$

where

P = selling price of the new product

x = break-even point

FC = fixed costs

VC = variable costs

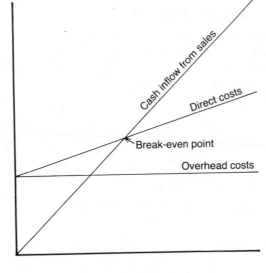

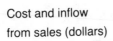

Sales volume (units)

FIGURE 16–2
Basic break-even chart

This break-even equation allows various prices to be used to determine various break-even points. These prices should be chosen based on experience and judgment, as well as the prices of competing products.

For example let us suppose that a firm developed a new non-fogging photochromic ski goggle. The fixed costs are $1,000,000, and the variable costs are $18.00 per unit. What is the break-even point at prices of $18.50, $20.50, and $32.50?

At a price of $18.50 per unit, the break-even point is:

$$18.50 \ (x) = 1,000,000 + 18.00(x)$$
$$x = 2,000,000 \text{ units}$$

At a price of $20.50 per unit, the break-even point is:

$$20.50(x) = 1,000,000 + 18.00(x)$$
$$x = 400,000$$

At a price of $22.50 per unit, the break-even point is:

$$22.50(x) = 1,000,000 + 18.00(x)$$
$$x = 222,223$$

From this analysis, the firm can estimate demand for the new product at the prices being considered to see which price should be charged to accomplish the new product's objective.

Price Determination Model

Closely related to break-even analysis is the price determination model, which requires determining the price/sales/quality relationship that will maximize profits.[9] Let us assume the results of a firm's test market indicated the following price/demand relationship:

Price	Estimated Demand (units)
$ 30	4,100
60	3,200
90	2,300
120	1,400
150	500

This linear relationship can be expressed in a more general model:

$$Y = a + bx$$

---

[9]For a discussion and application of this approach, see Donald V. Harper, *Price, Policy, Procedure* (New York: Harcourt, Brace and World, 1966), 143–48.

where

$$Y = \text{quantity}$$

$$x = \text{price}$$

$$a = \text{intercept (value of } Y \text{ when } x = 0)$$

$$b = \text{slope of the line}$$

By applying the market test data in the general model, it can be seen that when the price (x) is zero, the demand (Y) is 5,000 units. The slope (b) of this equation is -30. Substituting the figures in the general model, the demand for the firm's new product can be expressed by the following equation:

$$D = 5,000 - 30P$$

where

$$D = \text{new product demand (units)}$$

$$P = \text{price}$$

Since the fixed costs for this product are $75,000, and the variable cost per unit is $30, the total cost (TC) can be expressed as:

$$TC = 75,000 + 30D$$

Correspondingly the revenue resulting from selling D units at a price of P is:

$$TR = (P)(D) = P(5,000 - 30P)$$

Expressing the total costs in terms of unit price and demand gives:

$$TC = 75,000 + 30D$$
$$= 75,000 + 30(5,000 - 30P)$$
$$= 225,000 - 900P$$

This expression $(225,000 - 900P)$ for total costs indicates that if the price is zero, 5,000 units will be sold. Since each unit of the product has a variable cost of $30, the total variable cost of $15,000, when added to the fixed cost of $75,000, gives the first term of the equation—$225,000. In addition since each dollar increase in price will produce 30 fewer units sold, a savings of $900 (demand [30] $\times$ variable cost [$30]) in cost will result for each dollar increase in price. This is indicated by the second term in one equation.

Profit is equal to total revenue minus total cost:

$$\text{profit} = TR - TC$$
$$= P(5,000 - 30P) - (225,000 - 900P)$$
$$= 5,900P - 30P^2 - 225,000$$

Since the objective is to maximize profit, the optimal price can be found by taking the first derivative of price in this equation with respect to profit, setting the resulting expression equal to zero, and solving for the price that maximizes profit as follows:

$$\frac{d \text{ profit}}{dP} = 5{,}900 - 60P = 0$$

$$P = \$98$$

At this price, the quantity that will be sold is:

$$D = 5{,}000 - 30 \ (P)$$
$$= 5{,}000 - 30(98)$$
$$= 2{,}060 \text{ units}$$

Operational Approach to Demand

Another very useful approach to evaluate demand is the use of three sales volume estimates at each of the new product's prices under consideration.[10] These estimates are obtained from managers in the firm. The three demand estimates at three different prices are the most likely estimate of sales (Q_m), the most pessimistic estimate of sales (Q_p), and the most optimistic estimate of sales (Q_o). From these estimates and their probability of occurrence, the largest expected volume (Q_e) can be determined through the following formula:

$$Q_e = \frac{Q_p + 4 \ Q_m + Q_o}{6}$$

The estimate of the variance of this volume can be calculated through the following formula:

$$\sigma^2 = \frac{(Q_o - Q_p)^2}{36}$$

By repeating this estimating procedure, three demand curves can be obtained as indicated in Figure 16–3. Each of these demand curves can be used to make the final price decision. $(D_o$ is the optimistic demand; D_m is the medium demand; and D_p is the pessimistic demand.)

Price Sensitivity Measurement

Price sensitivity measurement (PSM) is still another way to analyze consumer's response to the new product's price.[11] This technique examines price perception by determining consumer resistance over a range of price/quality percep-

---

[10]This approach is presented in detail in Bill R. Darden, "An Operational Approach to Product Pricing," *Journal of Marketing* 32 (April 1968), 29–33.
[11]This approach is discussed in Kenneth M. Travis, "Price Sensitivity Measurement Technique Plots Product Price vs. Quality Perceptions," *Marketing News*, 14 May 1982, 6.

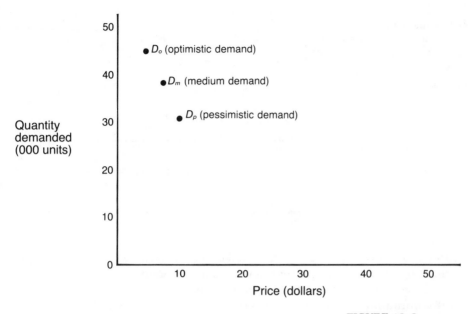

FIGURE 16–3
Subjectively estimated demand curves

tion levels. When confronted with different prices a consumer will compare the price with a standard or acceptable price based on experience with other products in the category.

The price sensitivity measurement technique involves determining the absolute threshold range between the low price where quality becomes questionable and the high price where product substitution is considered. This is accomplished by having respondents answer four questions using a rating scale with 15 to 30 prices spaced equally apart. These four questions measure prices respondents consider too cheap and too expensive with respect to quality at either end of the spectrum. The cumulative distribution of the responses is plotted to indicate the point of marginal expensiveness (where the number of people regarding the new product as too expensive is equal to the number of people who regard it as too cheap) and the point of marginal cheapness (where the number of people regarding the new product as too cheap is equal to the number of people regarding it as too expensive). The range between the point of marginal expensiveness and point of marginal cheapness is the range of acceptable prices, within which the new product should be priced for maximum sales (see Fig. 16–4).

This price/quality relationship can scarcely be overstated. It can be easily illustrated by recalling your reaction at the last good sale you attended. Did you not examine the items very, very carefully before purchasing, because you felt that something must be wrong with the merchandise to allow such low prices? These same doubts are harbored by potential buyers of new products. A buyer judges the quality of the new product based, in part, on the price charged. Too low an initial price can be as detrimental to sales as too high an initial price. A low price can cause buyers to perceive the product as inferior in quality. The price/quality relationship is very important to consider when

FIGURE 16–4

Use of price sensitivity measurement technique

Source: Kenneth M. Travis, "Price Sensitivity Measurement Technique Plots Product Price vs. Quality Perceptions," *Marketing News*, 14 May 1982, Section 1, p. 6.

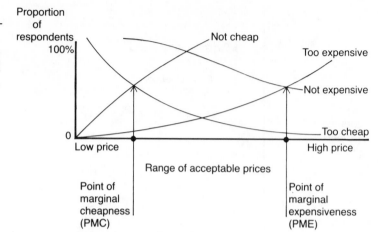

introducing a new product in the market, particularly when simultaneously evaluating the product's present as well as future competition.

Since a buyer's reaction to various levels of any stimulus, such as price, is discontinuous, only price differences greater than a specific percentage threshold or amount are perceived as being truly different. This noticeable difference is directly related to the overall price of the product category. For example the noticeable difference for an automobile costing $5,000 might be $300, but for an industrial machine costing $31,000, it might be $5,000.

Finally buyers also have fair-price reference points for product categories. This means that they view a new product's price in terms of a reference price that they consider fair. If the new product's price does not match this reference price, no purchase will occur.

COMPETITIVE CONSIDERATIONS

Aside from cost and the consumer, the final important factor that must be considered in the pricing of every new product is competition. Competition must be considered in terms of the price of existing products as well as possible new products. Every product that even remotely fills the same need as the new product should be evaluated for its ability to achieve sales. In this analysis it is helpful to classify each product with respect to its degree of competitiveness to the new product. One classification scheme uses three major categories: direct competitors, near competitors, and indirect competitors.[12] Direct competitors are those products that offer direct substitutes for the new product. They are the ones to be the most concerned with in establishing the price of the new product because they offer the consumer the best price comparisons. If the price of the new product is above the price of directly competitive products, then the new product must have enough differentiation to indicate its superiority and value and therefore the need for a higher price.

---

[12]This classification is developed in Barrie James, "A Contemporary Approach to New Product Pricing," in *Pricing Strategy*, eds. Bernard Taylor and Gordon Wills (Princeton, N.J.: Brandon/Systems Press, 1969), 521–33.

On the other hand, if the new product is priced too much below a direct competitor, consumers may appraise the offering as being inferior; in addition, the competitive product's price may be lowered to match the new product's.

Near competitors are products that partially meet the particular need of the buyer. Since they do not directly compete with the new product, their respective prices are not as comparable. Their prices should be evaluated, however, as they do provide a frame of reference for the buyer. In addition the buyer might purchase these products and use them as substitutes for the new product.

Individual competitors are the least competitive to the new product. Since they do not directly fill the same consumer need but merely vie for a share of the buyers' purchasing power, their price is not as important.

Regardless of the type of competitors available, the price of any new product is very susceptible to competitive reaction. This is the variable in the new product's marketing mix that is most easily duplicated. Since competition can react quickly to the company's price, it is important to analyze competitive reactions and outcomes of various actions.

Bayesian analysis is a useful technique for assessing the reaction of competitors to the price of the new product. Through this approach management can evaluate various pricing alternatives and assess the effect on profits of possible competitive reactions. The technique requires that management assess the profits for each of the possible competitive reactions and then assign the probability that the competitors will react in this manner. Expected payoff can be determined by multiplying the profit with the probability of occurrence for each possible reaction, as is indicated in the following example.

R.T.K. Enterprises has developed a new tire pump that weighs only four ounces. The pump is designed for use on both automobile and bicycle tires and has a built-in pressure gauge to eliminate over- and under-inflation. The company wants to determine whether the new product should be priced at $8.95 or $12.95. The firm has one major competitor who can effectively react to the new product and its price.[13] The new product manager at R.T.K. Enterprises determines that this competitor can respond to the new product's price in one of the following five ways:

1. Take no action on price of the new product
2. Develop a new, higher-priced competitive product
3. Develop a new competitive product and price it the same
4. Develop a new, lower-priced competitive product
5. Decrease price of somewhat similar competitive product in market

The expected profit for each of these possible reactions has been determined based on various cost and demand estimates (see Table 16–2). The new product manager also feels that there is a 30 percent chance that the competitive company will not respond to the new product if it is priced at $8.95 and a 20 percent chance that it won't respond at $12.95. Table 16–2 indicates that

---

[13]Of course this analysis can easily be expanded to more than two prices and one competitor. The methodology is the same, only expanded. With many computer software packages having a subroutine available, the number of options on both the prices to be analyzed as well as competitors is only limited by management's ability to assess profits and probable reactions.

TABLE 16–2
Bayesian pricing chart

	Profits of Competitive Reactions	Probability of Reaction	Expected Payoff
	A: $150,000	.3	$45,000
	B: $90,000	.4	$36,000
$8.95 Price	C: $110,000	.1	$11,000
	D: $60,000	.1	$6,000
	E: − $40,000	.1	− $4,000
	Total payoff		$94,000
	A: $180,000	.2	$36,000
	B: $110,000	.4	$44,000
$12.95 Price	C: $80,000	.2	$16,000
	D: $50,000	.1	$5,000
	E: $10,000	.1	− $1,000
	Total Payoff		$100,000

there is a 40 percent chance that the competition will develop a higher-priced competitive product regardless of whether the new product is priced at $8.95 or $12.95. All other competitive reactions have a probability of 10 percent, with one exception—if the new product is priced at $12.95, there is a 20 percent chance that the competition will develop a new product with a similar price.

The expected payoff for each competitive reaction has been determined by multiplying the profit by its respective probability (see Table 16–2). The expected payoff for no response to the price of the new product, if the new product is priced at $8.95, is $45,000 ($150,000 × 0.3). Summing each of the five expected payoffs for each of the proposed prices yields the total payoff: $94,000 for a price of $8.95, and $100,000 for a price of $12.95. Given that everything else is equal, the firm might price the new product at $12.95, as this price yields the greatest total payoff. However, before setting the final price, the firm would have to take into account the cost and consumer considerations previously discussed and the relationship of other new product marketing mix decisions with the price.

NEW PRODUCT PRICE AND OTHER MARKETING DECISIONS

A carefully designed price for a new product is a virtual prerequisite for successful new product launching. Aside from being one of the many variables that determine the consumer's response to a new product, pricing decisions are inter-dependent on external forces and internal variables in the firm's marketing environment.

External forces, such as consumer demand, costs of other variables in the new product's marketing mix, domestic and foreign competition, government regulation of price differentials, labor and material costs, and sources of supply, all affect the quantity of the new product produced and the price received. These forces establish parameters for accurate pricing decisions. Acting alone or in combination, any of these factors may necessitate that only certain prices or price levels be considered.

Similar interdependence exists between the price of a new product and internal variables. In establishing the price of a new product, a marketing manager must concurrently consider the product itself and its distribution and promotion. No decision regarding the price of the new product can be made without evaluating the impact on the other elements in the product's total marketing mix.

The price of a new product is influenced substantially by various aspects of the product and product policies. For example the price greatly affects the quality of the new product and therefore its appeal. A high price relative to the cost of production enables quality features to be built in, which might differentiate the new product. A low price may mean that the impact of every quality feature must be carefully assessed in terms of its importance to the consumer and influence on cost. Similarly the importance of a product in terms of its end-use must be considered. For example, whether the new product plays only an incidental or major role of the cost and functioning of the final industrial product influences the degree of price flexibility. Finally the relationship of the new product to the entire product should be considered. The price established for the new product affects to some extent all other products in the line and therefore must be considered in establishing the new product's price.

The distribution, including the channels selected, the types of middlemen desired, and the profit margins required by these middlemen, also influences the price set for the new product. In order to secure aggressive middlemen and adequately compensate them for their services, the new product must be priced to allow an appropriate margin. In fact the profit margin can be used as an incentive to secure an integrated marketing effort on the part of all distributors, as well as to entice them to carry the product.

The importance and impact of channel members on the price of the new product cannot be overstated. Since distributing the new product is the focus of Chapter 17, at this point, only the impact of channel members in terms of pricing the new product will be discussed. Many small companies fail to fully consider channel options and the final price to the consumer. Channel members operate on margins or markups. Some operate on markup on cost

$$\left(\text{percent markup on cost} = \frac{\text{dollar markup amount}}{\text{cost}} \right).$$ This is the markup used by manufacturing companies as well. However companies often do not realize that most channel members work on a markup on selling price

$$\left(\text{percent markup on selling price} = \frac{\text{dollar markup amount}}{\text{selling price}} \right).$$ This is sometimes called *markup off list* (list price). Markup on selling price is always a lower percentage amount but can have a larger impact on the final price than markup on cost. For example suppose a channel member receives a product costing $1.00 and wants a $.50 markup (margin). The selling price would be $1.50. The percent markup on cost is:

$$\text{\% markup on cost} = \frac{\text{dollar markup amount}}{\text{cost}} = \frac{\$.50}{\$1.00} = 50\%$$

However the percent markup on selling price is lower:

$$\text{\% markup on selling price} = \frac{\text{dollar markup amount}}{\text{selling price}} = \frac{\$.50}{\$1.50} = 33\frac{1}{3}\%$$

Management must carefully consider the markups being used by channel members and the impact they will have on the final price the consumer pays for the new product. Three examples are indicated in Figure 16–5. The final selling price to the consumer varies from $1.20 for direct, to $1.80 when one channel member is involved, to $1.99 when two channel members are involved. Management must carefully assess the value added and need for a channel member and the resulting price to the consumer. At times a channel system cannot be used because the price that results is too large to obtain consumer purchases.

In order to evaluate various channel members and pricing strategies, it is often necessary to convert the markups from one base to another. This is easily accomplished through the following two formulas:

$$\text{\% markup on selling price} = \frac{\text{\% markup on selling price}}{100\% - \text{\% markup on selling price}}$$

$$\text{\% markup on selling price} = \frac{\text{\% markup on cost}}{100\% + \text{\% markup on cost}}$$

Applying these formulas to the previous example indicates how the conversion from one markup to another can occur without moving the actual dollar margin or base. For example a markup on cost of 50% is a markup on selling price of:

$$\text{\% markup on selling price} = \frac{50\%}{100\% + 50\%}$$

$$= 33\frac{1}{3}\%$$

Similarly a percent markup on selling price of 33⅓ percent is equal to a percent markup on cost of:

$$\text{\% markup on cost} = \frac{33\frac{1}{3}\%}{100\% - 33\frac{1}{3}\%} = 50\%$$

In addition to the channel members used and margins required, the amount and types of promotional methods are closely related to the price of the new product. Whether the bulk of the promotional responsibility lies with the middlemen or the manufacturer, the costs must be reflected in the price.

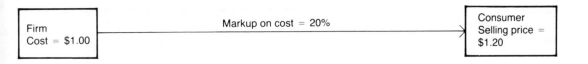

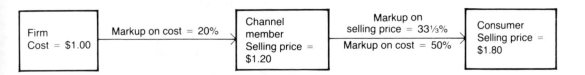

FIGURE 16–5
Channel members and the price

Since the price is related to the available promotion funds for the product, it is imperative to consider the promotional needs of the product when establishing its price.

Many consumer, intermediate-customer, and industrial-customer purchasing decisions are made with insufficient information regarding the available options, as well as considerable uncertainty about the attributes of the new product. The price should always be established in light of the buyer's conception of value, as it is possible to price a new product at such a low level that the value image of the product is lost. The initial price establishes the base from which any price changes are viewed and is therefore the most critical price established.

Table 16–3 summarizes the major factors affecting a new product's price.

OTHER PRICING OPTIONS

In addition to establishing the initial price of the new product, management must establish policies on other price features. This overall price structure is necessary before the new product is launched so that the total profit picture can be developed and channel members will know the terms of sales. These price features include geographic pricing decisions, discounts, and allowances and deals.

TABLE 16–3
Major factors affecting the new
product's price

Category	Major Factors
Cost considerations	Cost of production—start up
	Cost of production—future
	Volume anticipated
	Break-even points
	Amount of integration with other products
Consumer considerations	Utility to the buyer
	Comparable products
	Substitutable products
	Customary prices
	Buying habits
	Prestige position of product
Competitive considerations	Amount of competition
	Degree of competition
	Meeting, underselling, overselling, competition
	Actions of competition
	Market share and position of competition
Market considerations	Channels used and available
	Geographic distribution plan
	Distribution structure
	Promotion outlets available
	Service availability

Geographic Pricing Decisions

One of the most significant costs associated with marketing a new product today is transportation—moving the product from the firm to the final consumer or channel member. While the decisions regarding distribution are discussed in Chapter 17, from the standpoint of the new product's price, the price/distribution decision revolves around who is going to pay the shipping costs. If the company wants to account for shipping costs in the pricing structure, this can be accomplished by two general methods: f.o.b (freight on board) origin or delivered pricing.

F.o.b. origin pricing is used when the company marketing the new product quotes prices from the point of shipment. In other words the buyer pays all shipping charges and often selects the mode of transportation, chooses the specific carrier to be used, and handles any claims and damages. While this ensures the company marketing the product of the same profit regardless of where the product is sold, it also makes it more difficult to sell the new product farther away from the company's headquarters. Even more importantly channel members would rather have a delivered cost quoted to them when they are deciding whether to purchase the product. Using f.o.b. origin pricing can actually cause some sales to be lost.

The second method, delivered pricing (or f.o.b. destination), includes transportation in the price quoted to the buyer. In this case the company marketing the new product assumes the responsibility and cost of transporting the new product to the buyer.

Another geographic decision that needs to be made with respect to the new product is whether to use single-zone or multiple-zone pricing. In single-zone pricing all buyers pay the same average transportation cost, and the company marketing the new product receives a different net return from each sale, depending on the costs of transportation for that sale. The farther away the buyer, the less the net amount obtained. This is not the case in multiple-zone pricing, since there are different prices in different zones. The prices in the zone farthest away from the company marketing the new product are of course highest. Generally multiple-zone pricing allows the company much more flexibility in taking into account any geographic differences in price reaction and sensitivity. Of course accompanying this flexibility is more management responsibility, as transportation and costs must be carefully evaluated on a zone-by-zone basis.

Discounts

In addition to geographic pricing decisions, the company marketing the new product must decide whether to offer a reward to the buyer for payment of the invoice within a specified period of time—a cash discount for prompt payment. There are several advantages and problems associated with cash discounts. Probably the most important advantage of a cash discount is that it allows the cash of the company marketing the new product to turn over faster. This is particularly important for smaller firms and firms strapped for financial resources—as well as for all firms in times of high-interest rates. Another advantage in offering a cash discount is that it reduces credit risks and the costs of collecting overdue accounts.

However there are many problems associated with offering cash discounts. The biggest problem centers around the fact that large key accounts will take the cash discount even though the payment is made outside the payment period. Another problem is the loss of revenue. Many buyers will take advantage of any discounts offered, thereby reducing the profit of the company marketing the product. In some cases it is even more profitable for them to borrow money over the time period, if necessary, and not lose the discounted amount.

Generally it is better for a small company marketing a new product not to offer a cash discount unless it is a general industry norm and the buyers in the industry are encouraged by the discount to pay within the discount period.

Allowances and Deals

There are several different allowances and deals a company marketing a new product can use. These are usually in the form of either some type of trade allowances or promotional deal. A trade allowance is an allowance given to a channel member for performing some specified marketing activity. Since different channel members perform different activities, the allowance varies depending on the amount of effort involved. For example some channel members provide storage facilities for the company and therefore receive a storage allowance. Others do personal sales work, such as calling on other members in the distribution channel and perhaps setting up displays. Because of the costs involved, the legal problems that can occur, and the variety of allowances possible, it is usually better for the company marketing the new product to avoid offering trade allowances unless dictated by market conditions.

On the other hand, promotional deals are much easier to establish and control. Since promoting the new product is covered in Chapter 18, it is enough here to merely mention that the promotional allowances offered affect the price of the new product and the net return to the company. There are many forms of promotion deals available, such as cents-off, samples, coupons, rebates, premiums, contests, and two-for-one offers. Each of these must be carefully evaluated in terms of its impact on price and revenue and increasing the sales of the new product.

LEGAL ISSUES CONCERNING PRICE

One of the most complex and frustrating areas in pricing the new product is taking into consideration the legislation against restraint of trade, price discrimination, and deceptive pricing. Table 16–4 summarizes the provisions of the major acts affecting new product pricing.

Legislation Against Restraint of Trade

The Sherman Anti-trust Act of 1890, a landmark of federal legislation, established policy toward restraint of trade and monopoly in interstate and foreign commerce, thereby favoring the preservation of competition. The Sherman Act attacks two types of anticompetitive business behavior. Section 1 covers contracts, combinations, and conspiracies in restraint of trade or commerce. The acceptance of an invitation to participate in a plan that would restrain interstate commerce is sufficient to establish an unlawful conspiracy. For example, in *American Tobacco Co.* vs. *United States*, a simultaneous price increase by three major cigarette producers at a time of declining sales was sufficient to justify a conviction without direct evidence that the companies had actually communicated with one another.

Section 2 of the Sherman Act is directed at attempts to monopolize any part of interstate or foreign commerce. Its goal is to restore competition by preventing the monopolizing company from receiving the benefits of its illegal conduct and by rendering the company's monopoly power impotent. The law requires proof of the relevant market involved, since a monopoly or tendency toward monopoly must be of a certain product and geographic market.

TABLE 16–4
Provisions of major acts
affecting new product pricing

Sherman Act:

Section 1. "every contract, combination in the form of trust or otherwise, or conspiracy, in restraint of trade or commerce among several states, or with foreign nations, is hereby declared illegal. . . ."

Section 2. "every person who shall monopolize, or attempt to monopolize, or combine or conspire with any other person or persons, to monopolize any part of the trade or commerce among the several states, or with foreign nations, shall be deemed guilty. . . ."

Robinson-Patman Act:

Section 1 (a). "That it shall be unlawful for any person engaged in commerce, in the course of such commerce, whether directly or indirectly, to discriminate in price between different purchases of commodities of like grade and quality, where either or any of the purchases involved in such discrimination are in commerce . . . and where the effect of such discrimination may be substantially to lessen competition or tend to create a monopoly in any line of commerce . . ."

Federal Trade Commission Act:

Section 5. (a) (1) "Unfair methods of competition in or affecting commerce, and unfair or deceptive acts or practices in or affecting commerce, are hereby declared unlawful."

Source: *Trade Regulation Reporter* 4 (Chicago, Ill.: Commerce Clearing House, 1982).

It is a federal crime punishable by fine or imprisonment or both for any person or corporation to violate the provisions of the Sherman Act. While crimes under the Sherman Act were originally misdemeanors, in 1974 they became felonies. Corporations may be fined up to $1,000,000 for each price-fixing violation; responsible managers can be fined up to $100,000 each and jailed for up to three years. The two landmark cases in the Sherman Act—the Standard Oil Case (1911) and the Alcoa Case (1945)—increased the effectiveness of the antitrust policy by getting the "rule of reason" and considering per se violations.

A judgment under the rule of reason takes into account the alleged violation itself and its adverse effects on competition. A per se violation is one that is deemed to have a pernicious effect on competition and no defense of reasonableness is permitted. Any attempt by firms in the same market to jointly establish product or service prices or to change prices in a collaborative manner is a horizontal price agreement and considered price-fixing. When firms establish a joint price through collusion they are, in effect, behaving in a monopolistic manner.

The Department of Justice has indicated that practices that may represent horizontal price agreements will be closely scrutinzed and severe penalties will be assessed on firms and managers found guilty. Activities that have been investigated as potentially collusive price agreements include discussions between salespeople of competing companies concerning probable bids that each might submit on a new project, issuance of lists of anticipated price changes to competitors as well as to customers, sharing cost information among competitors, issuance of "codes of business ethics" that apply to industry members and that discourage price activities, and signaling future price changes through speeches and publicity releases. In addition to horizontal price agreements, there are vertical price agreements. These agreements, sometimes referred to as a *resale price maintenance* or *fair trade*, are agreements between manufacturers and distributors concerning the price at which the manufacturer's branded products will be resold. These agreements have been found to be per se violations of the Sherman Act.

For a period of time vertical price agreements were exempt from the antitrust laws concerning price-fixing. In 1937 Congress passed the Miller-Tydings Act, which allowed manufacturers to stipulate minimum resale prices for their branded products. It was felt that these agreements would promote a quality image for the product and provide an adequate distribution margin for distributor advertising and in-store promotion. However in 1975 Congress repealed the Miller-Tydings Act through the Consumer-Good Pricing Act. As a result of the repeal, vertical price agreements are currently per se violations of the Sherman Act.

Legislation Against Price Discrimination

Price discrimination is the practice of selling the same product to different buyers at different prices even though the cost of the sale is the same. The intended effect of price discrimination on competition may be between a manufacturer and his competitors, between a buyer and the buyer's competitors, or between customers of the buyer—in short, anywhere in the distribution channel.

The Robinson-Patman Act of 1936 was passed to prohibit price discrimination when such practices will substantially lessen competition and lead to monopolistic control. The primary objectives of this act were to prevent unscrupulous suppliers from attempting to gain an unfair advantage on their competitors by discriminating among buyers, and to prevent unscrupulous buyers from using their economic power to obtain discriminatory prices from suppliers to the disadvantage of less powerful buyers. The act permits price differences when the goods being sold are damaged and when the seller is closing out a particular line of merchandise in "good faith". A seller's price discrimination that injures competitors is termed a *primary-line injury.*

The leading case in interpreting this primary-line injury was the Supreme Court's *Utah Pie* decision in 1958. Although the Utah Pie Company had 66.5 percent of the frozen-pie market in the Salt Lake City area, it was still a relatively small company, with sales confined almost exclusively to Utah. Utah Pie's major competitors were Pet Foods, Carnation, and Continental Baking—all large, national firms. These national companies wanted to increase their market share in Utah Pie's market area and, consequently, lowered their prices in Utah while maintaining prices elsewhere. The Supreme Court, in finding for Utah Pie, indicated that a reasonable possibility of injury to competition can exist even though the volume of sales is increasing and competitors continue to make a profit.

The Robinson-Patman Act was also designed to reduce secondary-line injury—to reduce the advantage based on economic power not enjoyed by smaller competitors. The majority of cases involved with the issue of competitive effect of price differentials are at the customer level. The landmark case in this area is the *Morton Salt Company* case of 1948. Morton Salt sold its best brand of table salt on a per-case price schedule that ranged from a high of $1.60 for small purchasers to $1.35 for 50,000-case purchases in any consecutive 12 months. Only five national customers (large retail grocery chains) purchased in volume sufficient to receive the $1.35 price. The Supreme Court ruled that the difference in price gave rise to adverse competitive effects.

The two major defenses to charges of price discrimination under the Robinson-Patman Act are referred to as *good faith meeting competition* and *cost justification*. The good-faith defense can be used when a manufacturer lowers prices in one particular geographic area, or for one group of customers, in order to meet an equally lower price offered by a competitor. The cost-justification defense allows a manufacturer to charge different prices to buyers as long as such differences reflect the actual costs of serving the buyers.

The Robinson-Patman Act does allow quantity that is either noncumulative or cumulative in nature. Noncumulative quantity discounts are discounts that provide buyers with a price break at a specific order quantity for each order placed. These discounts must reflect real economic savings and must be available to all buyers. They can be challenged as illegal when only a very few buyers would be of sufficient size to take advantage of the largest discounts, as in the *Morton Salt Company* case.

Cumulative discounts are granted on purchases made over a period of time, usually one year. These discounts have proven difficult for firms to justify because they do not necessarily reduce the number of transactions or shipments and do not necessarily result in cost savings. Rather these discounts tend to tie buyers to a producer, making competition more difficult. These discounts also tend to discriminate against small buyers who would be unlikely to be able to take advantage of high-volume discounts.

Legislation Against Deceptive Pricing

Efforts to legislate unfair and deceptive business practices began with the establishment of the Federal Trade Commission (FTC) in 1914. The Federal Trade Commission Act prohibits unfair methods of competition in commerce. Pricing misrepresentations—particularly such practices as false claims of price reductions, sham lists of pre-ticketed prices, and "bait and switch" advertising—have been subject to careful scrutiny.

The FTC has taken action against manufacturers and distributors who have misrepresented the true nature of their prices. Probably the most deceptive pricing practice is the "bait and switch" tactic, in which the advertiser promotes a very low price for a particular item to attract customers to his or her store. Once customers arrive, the seller tries to "switch" them to a higher-priced item by not having the advertised item in stock or playing down the quality of the advertised item.

Another area carefully policed by the FTC is bargain offers based upon the purchase of other merchandise. This type of promotion often takes the

TABLE 16–5
Summary of legal issues and
corresponding antitrust law

Legal Issue	Antitrust Law Applicable	Type of Antitrust Violation
Horizontal price agreements	Sherman Act	Per se
Vertical price agreements	Sherman Act	Per se
Price discrimination	Robinson-Patman Act	Rule of reason
Deceptive pricing	Federal Trade Commission Act	Rule of reason

form of "Buy one—get one free." Often the firm making the offer is not offering anything free, since the consumer must purchase an article in order to receive the "free" item. If the seller increased the regular price of the article that the customer must purchase to get the "free" merchandise, deception may be involved. Whenever a "free," "2-for-1," or "half-price" sale is made, the seller must clearly state all terms of the offer.

The landmark case in this area of pricing is the *Mary Carter Paint Company* case. Mary Carter Paints had for many years been advertising that a buyer would receive a can of "free" paint with every can purchased. The FTC charged that since the company had never sold single cans of paint, thereby establishing a price per can, that it had actually priced one can to cover the cost of two cans. Consumers were, therefore, deceived.

The legal environment affecting new product pricing is substantial. Table 16–5 summarizes the various legal issues and their corresponding antitrust laws. Since there are yet few completely clear guidelines for developing the price of the new product and its price list, management must work closely with the legal department in establishing the initial price and schedule so that no laws are violated.

SUMMARY

The price for the new product is one of the keys to its successful introduction on the market. Establishing appropriate objectives and policies is necessary to implement a total new product pricing approach. Objectives can generally be classified into three major categories: profit; sales; and other, such as image, product line, early cash recovery, and meeting competition. Whatever objectives are adopted, a sound pricing policy must be formulated to provide the guidelines for price establishment. Two general pricing policies—skimming and penetration—are useful, depending on the specific market and product situation.

. Three fundamental considerations in establishing the price for the new product are cost, consumer, and competition. The cost of making and distributing the new product serves as the floor for the price. In all pricing decisions, the response of the consumer to the price is, of course, a basic issue. This demand for the new product can be tempered by the action of competitors whose possible reactions must be assessed.

Taking into account these three considerations, the price for the product must be established in view of the other marketing variables of the new product. This will allow a total product to be successfully introduced into the market.

SELECTED READINGS

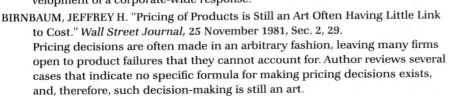

ARAGON, GEORGE. "Mandatory Price Controls: Managing a Response." *Business Economics* 15, no. 3 (May 1980): 25.
Argues that price controllers cannot properly anticipate the full range of consequences on their policies. Therefore management responsibility extends to the proper understanding of price control programs and the coordinated development of a corporate-wide response.

BIRNBAUM, JEFFREY H. "Pricing of Products is Still an Art Often Having Little Link to Cost." *Wall Street Journal*, 25 November 1981, Sec. 2, 29.
Pricing decisions are often made in an arbitrary fashion, leaving many firms open to product failures that they cannot account for. Author reviews several cases that indicate no specific formula for making pricing decisions exists, and, therefore, such decision-making is still an art.

CIGLIANO, J. M. "Price and Income Elasticities for Airline Travel: The North Atlantic Market." *Business Economics* 15, no. 4 (September 1980): 17.
Estimates price and income elasticities for airline travel across the North Atlantic. Differences in price elasticities are noted when markets are subdivided into United States/Europe and Canada/Europe groupings so that segmenting demand according to demand elasticities could lead to higher revenues.

DELLABITTA, A. J., MONROE, K. B., and McGINNIS, J. M. "Consumer Perceptions of Comparative Price Advertisements." *Journal of Marketing Research* 18, no. 4 (November 1981): 416–427.
Analyzes the issue of comparative price advertising, using a complex experiment and replication to examine some of the research questions. Public policy recognizes that comparative pricing may lead to some consumer misperceptions, and the authors review the regulatory setting.

DOLAN, ROBERT. "The Panic of the 1980's: It's Pricing." *Sales and Marketing Management* 124, no. 8 (9 June 1980): 47.
Discusses how competition is getting tough in the 80s economy, making pricing wars become more common. Reviews problems and options.

FOX, H. W. "Setting Prices According to Consumer Response Determinants: A North American Perspective." *Quarterly Review of Marketing* 12 (Summer 1980): 9–15.
States that the North American market can be entered based on consumer profile methodologies. Presents several case examples.

GABOR, ANDRE. "How to Price." *Management Today* (January 1979): 54.
Most managers claim to set their prices by adding some percentage to costs, but the author contends that this is an internally generated system that does not include external considerations and market conditions.

HATTEN, MARY LOUISE. "Don't Get Caught with Your Prices Down." *Business Horizons* (March 1982): 23.
Contends that cost overestimation and possible subsequent price cuts can actually result in high levels of consumer goodwill and even loyalty—both of which have future market value.

"How Price Tactics Feed Inflation." *Business Week,* 10 March 1980, 36.
Reports that corporations are initiating surcharges and markups to protect their margins. However, their tactics fuel inflation, which may erase margins altogether.

JAGPAL, H. S. "Pricing when Sales and Costs Are Uncertain." *American Marketing Association Proceedings,* 1981, 17–19.
Pricing under uncertainty is a classic marketing problem. Record reviews some of the guidelines for making just such a decision.

JOSAITAS, BOB. "Explore New Export Markets via Entry Pricing." *American Import-Export Bulletin* (July 1981): 54.
Contends that international markets can be breached by correct pricing decisions and that exploratory price sensitivity analysis done by a flexibility organized firm can lead to greater sales enhancement.

KRAUSHAR, PETER. "How to Research Prices." *Management Today* (January 1982): 50.
Asks what price the traffic will stand. Argues that many managers either do not understand this question or fail to answer it correctly. These managers may not be familiar with the research and examples that indicate sales and profits can be increased by raising prices.

MAURIZI, A. P., MOORE, R. L., and SHEPARD, L. "The Impact of Price Advertising: The California Eyewear Market after One Year." *Journal of Consumer Affairs* 15, no. 2 (Winter 1981): 290–300.
Reviews how price can affect new product introductions that have a new technological edge. Hindsight is a powerful learning tool.

URBAN, P. A. "Pricing Products to Gain Competitive Edge." *Marketing Times* 11, no. 3 (July–August 1981): 43.
Reviews competitive pricing issues in today's economy.

Distributing the
New Product

Firms often overlook distribution channel decisions when they develop and launch a new product. It is assumed that channel systems are readily available to access target markets. However, with increased competition in new product development, new channel systems are often needed to differentiate the new product. A new channel system can provide the means to stimulate product growth when a product has apparently reached saturation in its product life cycle. L'eggs pantyhose provided a good illustration of this strategy by using a compact display for its product in supermarkets—a channel that had not previously yielded significant sales.

Distribution decisions for the new product should be based on the firm's market segments and marketing objectives. The complexity of the channel strategy may be increased by the newness of the product. New products that are discontinuous or that require high learning may also require unique channel strategies.

For example, in 1959, Mead Johnson and Co. introduced Metrecal, a liquid diet product that could be categorized as a learning product. Mead Johnson identified its segments as those consumers who must diet for health reasons. The marketing objectives were to use the company's medical experience to market the new product and to use existing pharmaceutical distribution channels to reach the target segment and the product's objectives. Mead Johnson was able to dominate the market with this strategy until Pet Milk introduced Sego, a liquid diet product that was marketed through traditional

FIGURE 17–1
Channel of distribution decision process

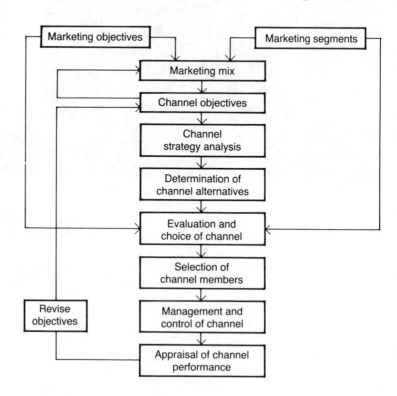

food channels. Pet Milk proceeded to capture 51 percent of the liquid food market, compared with 46 percent for Mead Johnson.

Figure 17–1 illustrates the process used for making channel strategy decisions. For each new product introduced, management should carefully consider the elements in this process. Thorough evaluation is particularly important in those situations where the product is new to the firm or is discontinued with little or no established channel experience in the industry.

Mead Johnson chose to minimize risk by marketing this high-learning product through those channels with which it had the most experience. This decision resulted in loss of market share due to competition and Mead Johnson's inability to adjust its marketing mix strategy.[1]

CHANNEL OBJECTIVES

Distribution channel objectives should specify the desired performance, the services needed, the control the firm wishes to retain, the financial support given to channel members, and any other operational concerns. These objectives are established based on a clear understanding of the market segments defined for the new product.

CHANNEL STRATEGY ANALYSIS

Table 17–1 summarizes the analysis required to determine feasible channel alternatives. This analysis should consider the directness of the channel, selectivity of channel members, selection of middlemen, and the number of channels to be used.

Degree of Directness

Channel strategy formulation begins with a determination of the directness or length of the channel. This decision must be based on market conditions, product attributes, middlemen requirements, company characteristics, and environmental factors.

Market conditions　Target market objectives are important in determining the length of the distribution channel. First, management must consider the geographic density of the target market(s). If most of the end-users are highly concentrated in one or a few geographic areas, then the firm may use a more direct channel to reach them. The more dispersed the end-users are, the more need there is to use middlemen, which would increase the cost of reaching end-users. Second, management must consider the number of potential end-users for the new product. If the number of end-users is unlimited, such as in the case of certain consumer packaged goods, then management would con-

---

[1]For more information, see "The Rise of Instant Skinny," *Sales Management* (4 November 1969): 52–54, and "Tying Up a Market by Tying in Fashion," *Printer's Ink* 287, 8 May 1964, 39–41.

TABLE 17–1
Channel strategy analysis

1. Degree of directness depends on:
   market conditions
   product attributes
   middlemen requirements
   company characteristics
   environmental factors
2. Degree of selectivity can range from:
   intensive
   selective
   exclusive
3. Criteria to select middlemen:
   reputation
   services provided
   degree of cooperation
4. Number of channels and their uses:
   one channel—one target market; multiple target markets
   multiple channels—one target market; multiple target markets

sider middlemen necessary. If there are only a few end-users, as in the case of specialized industrial machines, then management would consider a more direct channel of distribution.

Exceptions to using long channels occur when the market is geographically concentrated or dense and where there is a large number of end-users (as in the case of Avon, Fuller Brush, Electrolux, and a number of encyclopedia companies). These manufacturers have elected to market their products through a direct, door-to-door distribution.

Product attributes   The directness of the channel is affected by the nature of the new product. If the new product is perishable or a fad item, then the firm must use a more direct channel to avoid losses due to product spoilage or sudden shifts in consumer tastes (fads). Bulky new products also require more direct routes to the end-user because the costs of shipping and storage would be prohibitive. New products such as jewelry that appeals to very exclusive target markets have been traditionally marketed from the manufacturer directly to retail stores.

Middlemen requirements   Different levels in the channel of distribution provide important services to the firm introducing the new product. Given channel objectives and target markets, management must carefully consider what each level in the channel could offer as value-added or cost benefits to the new product. If there are no benefits in using wholesalers, then they should not be considered when setting up the channel. Instead more direct means could be used, and the savings could be passed on to other channel members. Such savings can be used by the manufacturer to achieve more control over price or promotion or as an incentive for more cooperation from middlemen.

Company characteristics   The channel strategy is influenced by characteristics specific to the firm introducing the new product. Factors such as size, financial strength, previous channel experience, product mix, and marketing policy will affect the channel decision. Larger firms have more flexibility and power in obtaining cooperation from channel members. A new firm would be limited in its choice of channel members because it would lack financial strength, experience, and a wide product mix. Smaller firms must use commissions to persuade channel members to store and deliver the new product. The larger firm with wider product lines and more experience would be able to provide its own storage and shipping and hence deal more directly with its end-users.

Environmental factors   When economic conditions are depressed, firms seek the least expensive means to move their products to the market. This usually results in shorter channels and fewer services. The firm introducing the new product may carefully analyze competitive techniques for distributing its products, particularly if its new products must be located in the same stores as competing products so that the consumer can compare alternatives. Legal regulations may also affect channel directness.

Degree of Selectivity

Once the decision about the number of levels in the company's channels has been made, management must decide how selective the distribution channel should be. This is generally a question of degree and is sometimes precluded by the nature of the product. When the strategy is to distribute to as many outlets as possible, the process is called *intensive*. Examples of intensively distributed products are cigarettes, aspirin, and candy. The firm introducing a high-learning product would be foolish to consider intensive distribution until the product is firmly entrenched in the growth stage of its life cycle or has attained high awareness by consumers. For high-learning products, a more selective or even exclusive distribution policy would be recommended because middlemen are hesitant to accept the new product until forced to do so by strong demand. Selective or exclusive distribution may be necessary when the firm is trying to develop or maintain an image of quality. The exclusive distribution strategy requires the selection of middlemen who have a reputation that conveys a quality image.[2]

Type of Middlemen

When considering the type of middlemen in the channel, management must emphasize the mix of services provided by each in view of the objectives in introducing the new product. The functions of a particular middleman must be coordinated with those of the manufacturer.

A small or young business would generally lack storage, transportation, and equipment—or even marketing expertise. Middlemen provide a broad variety of services and will, in many instances, provide a fledgling company with badly needed advice on how to market the new product.

The reputation of the middleman may be critical in achieving certain marketing objectives. For example a firm might want to place its new product

---

[2]For more information on selectivity, see Edwin H. Lewis, *Marketing Channels: Structure and Strategy* (New York: McGraw-Hill Book Co., 1968), 85–88.

with certain retail stores that have a reputation for carrying quality products. Some middlemen also have the knowledge or expertise to market the new product in a way that achieves the firm's marketing objectives. Head skis, for example, selected specialty ski shops that used personnel with professional knowledge of ski equipment. Cooperation from middlemen to market the new product in a consistent manner is important when trying to establish an image. Inconsistent efforts will slow the adoption process and give an advantage to the competition.

Number of Channels

The decision to use one channel or multiple channels differs from selectivity, which affects the number of middlemen in each channel. An example of the use of multiple channels is a clothing manufacturer who markets directly to the end-user through a factory outlet store, as well as distributing the products through retail clothing stores. This same clothing manufacturer also uses manufacturer's agents in certain sparsely populated areas but, at the same time, sells directly to large retail stores.

Many large firms use a multiple approach when they market the same product to different target markets. A calculator manufacturer distributes to large accounts in the industrial market using its own sales force, but uses manufacturer's representatives for smaller accounts and wholesalers for the household market. The decision of the number of channels to be used will depend on the company's objectives and target markets.

CHANNEL ALTERNATIVES

Figure 17–2 shows typical configurations of channel strategies, whose purpose is to select those alternatives that are best for the new product. This process should always consider the following:[3]

Financial investment needed Certain channel patterns require substantial financial commitments, such as promotion support for channel members, inventory, credit, and transportation. Those channel alternatives that place an unprofitable financial burden on the manufacturer should not be considered.

Timing Some channel patterns may need a significant length of time for training and development before they can perform efficiently. Established channels would require less time but may, in the long run, provide little differentiation from the competition. Thus compromises on timing and distinctiveness may be necessary.

Strengths and weaknesses of manufacturer Those functions the manufacturer handles well should not be carried out by the channel members. Channel alternatives should be screened so that channel strengths complement the manufacturer's weaknesses and vice versa.

---

[3]For further discussion, see David W. Cravens, Gerald E. Hills, and Robert Woodruff, *Marketing Decision-Making: Concepts and Strategy* (Homewood, Ill.: Richard D. Irwin, Inc., 1976), 539–45.

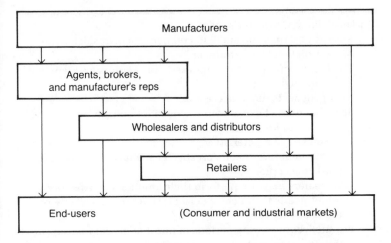

FIGURE 17–2
Alternative distribution
channel patterns

Bargaining power   The ability of the manufacturer to control channel members is an important consideration in the selection process. Typically channel members are most concerned with profit margins and the new product's demand potential. Hence high margins and strong support with advertising or other means of promotion can give the manufacturer a strong bargaining position with its channel system.

One of the most widely used channels for a new product requiring market penetration into new market segments is the manufacturer's representative (rep). Business start ups (with inventions) or firms that have diversified by developing a new product generally do not want to invest in a permanent sales force. These occasions require a careful evaluation of the possible use of the manufacturer's rep.

Table 17–2 provides a list of criteria that may be used to determine the feasibility of using a manufacturer's rep as a channel alternative to reach retailers and users. These criteria, which are categorized as being market related, product related, and company related, are discussed in the following paragraphs.

Market-Related Characteristics

Many new products may have a small market, particularly in the introduction stage of the product life cycle. During this period a permanent sales force would be too costly—especially when training, salary, benefits, and other cost factors are considered. Since the manufacturer's rep works on a commission basis and carries complementary products, the company benefits by limiting its selling costs as a direct function of sales. The representative can cover distant markets and take orders that are not frequent or seasonal but require continuous representation. New prospects can also be enlisted through the representative, who will be looking to expand his or her customers in order to increase profits.

Product-Related Characteristics

Often the nature of the product necessitates the use of a manufacturer's rep instead of any other alternative. A product that requires a prestige image and

TABLE 17–2

When to use a manufacturer's rep

Market-Related Characteristics
- Market potential too small to compensate own sales force
- Market distant from producer
- Product is infrequently purchased but requires frequent calling on customer
- New prospects must be reached
- Seasonal market exists with the need for continuous representation

Product-Related Characteristics
- New manufacturer needs prestige of agent to aid in gaining market acceptance
- Wholesalers may not perform their functions satisfactorily
- Product needs technical specialists that are unavailable or too costly to train
- Product requires long periods of negotiation before transaction is completed

Company-Related Characteristics
- Has insufficient financial resources for a direct sales force
- Product line insufficient to compensate a direct sales force
- Item produced requires a different market and distribution channel than current markets and channels
- Wants to expand distribution to territories that have low sales potential
- Perceives too many administrative problems with a direct sales force
- Wants to shift burden to other channel members
- Wants quick distribution

selling approach can take advantage of the existing reputation of a particular rep. A manufacturer's rep is also effective when the product requires special handling or technical support that is too costly to hire or train. A rep may also be appropriate when long periods of negotiation occur prior to completion of a transaction. This situation is common in industrial markets and the computer industry, particularly when the manufacturer is trying to penetrate completely different market segments.

Company-Related Characteristics

The company-related characteristics listed in Table 17–2 are particularly significant for a smaller company when resources are limited and the product line is too narrow to take advantage of any cost efficiencies. For a one-product company, the selling costs necessary to penetrate selected markets could be prohibitive.

For a small firm or a firm attempting to penetrate a brand new market, creating a permanent sales force can be risky as well as costly. The time involved in training, establishing contacts, and making the actual sales calls could be too long, giving competition an opportunity to move into the market more quickly and efficiently.

As a new product begins to reach the growth stage in its product life cycle, the distribution system will have to expand into territories that have lower sales potential. In order to maintain its market share, the firm will have to penetrate all of these market segments. The manufacturer's rep provides a feasible and cost-effective mechanism for this purpose.

Manufacturer's reps or agents provide an important service for manufacturers in many product categories. They can be a valuable means for achieving market sales goals for a new product. The criteria presented here can be applied to each new product in order to discern the feasibility of employing manufacturer's reps or agents.

EVALUATION AND SELECTIONS OF CHANNEL MEMBERS

The decision on the channel of distribution requires comparison of the feasible alternatives based on a set of relevant criteria. Table 17–3 illustrates a technique that management can use to determine which configuration should be used. In this example a small corporation manufactured a new liquid grill cleaner for the institutional market. This product has features which will revolutionize grill cleaning: it is odorless, will operate on a hot grill, converts grease to a dry ash, and is inexpensive. Three alternative channel strategies

TABLE 17–3
Method for evaluating channel alternatives

Factors in Order of Importance	Minimum Pass Level	Channel Alternatives		
		1	2	3
		Manufacturer Mfr's. Reps Wholesalers Restaurants	Manufacturer Mfr's. Reps Restaurants	Manufacturer Restaurants
1. Amount of investment needed to meet objectives	.4	.5 = P	.6 = P	.3 = F
2. Profit	.5	.8 = P	.8 = P	—
3. Reach of target market	.4	.8 = P	.7 = P	—
4. Impact on future new product development	.3	.8 = P	.5 = P	—
5. Flexibility to changing customer needs	.4	.4 = P	.8 = P	—
Ranking		2nd	1st	3rd

P = Pass
F = Fail

are being considered. Alternative 1 is to distribute the product through manufacturer's representatives, who will sell to wholesalers, who in turn will sell to restaurants. Alternative 2 is to sell the new product through manufacturer's representatives directly to restaurants. Alternative 3 is to sell directly to the restaurant.

Table 17–3 lists each relevant factor or criterion for the channel decision by order of importance. Management must analyze these criteria based on experience, judgment, and the firm's marketing objectives.

Alternative 3 fails to pass the most important criterion: the amount of investment needed to establish the channel. Alternatives 1 and 2 pass the minimum requirements for all five factors. Based strictly on the values assigned to each factor, alternative 2 would be chosen. However, since alternatives 1 and 2 are so close to each other, management must consider other factors, especially since some judgmental error is inherent. Because the price must be kept low—and because wholesalers provide little cost benefit—it may be more advantageous to distribute the new product through manufacturer's representatives to the restaurants (alternative 2). All payments to the manufacturer's representatives for service will be in the form of commissions. Delivery of the product will be directly from the manufacturer, who will control production quality and keep distribution costs to a minimum. Considering all factors, alternative 2 would be selected.

It is also reasonable to consider multiple channels. For example direct sales to large restaurant chains may be feasible in conjunction with alternative 2.

Other techniques are also available for evaluating alternative channel decisions.[4]

SELECTION OF CHANNEL MEMBERS

The last step in the design of the channel of distribution is the selection of the channel members. If the firm has chosen to intensively distribute the new product, some effort should still be made to avoid selecting retailers or wholesalers who will damage the reputation of the manufacturer. Any negative outcomes from these middlemen could result in damage to the firm and the new product.

Figure 17–3 is a sample list of some major criteria to be used by management to select channel members. It is also possible to develop a technique, discussed earlier, where the criteria are listed by order of importance with a minimum passing index for each. The rating scale in Figure 17–3 may be administered as a questionnaire to the management or customers of the potential middlemen. The complexity of the selection process will be based on how critical the decision is for the company and how many middlemen will be given the new product to sell.

---

[4]Evaluation methods are described in Philip Kotler, *Marketing Decision-Making: A Model Building Approach* (New York: Holt, Rinehart and Winston General Book, 1971), 293–98, and Barton Marcus et al., *Modern Marketing*, rev. ed. (New York: Random House Inc., 1980), 442–47.

FIGURE 17–3
Manufacturer's rating scale for
selection of middlemen

```
┌──────────────────────────────────────────────────────┐
│  1.  Contacts with customers in target market           │
│      Excellent                              Poor         │
│         □   □   □   □   □   □   □   □                   │
│                                                          │
│  2.  Reputation in target market                        │
│      Excellent                              Poor         │
│         □   □   □   □   □   □   □   □                   │
│                                                          │
│  3.  Past performance with other new products           │
│      Excellent                              Poor         │
│         □   □   □   □   □   □   □   □                   │
│                                                          │
│  4.  Overall quality of services provided to manufacturer│
│      Excellent                              Poor         │
│         □   □   □   □   □   □   □   □                   │
│                                                          │
│  5.  Compatibility of marketing mix strategy of middlemen│
│      with manufacturer                                   │
│      Excellent                              Poor         │
│         □   □   □   □   □   □   □   □                   │
│                                                          │
│  6.  Ability to maintain control of product with middlemen│
│      Excellent                              Poor         │
│         □   □   □   □   □   □   □   □                   │
│                                                          │
│  7.  Probability of effective long-term working relationship│
│      Excellent                              Poor         │
│         □   □   □   □   □   □   □   □                   │
└──────────────────────────────────────────────────────┘
```

There is an implicit assumption that the manufacturer can choose whomever it wants to carry the new product. In fact, in many instances, middlemen—particularly large retailers or distributors—are reluctant to add new products, especially from an unknown firm. Regardless of whether the new product is a consumer or industrial product, the risk in handling new products from small or new firms greatly reduces the number of willing middlemen. Only by providing them with incentives, such as liberal credit or high margins, will the firm entice the middlemen to accept a new product. Large corporations with a good reputation will have a much better chance of having their new products accepted by middlemen. For example, if IBM introduced a new, high-risk product, it would probably have little difficulty in gaining acceptance from channel members, even if the establishment of a new channel system were required.

CHANNEL MANAGEMENT AND CONTROL

Once the channel systems have been designed and implemented, the responsibility for managing the channel must be assigned. This function includes providing operating policies and procedures, motivating channel members, and gathering and distributing control information.

Large manufacturers are usually able to assure leadership in the channel since they built it through previous success in their market. Generally the leadership power results from the large manufacturer's financial strength, which enables it to maintain superior product research and development. In addition a large company can purchase materials in large quantities, achieving economies of scale that afford high margins.

A small manufacturer may also have control of the channel of distribution, particularly when it has developed a good new product.[5] The small manufacturer with a desirable new product may be able to dictate how it should be sold. The only control mechanism available to the small firm if their wishes are not accommodated is to withhold the product from the middlemen. In many instances the small manufacturer must build this control as the new product gains acceptance by consumers. Strong consumer demand will permit the manufacturer to expand distribution and eventually increase its control of the channel system.

The larger the channel system, the further the manufacturer is from the end-user of the new product. In such cases, it is important for the firm to develop effective communication with the middlemen to ensure that its objectives are being met.[6] In addition it is also helpful to establish direct communication with the end-user to satisfy complaints, requests for information, and other product-related questions.

Channel management and control requires careful consideration of the necessary information to maintain channel efficiency, incentives for middlemen, image development, servicing, and guarantee of the new product.

APPRAISING CHANNEL PERFORMANCE

The appraisal of channel performance must be based on how effectively the channel meets the manufacturer's objectives established in the first step of the channel design. Cost analysis, market share penetration, and consumer satisfaction are a few criteria that should be assessed.

Cost analysis may be carried out by identifying sales and cost data for individual channel members. Determining profit margins and specific financial ratios will identify unprofitable accounts that can either be eliminated or improved.

If market share is used, some comparable historical data must be provided. Identifying comparative data will enable the manufacturer to discern weakness and quickly take corrective action. A company may use competition to determine the overall attractiveness of its own products relative to the available alternatives. Such questions as whether total industry sales are increasing or decreasing, whether the new product is having a negative effect on any competitor's products, and whether the market has stabilized or declined should be answered.

Surveys may also be conducted at various points in the channel system to determine customer satisfaction and/or complaints. End-user surveys can determine specific problems with product availability, service, retailer support,

---

[5]Robert W. Little, "The Marketing Channel: Who Should Lead This Extracorporate Organization?" *Journal of Marketing* 34 (January 1970): 34.
[6]Cravens, Hills, and Woodruff, *Marketing Decision-Making*, 557.

and other factors. In addition, where larger channels are employed, surveys of retailers may be conducted to determine the performance of wholesalers or manufacturer's agents. Survey data can be invaluable in identifying specific problems that can negatively affect sales.

Channel appraisal may result in revisions, such as replacement of intermediaries, change in objectives and channel strategy, or a complete redesign of the channel system. However major changes may seriously affect the success of the new product and should be given careful attention. Continuous monitoring and analysis of the channel system can only aid the successful launching of the new product.

SUMMARY

New product launches require careful analysis of all elements of the marketing mix, regardless of whether the product is continuous, dynamically continuous, or discontinuous. A good understanding of channel strategy must be developed so that objectives can be achieved.

The channel system decision process must begin with identification of marketing objectives and target markets. Marketing mix decisions (including the channel system) must be made in this context. The firm's objectives and target markets provide the framework for the determination of the channel strategy.

After the channel objectives are determined, a channel analysis should be conducted to identify channel system alternatives. This analysis includes consideration of the degree of directness, selectivity, types of middlemen, and number of channels. Channel alternatives are then evaluated to determine which system or systems will best meet objectives.

The channel system requires careful management and control by the manufacturer. Leadership must be provided by establishing good information and communication flows in both directions. If performance of one or more channel members deteriorates, it is the responsibility of the manufacturer to take remedial action. Those intermediaries who perform poorly based on measures such as cost analysis, market penetration, and consumer satisfaction may be eliminated or forced to improve performance.

The channel system decision process provides management with a) an understanding of the purpose and structure of a channel system, b) an approach for designing a new channel system if necessary, and c) a framework for managing and appraising channel performance.

SELECTED READINGS

DENHAM, F. RONALD, and SHELLEY, GABRIEL C. "The Integrated Distribution Concept: Management's New Approach for the 1980's." *The Business Quarterly* (Winter 1979): 30–36.
Argues that as enterprises have grown and matured, emphasis has shifted from strategies based on market expansion, through the addition of new products, to improvement of profitability through more effective distribution. This strategy is used as a means of meeting higher energy costs in an era of slower market growth.

ETGAR, MICHAEL. "Channel Environment and Channel Leadership." *Journal of Marketing Research* (February 1977): 70.

States that an important question for channel managers and marketing researchers is whether it is possible to identify general environmental factors leading to or conducive to the emergence of channel leaders in distributive channels. A proper identification of such factors may help manufacturers and dealers to design appropriate channel strategies to ensure greater channel success and to minimize conflicts within the channel. Using frames of reference developed by students of organizations and organizational sets, the author explores the relationship between selected environmental factors and the control structure in distributive channels.

FITZELL, PHILIP. "Distribution: How Welch Cracked The Soft Drink Market." *Product Marketing* 6 (June 1977): 19.

Relates how Welch Foods, a producer of grape products, entered the soft drink market with a newly developed grape-flavored soda that was distributed through an existing network of food brokers, rather than the traditional franchised soft drink bottlers used by Coke, Pepsi, and Seven-Up. Discusses the problems Welch had with food brokers who only serviced retail shelves once a month along with many other products. Soft drink franchisers, on the other hand, carry only this one item and service retail stores weekly.

LAMBERT, DOUGLAS, and ARMITAGE, HOWARD. "Managing Distribution Costs for Better Profit Performance." *Business* 30 (September–October 1980): 46–52.

Explores the opportunities for potential improvement in profits from effective distribution. This is particularly important during the early stages of the product life cycle. In order to achieve increased profit improvement, it is necessary to develop an accurate cost data base and an information system to measure the incremental costs of alternative strategies.

STERN, LOUIS W., and EL-ANSARY, ADEL I. *Marketing Channels.* 2nd ed. Englewood Cliffs, N.J.: Prentice-Hall, Inc., 1982.

Managerial focus with an emphasis on planning, organization, and control. First two parts explore theories of why and how channels emerged and analyze channel management. Remaining sections deal with the planning and design of channels, as well as evaluation and control. Also discussed channel management in the international market and service industries.

WEBSTER, FREDERICK E., JR. "The Role of the Industrial Distributor in Marketing Strategy." *Journal of Marketing* 40 (July 1976): 10–16.

In an attempt to get an in-depth look at the industrial distributor, a field study of distributors and manufacturers was conducted. Article discusses results of survey, giving special attention to both the role of the industrial distributor in manufacturers' marketing strategies and the changing relationship between distributor and suppliers. Discusses numerous management issues pertaining to the supplier/distributor relationship.

WEIGARD, ROBERT E. "Fit Products and Channels to Your Markets." *Harvard Business Review* 55 (January–February 1977): 95–105.

Gives examples of various possible combinations of markets, channels, and products, and outlines the conditions that give rise to what the author calls *multimarketing*. Stresses the importance of understanding how the system works so that the manager can deal effectively with the conflicts and problems that arise along the channel or in the laws under which business is operated.

Promoting the New Product

Regardless of the type of new product or service to be marketed, effort must be expended in promoting it. The complexity of the promotion mix decision is a function of many variables, most important of which is the "newness" of the product. The truly unique or discontinuous product (see Chapter 1) requires management to carefully consider all possible means to communicate the product, the funds available, company objectives, potential competition, market segments, distribution networks, and customer needs. The fact that the new product is unique presents problems to management in that there is little experience available to help make the promotional decisions.

Products that are dynamically continuous present less difficulty in promotion mix decision-making because substitutes and previous experience may be adapted to the new product. "Me-too" products or continuous innovations generally require the least additional effort in determining the promotion.

It is particularly important to recognize that the promotion mix will vary according to the stages in the product's life cycle. This chapter presents all aspects of developing the promotion mix for a new product relative to the internal and external variables that affect its design.

ORGANIZATION AND THE PROMOTION MIX

What the company chooses to do in terms of organizing the promotion mix may depend on factors that are relatively fixed. The firm's size, the nature of the market (relative importance of promotion in marketplace), and the diversity of the markets and product mix can determine the nature of the organization at the communication function. In addition the organizational structure for the company as a whole, including the new product development organization, may also be an important determining factor.

A number of alternative common organizational arrangements and reporting relationships may be considered for the communications/promotion function:[1]

1. Advertising can be a subfunction within the marketing or sales department.
2. Advertising can be separated as a major function, reporting directly to general management.
3. Advertising can be a subfunction within some other functional department.
4. Sales promotions, public relations, or both can be treated as joint functions or subfunctions of advertising, or else separated from advertising in terms of both organizational location and reporting relationships.

---

[1]David L. Hurwood and Earl L. Bailey, *Advertising, Sales Promotion and Public Relations—Organizational Alternatives* (New York: The Conference Board, 1968), 2.

If a firm is multidivisional, with each division autonomous because of unique manufacturing and selling operations, then it is conceivable that the promotion function may be separated or fragmented in various levels of the company. Effectiveness depends on minimizing the overlap or redundancy in this function.

ADVERTISING AS A SUBFUNCTION OF MARKETING

The tendency to place advertising directly under marketing or sales is strongest at the division level. The director or manager of advertising would report to a vice-president of marketing or sales as represented below:

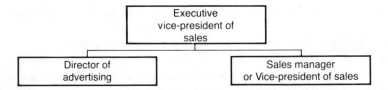

In this instance advertising is subordinated to marketing at a high level but also still reports to a member of top management who reports to corporate levels. Thus, because of the reporting relationship, management has some high-level control of divisional advertising activities.

ADVERTISING SEPARATED FROM MARKETING

When advertising is separated from marketing, the firm usually defines advertising and sales promotion as areas that service marketing where necessary. There is a close relationship, as well as constant contact, between advertising and marketing in this type of structure, as shown below:

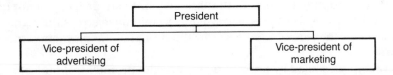

As mentioned earlier there are possible variations or combinations that describe the relationship of advertising and marketing within the organization. The effectiveness of the organization and relationships of advertising and marketing will always depend on the personnel, the experience of the firm, and the internal and external marketing variables mentioned earlier.

PROMOTION AND THE PRODUCT MANAGER

In Chapter 6 we discussed the responsibilities of the product manager, as well as other alternative product development organizations. The promotion mix is an important part of the decision-making process of the product manager, especially in consumer product markets.

Generally the product manager concentrates on the promotion plan as part of the market plan and works with the internal department (typically referred to as the advertising department or advertising/sales promotion department) and advertising agency to accomplish company objectives.

If the product manager's workload is heavy, he or she will have one or more people reporting directly on the advertising/promotion program. Another variation of this would be to have the complete advertising function under each product or brand manager, thus eliminating the need for an advertising manager. Elimination of this function can be a problem when coordination is necessary to reflect the appropriate image for the company and each of its products. If institutional (corporate-image) advertising is necessary, it would be difficult to identify responsibility internally in this type of organization.

Up to this point, we have assumed that the entire promotion mix is the responsibility of the advertising manager or director. Sales promotion can be combined with advertising or a separate function; however the same is true for public relations. In the industrial market it is more common to find sales promotion and/or public relations combined with the advertising function. The opposite is likely for large, consumer product firms, especially if the product mix is highly diversified.

THE PROMOTIONAL BUDGET

One of the most important and most difficult decisions is how much to spend on promoting the new product. Many managers are misinformed about the impact of the promotional dollar. They believe that more money expended for promotion will bring more sales. However promotion may reach a point of diminishing returns, so that each additional dollar expended returns less than a dollar in sales.[2] It is also possible to spend less on promotion than is needed to achieve the optimal return. Both these dilemmas are difficult—if not impossible—to resolve. Nevertheless every effort to minimize errors through careful analysis of all the variables inherent in the design of the promotion mix must be made.

Techniques used by firms to determine the promotional budget are percentage of sales, funds available to spend, and copying the competition.[3] These techniques are weak because they do not consider the nature of the market and the sensitivity of sales to the promotional budget. These weaknesses are enhanced when the product is discontinuous. For example attempting to decide on a budget based on the percentage of sales assumes that sales can be accurately forecasted without any knowledge of how much is to be spent on promotion. If the new product is continuous (no new learning required), then the percentage-of-sales technique may be used, since previous sales ex-

---

[2]Donald E. Sexton, Jr., "Overspending in Advertising," *Journal of Advertising Research* 2 (December 1971): 19–24.
[3]James Engel, Martin Warshaw, and Thomas Kinnear, *Promotional Strategy*, 4th ed. (Homewood, Ill.: Richard D. Irwin, Inc., 1979), 224.

perience with similar or substitute products would be assumed to be valid for the new product. Markets for continuous products also tend to be more stable and predictable.

When the product is discontinuous or dynamically continuous, management should initially conduct market segmentation analysis (see Chapter 12) to determine the size of the potential markets(s), competition, and the most effective promotion strategy to reach those segments.

Determining what is needed to reach the market segments is perhaps the most difficult aspect of setting the promotional budget. Figure 18–1 shows a checklist that assists management in the preparation of the promotional strategy. Management should carefully determine what strategy is to be taken and how much is to be spent on each promotional mix element. For example cents-off coupons—instead of free samples—may be mailed to consumers. Other similar decisions could be made so that less expensive strategies may be substituted for more expensive ones in order to satisfy budget constraints. At the same time, budget allocation can also be more carefully determined, since costs should be readily available for each item on the checklist. Most important is that the promotional budget be designed without guessing, but by applying available experience and scientific methodology.

NEW PRODUCT PUBLICITY

One of the most effective means to facilitate the introduction of a new product is to obtain free advertising, either in a trade journal for an industrial product or in newspapers or magazines for a consumer product. A news release explaining the company, its executives, and a description of the new product can provide much needed support in the initial phases of the product's life cycle.

Many trade magazines, as well as consumer periodicals, have new product sections. Usually a news release will contain the name of the company, a description of the product, and a phone number or address where a reader can obtain more information on the new product. Inquiries from news releases can, indirectly, be an important factor in marketing a new product, particularly in the design of the promotional mix. An innovative firm usually exploits opportunities for publicity as much as possible. Keeping the name of the firm before the consumer is the prime objective of advertising. It makes little difference how or what information source is used, as long as the consumer sees it, reads it, and remembers it in future purchase decisions.

Rules for Issuing New Product News Releases

The general framework of the news release should include certain specific items regardless of the information source or the nature of the product. These items are as follows:

1. Identify the news release as such when mailing it to the editor.
2. Begin the release with the new product's brand name.
3. List the features of the new product, emphasizing the most important factors.

Sales Department

Yes	No		Company allocation
(check)			
☐	☐	Introduction of product to sales force	
☐	☐	Sales conventions	
☐	☐	Sales meetings	_____
☐	☐	Brochures and other mailings	_____
☐	☐	Sales force training or education	_____
☐	☐	Sales force incentives	_____
☐	☐	Contests	_____
☐	☐	Special quotas	_____
☐	☐	Revised compensation systems	_____
☐	☐	New sales force equipment	
☐	☐	Sample cases	_____
☐	☐	Miniature models	_____
☐	☐	Films	_____
☐	☐	Drawings, pictures, or paintings	_____
☐	☐	Printed matter	_____

The Dealer

Yes	No		Company allocation
☐	☐	Dealer meetings	_____
☐	☐	Meetings with dealers' personnel	_____
☐	☐	Special deals	_____
☐	☐	Display material	_____
☐	☐	Counter displays	_____
☐	☐	Window displays	_____
☐	☐	Signs	_____
☐	☐	Printed material (e.g., folders and booklets)	_____
☐	☐	Trade advertising	_____
☐	☐	Trade shows	_____
☐	☐	Films	_____

The Employees

Yes	No		Company allocation
☐	☐	Employee magazines	_____
☐	☐	Posting advertisements	_____
☐	☐	Mailings	_____
☐	☐	Meetings	_____

The Consumer

Yes	No		Company allocation
☐	☐	The Media	
☐	☐	Newspapers	_____
☐	☐	Magazines	_____
☐	☐	Radio	_____
☐	☐	Television	_____
☐	☐	Direct mail	_____
☐	☐	Catalogs	_____
☐	☐	Outdoor advertising	_____
☐	☐	Car or bus cards	_____
☐	☐	Novelties	_____
☐	☐	Coupons	_____
☐	☐	Free samples	_____
☐	☐	Printed material	_____
☐	☐	Contests	_____

FIGURE 18–1

Checklist for preparing promotional strategy

4. Condense the written news release to 100 to 150 words to increase its chances of acceptance by the editor.
5. Include a high-quality, glossy photograph of the product. In some instances the firm may have to pay a nominal fee to defray the cost of printing the photograph.
6. Issue separate releases for each model and size to maximize the exposure of the new product. If possible, issue separate releases for each major application of the new product.

It is important that the new product decision-makers maintain a list of potential sources for news releases. Inquiries from potential consumers based on readership of news releases should also be carefully documented, since these consumers are likely to represent the initial market segment in the product's introductory stage.

USING AN ADVERTISING AGENCY

An advertising agency can provide many important promotional services to the firm introducing a new product. Advertising agencies exist in varying sizes and have different functions and specialties. The largest agencies provide worldwide service in all possible promotional decision-making areas. Traditionally the advertising agency has been perceived as an independent business organization composed of creative and business people who develop, prepare, and place advertising in advertising media for sellers seeking customers for their goods and services.[4] Today advertising agencies perform many services related to the development of the promotional mix, such as market research, package designing, and sales training. A small firm introducing a new product requires these expanded services, since it could not perform them economically on its own. The marketing experience of the advertising agency may also be utilized in the development of the market plan. Essentially the agency follows the procedure given in Figure 18–2 in the formulation of the promotional plan. This figure also identifies major services performed by the advertising agency at each stage in the planning process.

Evaluation of Agency Services

Whether the advertising agency is used exclusively for all promotional activities or whether it is used to complement the firms' own advertising department, it is important to evaluate the services an agency can provide. Although the total set of services provided by any advertising agency will vary, the major services are as follows:

1. Copywriting
2. Art techniques
3. Packaging design
4. Photography
5. Typography

---

[4]Frederick R. Gamble, *What Advertising Agencies Are—What They Do and How They Do It*, 4th ed. (New York: American Association of Advertising Agencies, 1963), 4.

6. Reproduction
7. Printing
8. Program planning
9. Public relations
10. Public speaking
11. Media analysis and selection
12. Market research
13. Space allocation
14. Trade shows
15. Sampling
16. Sales promotion
17. Others

The firm selecting the agency for a new product introduction should carefully determine which services are needed and whether some could be completed effectively within the firm's own advertising department. The advertising departments of most firms are weak in certain areas and require an advertising agency to fill the gaps. Duplication should be avoided to minimize costs and optimize the use of existing internal resources. It may be beneficial for the firm to evalute the strengths and weaknesses of the advertising depart-

Competition	Analysis of client's product or service	Advantages and Disadvantages
Location Seasonality Competition	Analysis of potential market segments	Market acceptance Industry conditions Economic conditions
Size Profitability	Analysis of characteristics of members of distribution channel	Competition Services Reputation
Reputation Influence Circulation	Analysis of available media that can best carry the message to all members of channel	Quality Location Costs
	Formulation of definite plan and presentation to client	
Writing, designing, illlustrating forms of message	Execution of promotion plan	Contracting for space, time and so on
Monitoring insertions, displays, and the like		Auditing, billing for services
Market research Analysis of sales by market segment	Feedback on promotional effectiveness	External services (e.g., Starch and A. C. Nielsen)

FIGURE 18–2

Formulation of the promotional plan by the advertising agency

ment on each of the services listed. If the agency is strong in the areas in which the advertising department is weak, then an ideal situation would exist for developing promotion strategy.

Selecting the Advertising Agency

Before choosing the advertising agency, management should thoroughly communicate their requirements, including a description of the new product markets, budget restrictions, competition, and other merchandising decisions, to the firm's own advertising department.

If possible, the agency chosen should have had some experience in promoting the type of product introduced. If the new product is dynamically continuous or continuous, it should be relatively easy to identify agencies that have had experience with similar products. Strengths demonstrated in campaigns for competitive products should provide enough information to judge the ability of any agency to handle the new product. However, if the new product is discontinuous, the choice of the agency will be more difficult, since no one will have had any previous experience with it. One means of judging or screening agencies at this point might be on the basis of the agency's ability to handle discontinuous new products in the past. If they have successfully introduced products of this type before, it may be an indication of their overall ability and creativity.

Once the initial screening has been completed, each agency should be invited to make a formal presentation to the marketing managers so that a final selection can be made. In order to help select the agency, a checklist is provided in Figure 18–3. For each item in the checklist, management should evaluate the agency and its presentation by assigning some scale value—for example, 1 to 7. Scoring of each criterion on the checklist is subjective but will provide some basis for making the final selection. More than one member of

Item	Value
1. Location of agency	_____
2. Organizational structure of agency	_____
3. Public relations department services	_____
4. Research department and facilities	_____
5. Creativity of agency staff	_____
6. Education and professional qualifications of agency's top management	_____
7. Media department qualifications and experience	_____
8. Qualifications and experience of account executives (if identifiable)	_____
9. Interest and enthusiasm shown toward firm and new product	_____
10. Copywriter qualifications and experience	_____
11. Art director's qualifications and experience	_____
12. Recommendations by other clients	_____
13. Experience and success with new products	_____
14. Ability of agency to work with company advertising department	_____
15. Extra services provided	_____
16. Accounting and billing procedures	_____
17. Overall formal presentation	_____

FIGURE 18–3
Checklist to select advertising agency

the marketing staff should be present at the formal presentation and each should be given an opportunity to evaluate the agency on the items in Table 18–1 (see page 372).

The relationship between the agency and the company should be honest and open so that problems may be minimized and optimal results provided for the new product introduction.

PROMOTION AND THE PRODUCT LIFE CYCLE

In Chapter 1 the product life cycle concept was introduced to illustrate the relevance of varying marketing strategy over the life of a product. The changes in marketing strategy are not only a function of the product's life cycle but are also dependent—particularly in the introduction and growth stages—on whether the new product is continuous, dynamically continuous, or discontinuous. During the other stages, the new product's promotional strategy is no longer dependent on the newness of the product because product concept knowledge would have been achieved across the potential market.

The Consumer Product

Figure 18–4 illustrates the general promotion strategy that should be incorporated into each stage of the product life cycle. Some differences are noted for high- and low-learning (continuous and discontinuous) consumer products during the introduction and growth stages. Industrial products are discussed later. It is evident that the high-learning or dynamically continuous or discontinuous type of products have longer introductory stages and hence are more likely to fail. Promotional strategy for these types of products must be carefully developed to avoid losses and ultimate failure. The biggest problem for the high-learning product is educating the consumer to the product's utility. Appeals must be based on the product concept rather than the brand name until there is sufficient awareness among the potential users to warrant a more aggressive brand identification campaign.

As the product reaches the growth stage, it is likely that competition will enter the market and, in the case of the high-learning product, take advantage of the fact that the consumer is now aware of the product and the needs it satisfies. The firm initiating the new product must be prepared for competitive introduction and should have its campaign ready for such changes in market conditions.

During the maturity stage, promotional strategy should appeal directly to important market segments. Special promotions and price deals may begin to be part of advertising copy. Emphasis on dealer promotions is necessary to ensure favorable treatment for shelf space, location, and the like.

As the product moves into the saturation stage, the firm should begin to research ways to reposition the product or use promotion strategy to convey new uses or to reach new target markets. Special deals to channel members will be needed to maintain trade loyalty. If the firm is not able to stimulate growth in sales, a decision must be made about whether to continue marketing this product or to withdraw it. If management decides to continue, the

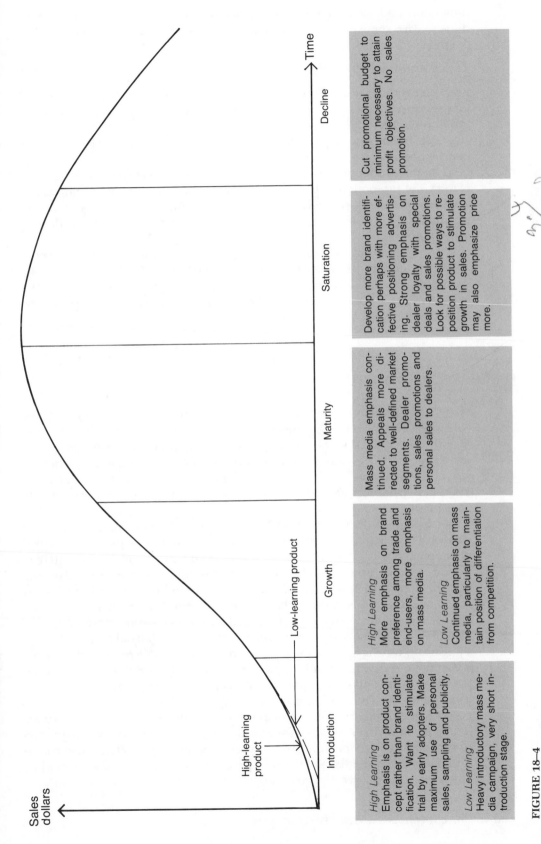

Sales dollars

High-learning product

Low-learning product

Introduction   Growth   Maturity   Saturation   Decline   → Time

High Learning
Emphasis is on product concept rather than brand identification. Want to stimulate trial by early adopters. Make maximum use of personal sales, sampling and publicity.

Low Learning
Heavy introductory mass media campaign, very short introduction stage.

High Learning
More emphasis on brand preference among trade and end-users, more emphasis on mass media.

Low Learning
Continued emphasis on mass media, particularly to maintain position of differentiation from competition.

Mass media emphasis continued. Appeals more directed to well-defined market segments. Dealer promotions, sales promotions and personal sales to dealers.

Develop more brand identification perhaps with more effective positioning advertising. Strong emphasis on dealer loyalty with special deals and sales promotions. Look for possible ways to reposition product to stimulate growth in sales. Promotion may also emphasize price more.

Cut promotional budget to minimum necessary to attain profit objectives. No sales promotion.

FIGURE 18–4
Promotional strategy during stages of consumer product life cycle

promotion budget will be cut back. Efficiency should allow the firm to continue to earn some profits or contribute some revenue to fixed costs.

The Industrial Product

Because products also follow a life cycle pattern, their promotional strategy is just as critical to their success. For the industrial product, the emphasis during the introductory stage is on trade shows, publicity, and personal selling. If it is a high-learning product, in-house seminars, films, or other training devices may be necessary to educate users.

As the product moves into the growth stage and competition enters the market, the promotional strategy will be mostly personal selling to emphasize service to the user—particularly delivery and maintenance. The reputation of the innovative firm is also important to ensure its continued success despite competition.

In the maturity and saturation stages, the firm will continue to develop its loyalty among users. Just as in the consumer market, the firm will be seeking to expand sales through the development of new uses or new target markets. Figure 18–5 summarizes the promotion strategy for a typical new industrial product.

POSITIONING WITH PROMOTION

Many new products do not offer enough novelty to make switching by the consumer worthwhile. Consumers may try the new product but will frequently switch back to the old one, since the old product was probably very satisfactory, and the new product simply did not offer enough uniqueness to warrant a permanent switch. This is often the reason for product failures. Thus new products should be positioned through promotional strategy to have novel attributes that are easily perceived by consumers.

Consumers often have difficulty remembering the product being sold but can recall everything else about the advertisement. One explanation for this phenomenon is that much advertising is burdened with creative and interesting formats that have nothing to do with the product. Marketers should ensure that the consumer learns the brand first. The brand name should be clearly stated or shown where appropriate to increase brand recognition.

There are many examples of ads that have failed to establish brand name memorability. In one television commercial Hermione Gingold appears singing a jingle and waving a can of the product in each hand. The result of the waving is that no one viewing the advertisement can actually read the brand name on the cans. At the end of the commercial the brand logo appears on the screen and disappears. The result is poor brand identification.

A second major error in new product commercials is sensory overload. In this instance the advertiser tries to say too much about the product in the 30-second or 60-second spot. These types of commercials often cause confusion and a negative image of the new product. Simpler information would be much more persuasive and credible.

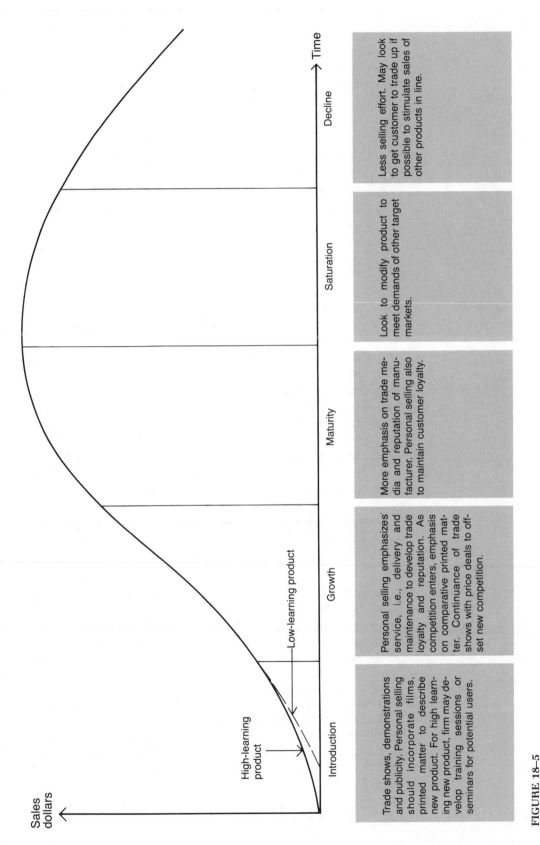

FIGURE 18–5
Promotion strategy during stages of industrial product life cycle

TABLE 18–1

Ways to improve commercials that introduce new products

1. Nothing can surpass the unique selling proposition or product attribute/benefit. Lacking this, however, it's important to establish areas of superiority and advantage over the competition.

2. Keep the new name up front. Don't be afraid to repeat, reinforce, and make meaningful that new name via copy, theme, story line, legends, logos, jingles, zooms, or other devices.

3. Put the word "new" up front and repeat it.

4. A distinctive, individualistic personality and image for the new brand can be of great help.

5. Position the product clearly (avoid ambiguities).

6. Demonstrate the product—show how it works and what it can do.

7. Emphasize rewards/end benefits (taste, consumption, enjoyment, relief). If the format is problem/solution, emphasize the solution.

8. Keep the message simple. Focus on a major, vital idea. Keep the exposition straightforward. Often, a single, continuous story line is preferable to vignettes.

9. Generate consumer identification. Portray the peer group and desired lifestyle of the prospective user. A recognizable, logical setting can be beneficial.

10. "Devices" should be supportive and relevant. For example, celebrity presenter should be appropriate, compatible, and a trustworthy authority on the product. Humor should be used cautiously.

11. Consider the 60-second format. Your message may be too ambitious to compress into 30 seconds.

12. Pretest to avoid getting the bad news later and to determine a corrective course.

SOURCE: "Goal of New Product Commercials Is to Make Brand Name Memorable," *Marketing News*, 23 January 1981, 5. Reprinted from *Topline*, December 1980 newsletter, McCollum/Spielman & Co. Inc., Great Neck, N.Y.

Research indicates that new product commercials are about 70 percent as successful as established product commercials. This ratio seems to hold true in about every product category. New brands have a much more difficult time developing identification when clutter exists in the commercial. This research suggests that, for a new product, advertising is most likely to fail at making the brand name memorable. Table 18–1 lists 12 ways offered by the researchers to improve commercials for new products.[5]

High brand recognition may occur through either high media frequency or low media frequency, but with high content efficiency. This last strategy entails making the new product the central element in all the advertising,

[5]"Goal of New Product Commercials Is To Make Brand Name Memorable," *Marketing News*, 23 January 1981, 5.

giving careful thought to what is being communicated. One good test of whether the advertising meets this objective is to substitute a different product for the one being advertised, still maintaining the same conceptual and visual elements. If this can be done, then the level of content efficiency is too low and new advertising copy should be developed.

The success of advertising a new product and obtaining high consumer recall is also dependent on how well the firm has defined its market segments, especially in relation to competition. The positioning of the new product through advertising has resulted in major success in highly competitive markets. A firm frequently has a wider range of positioning options for the new product than its competition.

Direct Comparative Advertising

It is possible to directly confront competition by comparing the new product with that of the leader in the market. This is done by placing the new product alongside the leading product in advertisements. At one time it was felt that this strategy might only enhance the advantages of the leading brand. Much of this fear was based on a Starch research study done in 1969, which concluded that out of 1,800 television commercials aired during prime time, an average of only 15 percent of the audience could recall a new product's brand name. Of this group, 8 percent credited the commercial to the new product's competition.[6] In addition similar research in 1970 and 1971 obtained similar findings but with increased misidentification.

More recent advertising strategy, however, is based on the belief that the direct confrontation approach can be successful. One advertising agency executive indicated that when the product is introduced in a category dominated by one or two well-established competitors, the positioning strategy should be head-on with the dominant brand.[7] Recent increases in the use of comparison advertising seems to indicate that these fears have diminished. Gillette, for example, positioned Earth Born directly against Clairol's Herbal Essence, Johnson's Baby Shampoo, and Head and Shoulders. In the industrial market, successful head-on collisions have occurred with Xerox entering the computer field and IBM introducing a copy machine. However, even with these successes, the general consensus is to utilize a variation of comparison advertising.

A study conducted by Ogilvy & Mather revealed the following about comparative campaigns:[8]

1. They offer no advantage to the advertiser.
2. They do not increase brand identification.
3. They increase consumer awareness of competitors.
4. They decrease credibility of advertising claims.
5. They result in increased brand misidentification.

The products used in this survey were in the drug, health, and beauty category.

[6] E. B. Weiss, "Direct Comparative Advertising," *Advertising Age*, 8 January 1973, 38.
[7] Lawrence Wolf, "How to Make Your New Product Advertising Work Harder," *Advertising Age*, 21 October 1974, 58.
[8] "Comparative Ads Ineffective: Ogilvy & Mather Study," *Advertising Age*, 13 October 1975, 16.

The conclusions on comparative ads are not definitive. Advertising for new products should be carefully designed and developed, particularly if a strategy of comparative advertising is undertaken.

Varying Direct Competitive Positioning

Most new products are introduced by varying the head-on approach. In the 1960s Avis represented themselves as being "number 2" in the industry but stated they would try harder to displace the dominant company. Coca-Cola introduced Tab by using the theme *Make Room for Number One*. B. F. Goodrich launched a recent campaign to educate the consumer in the difference between Goodyear and Goodrich.

Greater variations have also been achieved by firms seeking to carve out a new market. Vicks' Nyquil faced a very competitive market that was saturated with different brands of cold remedies. Instead of trying to introduce a new product to meet the competition head-on, Richardson-Vicks, Inc. decided to position this new product by promoting it as a nighttime cold remedy only. The product is in liquid form, which also enabled the consumer to identify it easily and to recognize its purpose as a cold remedy to be taken before bed. Vicks has since introduced a daytime, liquid cold remedy for adults, to compete directly with other products in this class, and also a children's version aimed at widening the use of this product category.

A firm may find that a new product has not achieved the success anticipated and may withdraw it from the market and reposition it using much advertising to enter a new market. Sealtest's Light n' Lively ice milk was initially positioned by using an advertising theme stating that the product contained less fat and would be lower in calories. This diet image was successful for a short time until competition began to enter the market, thereby reducing Light n' Lively's market share. Sealtest realized that the market for a diet ice milk was limited but could be expanded to compete with ice cream if consumers would believe that ice milk tasted good. After extensive research, Sealtest repositioned Light n' Lively with new, bright packaging and flavors and a large advertising campaign that emphasized the good taste of the product. To enhance the creamy taste of ice milk, Sealtest also added milk solids. The results of the repositioning were very successful and the product achieved a substantial increase in market share.

Industrial product advertising has generally lagged far behind consumer product advertising largely because many members of top management look upon advertising as a dispensable fringe contributor to sales. Thus management tends to minimize advertising. Quantitative data on the effectiveness of industrial advertising is generally limited to coupon returns, number of inquiries, and long-range volume trends. A study conducted on industrial advertising of 100 industrial firms in 408 trade papers and periodicals concluded the following:[9]

1. Sales increases followed advertising increases.
2. Sales increases prodded by increased advertising were seldom experienced in full in the same year.

---

[9] "Special Report," *Printer's Ink* 269, (9 October 1959) 25–31.

3. Curtailment of advertising decreased sales with increasing momentum.
4. A company must increase its advertising at least as much as the overall industry to retain its market share.
5. Short-term advertising was generally ineffective.
6. Advertising capacity should be increased during business down-swings to gain competitive advantages.

A small manufacturer of gas and fluid couplings had difficulty marketing a new product that had an automatic shut-off feature and greater adaptability than its competition. But because the product was perceived by customers to be the same as other products, a print campaign was conducted to promote the product as different. Each ad contained copy on functional benefits and quality by using two cartoon mice who explained the facts humorously. The results were significant sales increases—62 percent in the first eight months.

RCA and General Electric Co. had little success in attempting to meet IBM head-on in the computer market. Both of these manufacturers were forced to dissolve or sell their computer interests because their positioning strategy for matching IBM was unsuccessful. Yet Honeywell was successful in its bid to position itself as the "other computer company."

The overall effect of a positioning strategy may be the critical factor in determining the success of a new product. This strategy must be based on research and market segmentation analysis to ensure that the position sought does not belong to anyone else.

Push Versus Pull Strategy

The decision to use a push or pull advertising strategy is dependent on the acceptance of the product, particularly among channel members. A push strategy would entail an advertising campaign to support and aid sales at the retail level. When channel members accept the new product, the promotion strategy should be to inform the consumer through media, point-of-purchase displays, and coupons, with the purpose of increasing turnover for the retailer. Cooperative advertising may also be used to encourage media advertising on a local level. This practice is particularly evident in the food industry. Special promotional allowances may also be used to encourage channel acceptance (for example, offering a free case with every ten cases bought).

When the new product is unique or discontinuous, retailers may be reluctant to take the risk and use valuable shelf space for an untried product. This problem is one of the major causes of new product failures. If the retailer and other channel members cannot be induced to carry the product, then the probability of success is reduced significantly. In order to solve this problem, manufacturers have resorted to appealing directly to the consumer so that the consumer will pressure the retailer into carrying the product. This strategy can be very expensive since it is often necessary to use mass media during prime time to appeal to the largest audiences. Repetition and well-designed copy must be used to attract the consumer's attention, as well as to make the consumer request the product.

Probably the most successful recent pull strategy was conducted by Pure 1 chicken. This product was faced with a saturated market as well as retailers who were reluctant to change suppliers or add a new brand of chicken. The only direction Pure 1 could take was to directly promote the product to the consumer with a unique advertising plan to facilitate retailer acceptance. Consumers were puzzled by the unique advertising copy and were curious enough to ask their retailers why they did not carry Pure 1. The word-of-mouth activity was effective, enabling Pure 1 to take a solid position in the chicken market.

Matching Media with Target Markets

To ensure advertising success, the firm must use all available information on market segments (see Chapter 12). The profiles of target markets enable management to match the media's audience with the target market profiles. In most cases demographic data are readily available describing the audience of various media. Nielsen Product Media Ratings or Target Group Indices are two sources that provide demographic characteristics for both products and media (print as well as television).

Profile-matching has been used by psychologists to compare characteristics of an individual represented by a set of test scores with those of another individual. The technique allows the researcher to compare the configurations of two sets of test scores, then combine them into a single index.[10]

Profile-matching has been used in marketing to determine whether the congruency between self-image (profile) scores and product or store-image (profile) scores has any effect on purchase decisions. Results indicate that consumers are attracted to stores or products that reflect the image they have of themselves.[11] This same methodology has numerous applications in marketing, particularly in matching media audiences with target market profiles. The technique used for matching is a measure of distance between two profiles of demographic characteristics.[12] The goal is to determine which media has the closest audience characteristics to the profile of market characteristics. The formula used is as follows:

$$D_{12}^2 = \sum_{j=1}^{k} (x_{j1} - x_{j2})^2$$

where

d_{12} = distance between two sets of data
(e.g., the market and media characteristics)

j = any k number of demographic or other defined variables

x_{j1} = score of market variable j

x_{j2} = score of a medium's audience on variable j

---

[10]Lee J. Cronbach and Goldine C. Gleser, "Assessing Similarity Between Profiles," *Psychological Bulletin* 50, no. 6 (1963): 456–473.

[11]See Ira J. Dolich, "A Study of Congruence Relationships Between Self Images and Product Brands," *Journal of Marketing Research* 6 (February 1969): 81–84, and Ira Dolich and Ned Shilling, "A Critical Evaluation of the Problem of Self Concept in Store Image Studies," *Journal of Marketing* (January 1971): 71–74.

[12]Jack Z. Scissors, "Matching Media with Markets," *Journal of Advertising Research* (October 1971): 39–43; also see Ronald E. Green, "Numerical Taxonomy in Marketing Analysis," *Journal of Marketing Research* 5 (February 1968): 83–94.

TABLE 18–2

Matching camera market profile with magazine readership profiles for occupation status

Occupation	Professional Managerial %	Professional Managerial Squared Difference	Clerical Sales %	Clerical Sales Squared Difference	Craftsmen Foremen %	Craftsmen Foremen Squared Difference	Other Employees %	Other Employees Squared Difference	Sum of Squared Differences
Camera Market Profile	22		15		9		18		
Better Homes and Gardens	20	(4)	14	(1)	5	(16)	13	(25)	46
Esquire	30	(64)	14	(1)	11	(4)	14	(16)	85
Playboy	26	(16)	17	(4)	11	(4)	18	(0)	24
Time	32	(100)	13	(4)	4	(25)	11	(49)	178
Reader's Digest	22	(0)	14	(1)	6	(9)	13	(25)	35

Tables 18–2, 18–3, and 18–4 show the matching of five potential print media profiles with the profile of the 35mm camera market. Thus a firm introducing a new 35mm camera may have some difficulty in deciding which print media to choose or how to allocate its print media budget. The problem here is simplified by selecting only five magazines and only three demographic variables. The process would also be appropriate in choosing specific television or radio programs based on similar matchings. For each magazine, distributions of the demographic variables of occupation, income, and age are determined from secondary sources. Tables 18–2, 18–3, and 18–4 identify and then match distributions for these variables in the 35mm camera market by taking the squared difference in each category and summing to get the total squared difference for each magazine and demographic variable. In Table 18–4, which shows the distribution of occupation, *Playboy* magazine provides the best match for the 35mm market, indicated by the low, total squared difference

TABLE 18–3

Matching camera market profile with magazine readership profiles for household income

Income	$25,000+ %	$25,000+ Squared Difference	15,000–24,999 %	15,000–24,999 Squared Difference	10,000–14,999 %	10,000–14,999 Squared Difference	5,000–9,999 %	5,000–9,999 Squared Difference	Less than 5,000 %	Less than 5,000 Squared Difference	Sum of Squared Differences
Camera Market Profile	5		22		33		29		11		
Better Homes and Gardens	5	(0)	25	(9)	35	(4)	23	(36)	12	(1)	50
Esquire	7	(4)	31	(81)	32	(1)	22	(49)	8	(9)	144
Playboy	4	(1)	25	(9)	34	(1)	27	(4)	10	(1)	16
Time	10	(25)	32	(100)	30	(9)	21	(64)	7	(16)	214
Reader's Digest	5	(0)	23	(1)	33	(0)	27	(4)	12	(1)	6

TABLE 18–4

Matching camera market profile with magazine readership profiles for age categories

Age	18–24		25–34		35–49		40–64		65+		Sum of
	%	Squared Difference	%	Squared Difference	%	Squared Difference	%	Squared Difference	%	Squared Difference	Squared Differences
Camera Market Profile	20		25		30		19		6		
Better Homes and Gardens	14	(36)	21	(16)	29	(1)	25	(36)	11	(25)	114
Esquire	27	(49)	28	(9)	26	(16)	14	(25)	5	(1)	100
Playboy	38	(324)	35	(100)	17	(169)	8	(121)	2	(16)	730
Time	19	(1)	23	(4)	30	(0)	19	(0)	9	(9)	14
Reader's Digest	13	(49)	20	(25)	28	(4)	26	(49)	13	(49)	176

value of 24. Other magazines in Tables 18–4 and 18–5 provide the best match for household income and age in the 35mm camera market.

The total squared differences for each demographic variable are summarized in Table 18–5, which also ranks the print media, with the lowest difference being the closest match. As indicated above, for occupation status, *Playboy* is the closest match to the 35mm camera profile. For income comparisons, Reader's Digest is the closest match; and for age, *Time* magazine is the closest match. Summing the total differences for all three demographic variables indicates that *Better Homes and Gardens* would be the best match for the advertising of the new camera. However there are some critical factors that should be discussed relative to this example. First we have assumed that the three demographic variables are the most important in buying this product. Second we do not know the relative weights of each demographic variable. If occupation and income were weighted the heaviest for the camera market, then *Playboy* and *Reader's Digest* would clearly be the top choices. Therefore the firm should carefully determine the relevant profile characteristics and weight the variables based on their own experience in this market.

The allocation of the budget for print media advertising can also use the information from profile matching. The top-ranked media would receive the

TABLE 18–5

Sum of squared differences for three demographic variables

Magazine	Total Squared Differences						Sum of Squared Differences	Rank
	Occupation	Rank	Income	Rank	Age	Rank		
Better Homes and Gardens	46	(3)	50	(3)	114	(3)	210	(1)
Esquire	85	(4)	144	(4)	100	(2)	329	(3)
Playboy	24	(1)	16	(2)	730	(5)	770	(5)
Time	178	(5)	214	(5)	14	(1)	406	(4)
Reader's Digest	35	(2)	6	(1)	176	(4)	217	(2)

highest proportion, and the lower-ranked media lesser amounts. Or lower-ranked media may be eliminated from consideration altogether.

The matching process may be used not only to compare magazine versus magazine and TV program versus TV program, but also for intermedia comparisons, such as magazine versus TV program. As the variables and the media involved increase, the calculations will become more difficult. Computer programs, however, may be easily developed to accommodate this simple algorithm.

Pretesting Advertising

One of the principal occasions when pretests are considered desirable by advertisers is when a new product or brand is introduced or is under consideration. The risk or uncertainty involved in this situation generally necessitates some pretesting.

To pretest either a single advertisement or alternative advertisements, it is important to establish a normal setting for conducting research. A normal setting for pretesting research is generally believed to possess three characteristics:[13]

1. Those individuals participating in the test are representative of the audience or market for the product.
2. The attention given to the advertisements tested is equivalent to regular advertising. The mere existence of the research can cause the respondents to pay more attention to the commercials.
3. The context of the test commercials is the same as the regular advertising. This implies use of the same medium, with the same form (color, content, and finish).

Developing a normal test environment is very difficult and can be extremely costly and time-consuming for the company. Thus pretesting may not be appropriate in every instance, since the advantage gained from the results could be negated by the costs. As a result of the problems in developing normal settings, most firms choose artificial contexts for pretesting, with the idea that this limited information will at least be valuable in company alternative approaches.

A variety of artificial testing methods can be used to pretest commercials in various media. For television ads, a group of consumers selected to match the target market would view regular programming with inserted test ads for the new product. They would then be interviewed and asked to complete a questionnaire, which would provide management with information on the most effective ad. Initial objectives would also be considered in evaluating the results.

For print media, individuals may simply be shown the ads and then asked to comment or give their reaction. A more effective technique is to leave dummy magazines or newspapers, in which the ads actually appear, with a sample of respondents. Again follow-up interviews would be used to analyze the effectiveness of each ad.

---

[13]Harry D. Wolfe et al., *Pretesting Advertising* (New York: The Conference Board, 1968), 12.

Like in the television medium, pretesting radio ads would involve presenting the ads as part of the recorded programming. As with television commercials, radio ads can be tested with or without programming.

Since the cost of the final television commercial can be extremely costly, some firms have used animated versions to determine specific reactions from customers. The animated versions avoid the cost of actors and photographers and long shooting sessions. However an animated version may not provide the type of realism necessary to determine the actual effectiveness of the commercial.

For new product introduction it is valuable for management to test alternative commercials in order to identify the approach that will best meet the firm's marketing objectives. When little prior experience exists, this procedure can prove to be informational and educational to marketing management.

SUMMARY

The decisions concerning promotion for the new product may be difficult and complex. The complexity of promotional mix decisions is dependent on the product's "newness" and management's experience. Where no internal expertise exists, the firm may choose to engage an advertising agency that will aid in making promotional mix decisions, ranging from setting the budget to designing the ads.

Management must be aware of the important criteria to be used when selecting an advertising agency. A checklist for selecting the agency can be used in the evaluation process.

This chapter has provided some insight into the significance of publicity in communicating the introduction of a new product. News releases are a means of gaining publicity for new products.

The product life cycle represents a valuable tool for studying and modifying the promotional mix strategy for the new product. The promotional mix strategy is also affected by the type of new product—that is, high learning versus low learning—particularly in the introduction and growth stages. The promotional strategies over the life cycles of consumer products and industrial products are illustrated in this chapter.

Marketers must position a new product either close to or distinct from competition in order to optimize its opportunities. Strategies for positioning a new product using promotion that may enhance the success of a new product introduction are as follows:

1. Using direct comparative advertising where the new product is shown in ads beside competitive products and comparisons are made.
2. Varying the positioning by seeking a different approach than competition and thus avoiding direct comparisons.

Two of the most difficult promotion decisions to make are the selection of media and subsequent allocation of the promotional budget. By matching media (both within the same medium and between two different media), the firm may optimize its allocation of the promotional budget. A distance formula may be used to match profiles of the market for a new product with media audiences.

SELECTED READINGS

BARTOS, RENA. "Ads that Irritate May Erode Trust in Advertised Brand." *Harvard Business Review* 59 (July–August 1981): 138–140.
  Provides evidence to illustrate how an increasing dislike of ads can lead to a decrease in loyalty to an advertised brand. Points out that a shift to more positive and entertaining advertising in Great Britain has led to increased public approval of the industry and toward the advertised brand.

BODDEWYN, J. J. "Advertising Regulation in the 1980's: The Underlying Global Forces." *Journal of Marketing* 46 (Winter 1982): 27–35.
  Argues that new media technologies, privacy and fairness issues, environmentalism, religion, changing economic conditions, the deregulation movement, and foreign regulatory initiatives will shape advertising regulation during the 1980s.

CANNON, HUGH M., and MERZ, G. RUSSELL. "A New Role for Psychographics in Media Selection." *Journal of Advertising* 9 (Spring 1980): 33–36.
  Proposes the use of psychographics for indirect matching of the target market with media users. This method is believed to be more effective than indirect matching, particularly with demographic profiles.

"Co-op Advertising." *Advertising Age* 52, 17 August 1981, S1–S16. Special section provides an in-depth discussion of how firms can use Co-op advertising more effectively. This particular issue may have a significant impact on new product introductions. This section presents the advantages of each mass medium for Co-op dollars and FTC guidelines.

ENGEL, JAMES F., WARSHAW, MARTIN R., and KINNEAR, THOMAS C. *Promotional Strategy.* 4th ed. Homewood, Ill.: Richard D. Irwin, Inc., 1979.
  Differs from many books in this area by building on a rigorous base of consumer behavior. Treats advertising, sales promotion, reseller stimulation, personal selling, and other communication tools as part of an overall promotion mix. Integrates theory with managerial issues and problems.

HURWOOD, DAVID L., and BROWN, JAMES K. *Some Guidelines for Advertising Budgeting.* New York: The Conference Board, 1972.
  A useful report for those responsible for advertising budgets or researchers who provide data for the budgeting process. Discusses many widely used budgeting methods and provides a framework for budgeting decisions and research.

KELLY, PATRICK. "The Ad Budget." *Marketing and Media Decisions* (January 1982): 53–55, 136–137.
  Summarizes a survey of key marketing/media executives in the top 200 brand advertisers regarding advertising budget. Argues that, besides being a team project with both company and agency specialists, budgeting is becoming a year-round activity.

KINCAID, WILLIAM M., JR., *Promotion: Products, Services and Ideas.* Columbus, Ohio: Charles E. Merrill Publishing Co., 1981.
  Provides an excellent conceptualization of the "tools" of promotion: advertising, personal selling, sales promotion, and publicity. The book is management-oriented, focusing on the management of the promotion function, rather than the creative aspects associated with its devices.

RAY, MICHAEL L. *Advertising and Communication Management.* Englewood Cliffs, N.J.: Prentice-Hall, Inc., 1982.

Provides a direct and simple discussion of the basics in the communication process. It covers the details of advertising planning and creative and media decisions that will provide managers with a clearer understanding of how to develop effective communication.

VII

Controlling and Managing the Product Line

Controlling the
New Product

T

he marketing program described in Chapters 15 through 18 is subject to many controllable and uncontrollable variables. This chapter discusses the importance of monitoring and managing these variables and their effect on the new product during the stages of its life cycle.

NATURE OF CONTROL

Control is the means by which management assures achievement of the objectives of the marketing plan and takes the necessary steps to bring the actual results closer to the desired results. Management must establish standards for comparing the actual results of the marketing mix strategy with the planned results. These performance standards are used to identify needed marketing strategy modifications. The control process must be active throughout the product life cycle.

Control and the Product Life Cycle

The amount of control in each stage in the product life cycle will differ for firms operating in different markets. Nevertheless every firm must establish a system for monitoring market characteristics to ensure success in meeting market objectives. Table 19–1 describes some of the elements considered to be critical information in monitoring and controlling the new product.

In the early stages management must be particularly concerned with how well the product performs physically and whether the market segments are correctly identified. During the growth stage, concern must be given to

TABLE 19–1
Control during the product life cycle

Stage of Product Life Cycle	Probable Control Problems
Introduction	Emphasis should be on developing a system to identify product engineering problems and market segments.
Growth Stage	Particular concern is given to the brand position, gaps in market coverage, new segments, and competition positioning.
Maturity	Repurchase rate, product improvement, market expansion, and new promotional techniques are the most critical concerns in this stage.
Saturation	Symptoms for product decline should be closely watched. Possible product improvements, new uses, and new market segments should be evaluated.
Decline	Information is needed to determine whether product should be withdrawn or dropped from product line.

competitive positioning and possible gaps in market coverage. The position of the new product and competitive brands will indicate how well the firm has achieved product differentiation.

As the product moves through the maturity and saturation stages, management must carefully monitor sales, profits, and costs. Efforts must be extended to find new uses, new users, or product modifications to stimulate sales. It is also during these later stages that a system must be designed to identify the critical turning point at which a product should be phased out. Marketing activities for the mature product are discussed further in Chapter 20. Product deletion is a critical decision and can be implemented at any stage in the product life cycle. Although they are costly, early product deletions may prevent more serious problems in the long run.

The performance standards will change at any stage in the product life cycle because of changes in the competitive environment. A monitoring and control system is important regardless of whether the new product is a high-learning or low-learning product.

THE CONTROL PROCESS

Figure 19–1 shows the process of establishing a control system for a new product. Each of the stages or control activities is interrelated and therefore must be coordinated when designing this system. Incorrect decisions at any stage may cause new product failure because these activities are so dependent on each other. For example, if the company fails to consider the growth rate of the industry in identifying the key marketing variables, it could assume that the new product is successful, since product sales are increasing. However

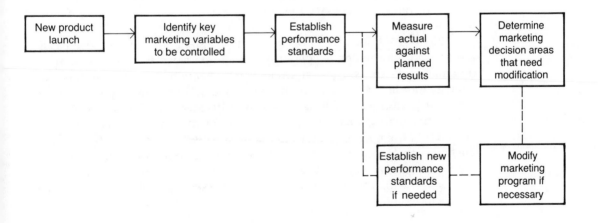

FIGURE 19–1
The control process for new products

this rate is less than that for the industry. A low rate of growth in product sales compared to industry sales may be indicative of poor product positioning.

Identifying the Marketing Activities to be Controlled

Management must determine which marketing activities are most critical to achieving a successful product launch. Particular areas should be identified, and then area managers should be asked for input about specific factors within their responsibility that they feel would significantly contribute to the success of the new product launch. Some of the most obvious areas to be controlled are sales, market share, profit contribution, marketing program costs, intensity of distribution, consumer awareness, and competitive activities. Each of these may be assigned priorities depending on the marketing objectives. The relevant variables may be more specifically defined by the area managers.

Establishing Performance Standards

Once the key performance areas have been determined, it is necessary to set standards. These standards should be in some measurable unit, such as 10 percent of market share at the end of first year after product launch.

It is important at this point to reflect on the strategic and marketing plans that establish objectives and the corporate mission (long term) as well as the activities and strategies (short term) that will take place within the framework of the strategic plan. These activities and strategies, within budget constraints, can be reflected in performance standards used for the control process.

For new products in the critical introduction stage, performance should be monitored in short intervals (monthly or quarterly). As the product moves on to another stage of the life cycle, the performance standards should change. If the new product is not meeting its standards, the problems must be diagnosed so that the necessary modifications can be made at an early point in the life cycle.

Figure 19–2 illustrates how the control system operates. It provides all of the needed information for performance comparisons as well as written documentation of the suggested strategy modification. The measurement and strategy modification decisions are discussed in the following paragraphs.

The performance standards in Figure 19–2 were established by the members of the management team responsible for the decision areas. It was expected that the new product would achieve a 10% market share by the end of the first year. At the end of six months, the new product was expected to reach 5 percent market share. These estimates were based on test market data, experience with other similar products, market research data, or management's judgment.

Measurement of Expected Versus Actual Results

The timing for measurement of expected and actual performance will vary depending on the newness of the product and management's knowledge of market conditions. In a high-risk market, where competition and consumer

	Six months			Twelve months		
Objectives for First Year	Expected	Actual	Corrective Strategy	Expected	Actual	Corrective Strategy
1. Achieve 10% market share	5%	3%	Increase advertising by 5%	10%	8%	High awareness but low market share may indicate poor product performance. Consider possible product modification based on carefully designed research.
2. Achieve 25% intensity of distribution at retail	15%	12%	Provide better margins and incentives to distributors and contact more wholesalers	15%	15%	Product location in store may be a problem. Continue plan to develop better relationship with retailers.
3. Reach 50% awareness in target market	30%	25%	Modify advertising appeal to target market	50%	60%	Advertising may need to express utility of product, as well as awareness.
4. Achieve marketing costs of 50% of sales	60%	60%	Maintain close cost controls, particularly with increases in advertising and higher margins to distributors	50%	60%	Determine highs and cost marketing items. Establish tighter cost controls and efficiencies with marketing expenses.

FIGURE 19–2

Example of monitoring new product launch in year 1

tastes are changing quickly, management may need to evaluate performance monthly. It is recommended—even for low-risk new products—that evaluations be considered on a quarterly basis.

In Figure 19–2 the actual results were less than expected for three of the objectives. Market share, intensity of distribution, and consumer awareness were below the standard established for the first six months. The lower-than-expected market share may be explained by the low distribution coverage, as well as the low consumer awareness. In many instances, the criteria for measurement and control are related, so that a change in one area for the marketing plan may have a positive impact on all the interrelated marketing mix variables.

In measuring the performance, it is necessary to identify sources of quantifiable information that will provide bases for comparison. Market share measurement requires some estimate of total industry sales. Secondary sources, such as trade data or government reports, provide some data on industry sales. In order to determine the intensity of distribution, shipping invoices, salespeople's reports, or wholesaler's records could be used. Consumer awareness may be more difficult to measure on a frequent basis. Even though this information can be derived from market research studies, the cost of such research may prohibit its occurrence more than once a year.

Marketing cost data should be readily available from accounting records. Each marketing cost item should be evaluated separately to determine the specific problems and the subsequent action.

Corrective Strategy

When actual performance deviates significantly from expected performance, corrective action is warranted. Figure 19–2 describes the corrective action at the six-month checkpoint and at the end of the first year. In order to correct the gaps in market share, management increased advertising by 5 percent. At the same time, it was determined that consumer awareness was below the desired standards, so the advertising plan should be modified to appeal more directly to the target market(s).

The intensity of distribution (12 percent) may have contributed to the low market share. In order to increase the intensity of distribution, greater incentives may be used to spur distributors to obtain new retail accounts. Perhaps higher margins to the distributors will be passed on to the retailer, thus developing more interest in carrying the new product.

All corrective actions for items 1 through 3 in Figure 19–2 will have an impact on marketing costs. The corrective action should cause an increase in sales, which is disproportionate to the increase in marketing costs, so that all objectives are satisfied. With corrective action taken for the first items, management must pay close attention to any changes in the ratio of marketing costs to sales.

Performance standards should not be rigid because the environment in which the firm operates is subject to change. Sudden changes in the environment, such as the energy crisis, would necessitate a change in performance standards. Thus the purpose of the control system is not merely to measure and evaluate performance differences from the plan but also to accommodate environmental shifts by modifying the standards. For example, in Figure 19–2, the actual market share at the end of six months was 3 percent—significantly lower than the expected level of 5 percent. The corrective action may be the solution to closing the performance gap, but it is also conceivable, given information about the environment, that the standard is unreasonable and should be lowered.

At the end of the twelve-month period, the actual performance improved somewhat but was still below standard for market share. In addition marketing costs did not decrease to the desired 50 percent of sales. Surprisingly consumer awareness had increased to 60 percent, exceeding the expected level. Management felt that advertising may not have sufficiently expressed the utility or benefits of the product, which may have had a negative effect on sales. At the end of the first year, the major problems appear to be inefficiencies in distribution and advertising strategy resulting in high marketing costs and low yield in reaching the desired market. Other reasons or possible problem areas should also be explored so that other corrective action may be implemented. For example sales people may need more training, or new marketing personnel may be needed to achieve marketing goals.

It is rare to find a situation where a firm has been able to design an optimal marketing plan for its new product. Marketing management is a process of human activities and decisions that attempts to forecast and react to changes in the competitive environment. The control process exhibited in Figure 19–1 provides a mechanism to reduce error in judgment, as well as to establish an adaptive system to ensure the success of the new product. The control process has significant responsibility in the new product management process.

PRODUCT ABANDONMENT: A CONTROL APPLICATION

All products eventually reach the point in their life cycle where they should be eliminated from the product line. This decision is often a difficult one for many firms to make because the product had been a part of the company for such a long time that it has acquired some sentimental value. In addition the impact of the elimination on other products in the product line must be considered. The relationship of the one product to others is often difficult to assess and complicates the elimination decision.

The product abandonment decision is becoming more critical today than in the past, primarily because of environmental circumstances, such as raw material shortages, high interest, unemployment, and shrinking markets. In the past a firm would retain a product longer because there were sufficient raw materials, and the product may have been meeting minimum goal requirements. This option is not likely to exist in most industries today, which should make the product abandonment decision a more active element in marketing strategy. Most firms market more than one product; this requires a systematic approach to introducing or eliminating products. Too many weak products can result in the thinning of resources among all existing products, which can adversely affect potentially successful products. Resources appropriated for weak products should be diverted to work more efficiently for successful products that could enhance the company's overall profit position.[1]

The resources absorbed by the weak product(s) may lead to delay in the development of profitable new products. Competitive strength may be reduced because of failure to provide an adequate system of controlling weak products.

Experience has shown that tremendous savings can result by eliminating marginal products. For example, Hunt Foods reduced its product line from over 30 to 3 over an eleven-year period, but still increased sales from $15 million to $120 million.[2] Another firm achieved similar success by eliminating 16 products with a total sales volume of $3.3 million, but increasing company-wide sales by 5 percent and profits by twenty times.[3] However these examples are limited to a few aggressive firms. Typically companies only undertake a pruning of the product line when a crisis or near crisis occurs, such as drastic reductions in profits or sales. In fact it is quite reasonable to conclude that new product planning and development activity far exceeds the managerial

---

[1]For a good overview of the product abandonment decision, see Philip Kotler, "Phasing Out Weak Products," *Harvard Business Review* 43, no. 2 (March–April 1965): 108–18.
[2]Ibid., 109.
[3]Ibid., 110.

activity devoted to the product abandonment decision.[4] Management must give more attention to assessing the contributions of such members in the product line to the profitability of all the company's products, new and mature.

Despite the obvious need for the careful analysis of weak products, sentiments and rationalizing often cause management to retain weak, mature products. Philip Kotler cites the following factors that contribute to this aversion to abandoning a product:

> Sometimes it is expected—or hoped—that product sales will pick up in the course of time when the economic or market factors become more propitious.
> Sometimes the fault is thought to lie in the marketing program, which the company plans to revitalize.
> Even when the marketing program is thought to be competent, management may feel that the solution lies in product modification.
> When none of these explanations exist, a weak product may nevertheless be retained in the mix because of the alleged contribution it makes to the sales of the company's other products. If none of these functions are performed by the weak product, then the retention rationale may be that its sales volume covers more than just actual costs, and the company temporarily has no better way of keeping its fixed resources employed.[5]

Even though these seem to be logical reasons for retaining a weak product, they may be excuses for vested or sentimental interests. While these rationales may be tenable in the short run, they will eventually fail to conceal the low profitability and contribution of the weak products. It is unfortunate that the inevitable is so often delayed, since the company tends to incur significant costs by retaining weak products.

Techniques for Product Deletion

Several approaches have been suggested for the deletion or abandonment decision. Early publications were almost completely subjective and provided suggestions as to what factors needed to be evaluated periodically in conducting a product line analysis.[6] Some numerical assessments were made by assigning weights to each criterion, but these assignments were based on managerial judgment.

The Kotler Control System

Philip Kotler's control system, illustrated in Figure 19–3, provides a significant contribution to the product deletion decision. The procedure is performed by

---

[4]See James T. Rothe, "The Product Elimination Decision," *MSU Business Topics* (Autumn 1970): 45, and Richard T. Hise and Michael A. McGinnis, "Product Elimination: Practices, Policies and Ethics," *Business Horizons* (June 1975): 25–32.
[5]Kotler, "Phasing Out Weak Products," 110.
[6]See D'Orsey Hurst, "Criteria for Evaluating Existing Products and Product Lines," in *Analyzing and Improving Marketing Performances*, AMA Management Report No. 32 (New York: American Management Association, 1959), 91. Also R. S. Alexander, "The Death and Burial of 'Sick' Products," *Journal of Marketing* 24, no. 2 (April 1964): 1.

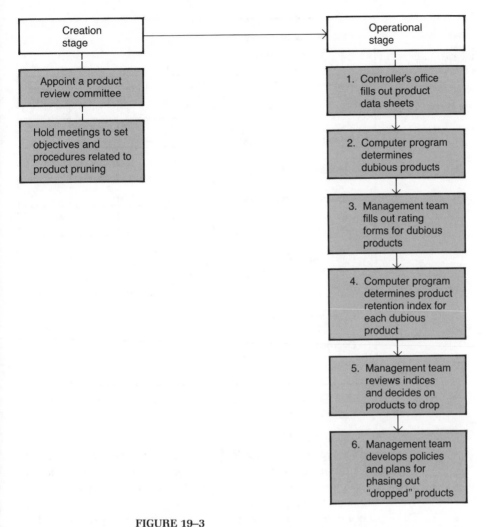

FIGURE 19–3
The Kotler control system
Source: Philip Kotler, "Phasing Out Weak Products," *Harvard Business Review* (March–April 1965): 110.

a team of managers from various functional areas and is formal and systematic, ultimately yielding an index that indicates the product's weakness.[7]

The creation stage of the Kotler system consists of two steps. The first is the development of objectives and procedures for reviewing the product line. The second involves meetings to define the factors that should be considered and procedures to be used in the analysis. The format and procedures implemented by this management team may vary, depending on the size of the company and the industry in which it operates.

The operational stage of the control system has six steps:

1. The controller's office prepares a data sheet for every product. (An example of this data sheet is shown in Figure 19–4.)

---

[7]Kotler, "Phasing Out Weak Products," 112.

FIGURE 19–4

Product data sheet

Source: Philip Kotler, "Phasing Out Weak Products," *Harvard Business Review* (March–April 1965): 113.

Product_____				
Date_____				
Criteria	Past Years			Current Years
	3	2	1	
Industry sales ($)				
Company sales ($)				
Physical volume				
Unit total cost ($)				
Unit variable cost ($)				
Price ($)				
Cyclical adjustment factor				
Overhead burden				
Comments:				

2. Identify which products should be further analyzed by considering such information as declining gross margin and declining sales relative to company and industry sales. A computer program can be written to scan these data to identify weak products.

3. Develop comprehensive criteria to evaluate the weak products identified in step 2. These criteria are determined from meetings of the management team and are incorporated into a rating form illustrated in Figure 19–5. Each criterion is also assigned a weight by management. Members of the committee must then agree on a rating of each question.

4. Calculate a product retention index based on the summation of the weights times the ratings for each criterion (see Figure 19–5).

5. The committee reviews each product retention index. Those products with indices below the minimum established by the committee may be eliminated, provided other justifications are also met. Marginal products must be carefully analyzed before a final decision is made.

6. Management must establish a procedure for phasing out the product. Problems such as replacement parts, sale of machinery and unfinished stock, and compensation for middlemen must be solved. Schedules for the phasing out procedure may be prepared by using critical-path programs, such as PERT.

Weight *(W)*	Criteria	Rating *(R)*	
1	What is the future market potential for this product?	———————————— .0 .2 .4 .6 .8 1.0 Low High	$W_1R_1 =$
2	How much could be gained by product modification?	———————————— .0 .2 .4 .6 .8 1.0 Nothing A great deal	$W_2R_2 =$
3	How much could be gained by marketing strategy modification?	———————————— .0 .2 .4 .6 .8 1.0 Nothing A great deal	$W_3R_3 =$
4	How much useful executive time could be released by abandoning this product?	———————————— .0 .2 .4 .6 .8 1.0 A great Very deal little	$W_4R_4 =$
5	How good are the firm's alternative opportunities?	———————————— .0 .2 .4 .6 .8 1.0 Very Very good poor	$W_5R_5 =$
6	How much is the product contributing beyond its direct costs?	———————————— .0 .2 .4 .6 .8 1.0 Nothing A great deal	$W_6R_6 =$
7	How much is the product contributing to the sale of the other products?	———————————— .0 .2 .4 .6 .8 1.0 Nothing A great deal	$W_7R_7 =$
		Product retention index*	————

*All *WR*'s are summed to obtain product retention index. The higher the index the greater the arguments for retaining the product.

FIGURE 19–5
Rating form
Source: Philip Kotler, "Phasing Out Weak Products," *Harvard Business Review* (March–April 1965): 114.

The Kotler Control Process provides a systematic procedure for eliminating weak products. It does have some weaknesses of its own, however. For example the management team responsible for determining criteria and weights also rates the criteria that may affect the committee's objectivity. Consensus may minimize this problem if committee members have equal power in the company.

Product Review and Evaluation Subsystem (PRESS)

The PRESS model to delete weak products was developed by Hamelman and Mazze and consists of four integrated parts: PRESS I applies standard cost accounting and marketing performance data; PRESS II, III, and IV are concerned with price changes, sales trends, and product interaction.[8]

PRESS 1: Cost analysis This model is based on a variable cost accounting procedure, which includes standard cost, unit price, and volume for the most recent period. Only those costs that specifically contribute to the production and sales of a product are included. Allocation of fixed costs is not included because of the difficulty in determining their impact on any one product and the fact that they are "sunk" costs and are not affected by the abandonment decision. A sample listing of costs for three hypothetical products is illustrated in Table 19–2.

In this model contribution margin is the primary criterion for comparing the several products in the line. However it does not provide any information about the resources required to attain that contribution. A more comprehensive measure should include an adjustment for the use of resources to give a comparison of return on investment rather than on earnings only. All costs, where easily attributable to each product, should be included in this analysis.

Each product is compared in the PRESS 1 model by using a selection index number (SIN). The formula determining the SIN is as follows:

$$SIN_i = \frac{CM_i/\Sigma CM_i}{FC_i/\Sigma FC_i} \cdot (CM_i/\Sigma CM_i)$$

where

SIN_i = selection index number for product i

CM_i = contribution margin for product i

FC_i = facilities' costs for product i

ΣCM_i = summation of contribution margin for all products

ΣFC_i = summation of facilities' costs of all products

Table 19–3 shows the SIN indices for products A, B, and C. It also includes the total contribution margin (unit contribution × units sold), product contribution margin as a percentage of the three products' contribution margin, cost of facilities utilized, and percent of the firm's resources used to produce each product. The products are arranged by the order of their SIN value. The lower the SIN number, the more reason to consider the product for deletion. A SIN number lower than 1.00 is a definite sign that the product is marginal. Thus Product C in Table 19–3 is a candidate for deletion, provided that there are substitutes, negative product image, or possible negative impact on other products in the line.

---

[8]For a complete description of the model, see Paul W. Hamelman and Edward M. Mazze, "Improving Product Abandonment Decisions," *Journal of Marketing* 36 (April 1972): 20–26.

TABLE 19–2
Standard costs, unit price, and volume data

Product	Unit Material Costs	Unit Labor Costs	Unit Variable Overhead Costs	Unit Sales Price	Unit Quantity Sold	Unit[a] Variable S and A
A	4.00	3.00	1.80	22.00	1,500,000	2.00
B	1.50	1.00	.50	12.00	2,000,000	2.00
C	6.00	3.00	5.00	29.00	200,000	4.00

[a]Includes advertising, salespeople and other administrative and selling costs that can be attributed to each product.

Source: Paul W. Hamelman and Edward W. Mazze, "Improving Product Abandonment Decisions," *Journal of Marketing* 36 (April 1972): 22.

PRESS II: Price volume relationships   PRESS II determines the effect of a price change on the marginal products identified by PRESS I. It is possible that a price increase of Product C may increase marginal contribution with little reduction in total units sold. This situation would improve the SIN index and perhaps give new life to the marginal product. Again, all other factors mentioned here must be considered before final judgment is made on the deletion of Product C.

One major problem with PRESS II is that sensitivity of demand to price increases cannot always be readily determined. Competitive actions may prevent the declining product from attaining a better position, thus limiting the probability for improving the failing product's marginal contribution.

PRESS III: Sales trends   It is possible that Product C in Table 19–3 is in the early stages of its life cycle and could conceivably become a good performer in the near future. The PRESS III subsystem forecasts future demand for marginal products by extrapolating from historical sales data using an exponentially weighted moving average.

Certain variables are assumed by PRESS III to remain constant. These are selling price, variable costs per unit, and selling and administrative costs. The

TABLE 19–3
Rankings of products

Product	Total Contribution Margin	% of Total Contribution Margin	Costs of Facilities Used	% of Costs of Facilities Used	SIN
A	16.8 M	52.8	400,000	47	59.32
B	14.0 M	44.1	300,000	35	55.57
C	1.0 M	3.1	150,000	18	.53
	31.8 M	100.0%	850,000		

Source: Paul W. Hamelman and Edward M. Mazze, "Improving Product Abandonment Decisions," *Journal of Marketing* 36 (April 1972): 23.

projections that result from PRESS III can be used for a reiteration of the PRESS I model to determine whether the SIN has improved.

PRESS IV: Product complementarity and substitutability It is likely that any product in a product line has some effect on the sales of another product in the line. The PRESS IV subprogram assesses the volume of sales of other products attributable to sales of the marginal product. This information may be provided by salespeople, product managers, and other members of management who are familiar with the product line. If the impact of the marginal product is considered minimal, then PRESS IV may be eliminated.

Since Product C would be deleted without PRESS IV, where some substitution of one product for another occurs, the PRESS IV program calculates a factor called *RESIN*, which is the original SIN adjusted for tie-in sales associated with the marginal product. If the marginal product, Product C, is deleted, it will be necessary to calculate a new SIN for Products A and B if there are any tie-in sales.

PRESS I provides valuable insight into the contribution of each product and gives management an index to evaluate marginal products. The major problem with the PRESS model exists in PRESS II, III, and IV, where a great deal of subjective evaluation is applied. For example, PRESS II's assumptions regarding the estimation of the sensitivity of demand to price changes and PRESS III's assumption of management's ability to forecast sales can result in weaknesses in the total PRESS program. PRESS IV requires careful analysis, since it may be difficult to estimate the sales associated with any particular products. Nevertheless the PRESS program is a valuable starting point for management to implement product deletions.

Strategic Intelligence System

The prior discussion on control of a new product focused on performance variables, such as market share and sales volume that were internal to the firm. It is also necessary in the control process to monitor and forecast changes in economic, social, technological, governmental, competitive, and physical factors that may affect a firm's long-term and short-term planning cycle. The firm's strategic intelligence system is concerned with monitoring these external environmental variables that could affect performance variables and the control process.

It is obviously impossible for any firm to attempt to change or control these external variables, but monitoring and forecasting activities related to these variables could provide valuable information to support a new product idea, a specific marketing program, or a change in long-term objectives.

A strategic intelligence system may range from a formal information system, such as a subscription service (of industry trends data), to an informal surveillance of the marketing environment (a review of possible new legislation in committee). Whatever the formality of the approach, the information represents an important basis for determining a course of action or identifying an opportunity.

Since the function of a control system is to determine performance (expected versus actual), strategic intelligence can provide important information for corrective action. An example of informal strategic intelligence substantiates the significance of such a system. A firm noticed that a competitor was modifying its selling efforts by eliminating some of its manufacturer's reps, who used to supplement its sales force in selected territories. Upon investigation it was found that the competitor was planning to replace the reps with its own sales force. When they learned of the new strategy, many reps dropped the product line and sought alternative suppliers. The competitor's slow reaction to the changeover enabled the firm to increase its distribution using these reps and to increase its market share.

More formal systems, such as industry trends provided by subscription service organizations, can also benefit a firm in controlling its marketing plan. For example research data on the wine and spirits market indicated a strong trend toward lighter spirits. Development of 80-proof whiskey and the reallocation of marketing efforts toward lighter spirits (such as gin and vodka) enabled the firm to increase its total sales.

The competitive analysis may be regarded by some as that external variable requiring the most attention. Sources related to what competitors say publicly about themselves, what competitors actually do, and what others say about the competitors can be easily evaluated or deduced.[9]

The key to any strategic intelligence system is how the information (formal or informal) is used. It is argued here that this information provides an extension of the internal performance evaluation provided by the control process and hence provides a more valid indication of appropriate performance standards or corrective actions.

SUMMARY

Implementation of marketing mix strategy in a new product launch requires a monitoring system to determine the effectiveness of the program. In order to achieve optimal profitability, a control system for the new product during its launch and at subsequent points in the product's life cycle is necessary.

The control process consists of the following steps:

1. Identify key marketing variables to be monitored.
2. Establish performance standards.
3. Measure actual results against planned results.
4. Determine marketing mix modifications.
5. Modify marketing program.
6. Establish new performance standards if necessary.
7. Repeat measure of actual and planned results at specified time intervals.

This chapter discussed how to design a control system and implement corrective marketing strategy.

One of the most difficult and complicated decisions faced by management is product abandonment. Two widely used methods for making the

[9]See David B. Montgomery and Charles B. Weinberg, "Toward Strategic Intelligence Systems" *Journal of Marketing* 43, no. 4 (Fall 1979): 41–52, and William E. Rothschild, "Comment," *Journal of Marketing* 43, no. 4 (Fall 1979): 53–54.

product abandonment decision are the Kotler system and the Hamelman and Mazze PRESS program. The Kotler system consists of six steps, beginning with preparation of product data sheets. Then the system identifies marginal products based on declining gross margin or sales. It develops criteria for evaluating a product and calculates the product retention index. The indices are reviewed by management, which finally makes the decision to phase out the product. The Kotler process provides a systematic procedure for making the product abandonment decision.

The Product Review and Evaluation Subsystem (PRESS) provides a computer analysis using standard cost accounting to determine the contribution of each product. PRESS I is the model that calculates the selection index number (SIN) to identify marginal products. PRESS II determines the effect of a price change on the demand and contribution margin of the marginal product(s). PRESS III forecasts the future demand for the marginal product to determine if the product's SIN can be improved. PRESS IV determines the impact of sales on all products in the product line if the marginal product(s) is deleted. A new factor called *RESIN*, which is the original SIN adjusted for tie-in sales due to the marginal product, is calculated.

The major limitations of all processes for product abandonment decisions are management's ability to identify and allocate costs to all products and to apply subjective criteria to the product's analysis.

Despite inherent weaknesses, a control process must be implemented by management to monitor new products, as well as products near the end of their life cycle. Insufficient control can result in inefficiencies, which can lead to low profits. Consistent new product failures or a continuation of low or negative profits is likely to force a firm into bankruptcy.

SELECTED READINGS

HULBERT, JAMES M., and TOY, NORMAN E. "A Strategic Framework for Marketing Control." *Journal of Marketing* 41 (April 1977): 12–20.
Outlines a strategic framework for marketing control. States that since the evaluation of performance against plan does not include unanticipated events during the planning period, it is necessary to provide a conceptual framework to take planning variances into account.

KOTLER, PHILIP. "Phasing Out Weak Products." *Harvard Business Review* 43, no. 2 (March–April 1965): 108–18.
One of the most widely referenced articles relating to the problem of control of the product mix. Details and discusses basic steps of phasing out products.

KOTLER, PHILIP, GREGOR, WILLIAM, and RODGERS, WILLIAM. "The Marketing Audit Comes of Age." *Sloan Management Review* 18 (Winter 1977): 25–44.
Describes how the marketing audit has become an increasingly useful tool for managers to evaluate and improve a company's marketing operations. Explains how it is used to assess the effectiveness of marketing activities against opportunities, objectives, and resources.

LEAF, ROBIN H. "Learning From Your Competitors." *McKinsey Quarterly* (Spring 1978): 52–60.

Argues that part of the control process should include an analysis of competitive strategy. Stresses that effective use of competitive analysis requires special organizational devices. Provides examples of three United Kingdom-based firms to illustrate how competitive analysis can be used to control and effectively develop marketing strategies.

MONTGOMERY, DAVID B., and WEINBERG, CHARLES B. "Toward Strategic Intelligence Systems." *Journal of Marketing* 43 (Fall 1979): 41–52.
States that with the increase in use of strategic planning it is necessary to avoid collecting meaningless data. A strategic intelligence system will provide a mechanism for assuring control over information and long-term marketing strategy.

"Report of the Committee on Cost and Profitability Analysis for Marketing." *Accounting Review* Supplement to 47 (1972): 575–615.
Provides a framework and recommended accounting practices for control of marketing costs. Focuses on differences between actual and planned total market size, actual and planned total market share, and price/cost per unit.

ROUSH, CHARLES H., and BALL, BEN C., JR. "Controlling the Implementation of Strategy." *Managerial Planning* (November–December 1980): 3–12.
Illustrates how serious problems in strategy implementation can be minimized through a strategic control system. Identifies six steps in the strategic control system illustrated in this article.

WORTHING, PARKER. "Improving Product Deletion Decision Making." *MSU Business Topics* (Summer 1975): 29–38.
Provides a summary of alternative approaches that can be used in the product deletion decision. The importance of this decision is emphasized and examples are provided to illustrate the decision process.

Managing the Product Mix and Mature Products

Previous discussions in this text have focused mainly on marketing decision-making relative to new products. The product development process from idea generation to commercialization requires careful market planning and decision-making in order to achieve the corporate objectives. Major marketing mix variables relevant to the introduction of any new product or service were discussed. Control systems necessary for market planning in the early and later stages of the product's life cycle were also presented. This chapter extends the discussion and analysis to the mature product and to relevant product line strategies and options that a firm may consider in order to maintain profitability and flow or market share.

THE PRODUCT LIFE CYCLE

It may be recalled from earlier chapters that each product can be viewed as passing through various stages during its market life. In each of these stages market conditions change, thereby necessitating appropriate adjustments in marketing strategy. This chapter focuses on the mature product, which is likely to be the longest stage in the product's life cycle.

Figure 20–1 illustrates some of the key market characteristics of various stages of the product life cycle. It can be seen from this figure that the more mature product is likely to involve more market, many channels, lowest price, lowest gross margins, many competitors, high incentives to consumer and trade, sophisticated segmentation, superior quality, and optimum capacity. Within these market characteristics management must defend the product against competition and look for new market segments, possible product modifications or spin-off products that can extend the product's life cycle and enhance profits. Innovations or efficiencies in production processes or marketing costs may also be considered for the mature product.

Chapter 19 presented control systems that can be used throughout the product life cycle. Continued monitoring of a product provides the basis for effective market strategy adjustments to maintain the goals and objectives set forth in the market/product plan. In addition this monitoring will inform management when the product is reaching maturity and whether it may need special attention.

During the late growth and early maturity stages, certain signs in market conditions will be evident. Increased competition will have likely squeezed profit margins and total profits. Some of the weaker competitors will begin to leave the marketplace if management of these firms finds other opportunities and directions for their product line. During transition to the mature market, supply will normally outpace demand, leaving the firm with surplus capacity. Price reductions by competitors, as well as an increased requirement of trade incentives for retailer support, will also become increasingly evident during this stage of the product life cycle. It is also likely that the product will not have been changed for a few years, which may be reason enough to consider change.

	Introduction	Growth	Maturity	Decline	Termination

Marketing

	Introduction	Growth	Maturity	Decline	Termination
Customers	Innovative/high income	High income/mass market	Mass market	Laggards special	Few
Channels	Few	Many	Many	Few	Few
Approach	Product	Label	Label	Specialized	Availability
Advertising	Awareness	Label superiority	Lowest price	Psycho-graphic	Sparse
Competitors	Few	Many	Many	Few	Few

Pricing

	Introduction	Growth	Maturity	Decline	Termination
Price	High	Lower	Lowest	Rising	High
Gross margins	High	Lower	Lowest	Low	Rising
Cost reductions	Few	Many	Slower	None	None
Incentives	Channel	Channel/consumer	Customer/channel	Channel	Channel

Product

	Introduction	Growth	Maturity	Decline	Termination
Configuration	Basic	Second generation	Segmented/sophisticated	Basic	Stripped
Quality	Poor	Good	Superior	Spotty	Minimal
Capacity	Over	Under	Optimum	Over	Over

FIGURE 20–1
Marketing characteristics during product life cycle
Source: John E. Smallwood, "The Product Life Cycle: A Key to Strategic Marketing Planning," *MSV Business Topics* 21 (Winter 1973): 31.

The key factor during maturity is effective market planning and strategy. Marketing skills during this stage will probably provide the edge needed to help the product enjoy "old age," even after many of its competitors have left the marketplace.[1] The strategic options available for a mature product may involve one or a combination of the following: new segmentation strategy, product redesign or modification, new uses for the product, or a change in marketing mix variables. Each of these will be discussed in the following paragraphs.

NEW MARKET SEGMENTS

During product maturity, management will usually find it necessary to use marketing research to identify significant customer segments, with attention focused on the differentiated needs of those segments. The effect of the pene-

[1]Donald K. Clifford, Jr., "Leverage in the Product Life Cycle," in *Product Planning*, 2nd ed., ed. A. Edward Spitz (New York: Petrocelli/Charter, 1977), 257–64.

tration of these new market segments is to extend the growth of the product beyond what might have been possible with the existing target markets.

The most common strategy to penetrate new market segments is to market a product in the international market or in some other market (e.g., introducing an industrial product into a consumer market). Monopoly is a children-to-adult board game that was modified and produced in many different languages in order to appeal to new international markets. Coke and Pepsi have recently made national news with their new market arrangements with China and the Soviet Union. Janitor-in-a-Drum is the most publicized example of taking an industrial product to the consumer market. In fact this particular strategy was not only quite successful, but the industrial experience of the firm, emphasized in packaging, branding, and advertising appeal (industrial strength), helped to differentiate the product from other consumer cleaning products.

The identification and expansion of a product into new market segments provides an alternative yet feasible strategy for a company faced with a declining market. The most common error made during the maturity stage is to attempt to emphasize growth through penetration of existing market segments. Although these efforts may be effective in the short term, by increasing sales and profits (through volume), the longer term offers serious problems for all firms in this particular market. Inevitably the competitors will not sit back if an aggressive growth strategy is taken by another firm. They will not be able to ignore the fact that one firm's growth strategy has come at the expense of market share and profits of other firms in this market.

The result of counter strategy by the competition is reduction of prices, increased marketing costs (advertising and trade incentives), and hence a lowering of profits for all of those firms competing in this market. A slower penetration/differentiation strategy supported by expansion into new market segments would have been much more successful in the long run for this firm.

It must be remembered that in the maturity stage the product has reached all of those consumers who might purchase it. If there are five major competitors in this market, all with about the same market share, an attempt at a growth strategy during the maturity stage, within the context of these same market segments, can only succeed at the expense of the market share of the competitors. It is possible that such a strategy could be developed with the purpose of shaking up the competitive environment or forcing the weaker firms out of the market. However it is extremely important for management to consider the counter strategies that the competition will develop. If they all have a reasonable market share it is unlikely that any one of them would drop out of the market. An analysis of these firms would be most appropriate to determine whether the product at hand supports other products in their product line or whether it is a cash cow or has some emotional meaning to them (e.g., first product ever introduced). Based on this information a more meaningful strategy can be developed.

Rather than face price wars, increased marketing costs, and lost profits, it generally makes sense for a firm to explore new market segments for the product. If the existing product cannot appeal to these new market segments, then it may be necessary to consider other alternatives, such as product modifications and spin-offs.

PRODUCT MODIFICATIONS AND SPIN-OFFS

Modifying a product to enhance consumer satisfaction can be used as an alternative to (or complement of) expanding into new market segments. It is also conceivable that a mature product may be the basis for spin-off products to take advantage of the historical success and image-building of the original product, as well as to flank the existing product with other alternatives that would broaden the appeal and market for the product mix.

Designer jeans represented an attempt by manufacturers to expand the market (appeal to new market segments) through product modification and spin-offs. As a result, sales of jeans rose significantly and reflected large profits for some of these manufacturers.

The large toothpaste market is presently going through some product modifications as firms attempt to gain a larger share. Beechum has introduced a third stripe on its Aqua-Fresh toothpaste. The firm is seeking support from the American Dental Association for their claim that the third stripe represents a formidable plaque-remover and thus differentiates it from all other toothpastes. In addition to this strategy other firms are turning to taste as a key consumer need. The development of gels in toothpaste by Procter & Gamble, Colgate-Palmolive, Beechum, and others has also been the basis for new direction in this market. The purpose of these modifications is to get a jump on the competition to increase market share in a mature market.

The A. T. Cross Company recently introduced a product modification to appeal to a market segment that was lost because of the increased cost of silver and gold. As the cost of these materials rose, the market segments also changed, leaving some consumers, who could not afford the expensive writing instruments, with no market alternative. Thus Cross introduced the Classic Black, a pen with the same internal mechanism as other pens but with an exterior finish of black-matte epoxy baked into a brass shell. The design, which is similar to all other Cross writing instruments, appeals to an executive male looking for a moderately priced product as an alternative to expensive gold and silver. Other spin-offs, such as a sleek, slender version of the Classic Black, are planned soon to appeal to the female market.

Sealtest Light n' Lively was an example of a product modification that was substituted for the original product. The Sealtest Ice Milk was approaching the later stages of the growth stage and was losing market share to competition. An investigation of this market indicated that consumers found ice milk to be low in flavor and texture. Sealtest enhanced the product's texture and flavor and reintroduced the product under the Light n' Lively name, which helped the firm repair their market position.

Some of these product modifications represent additions to the line and are considered spin-offs, such as the Cross Classic Black pen or Crest Gel, and others represent substitutions of the original product, such as Sealtest Light n' Lively. In either case, increased sales and profits are the ultimate motive.

An investigation of more than 152 spin-off situations revealed that the service or product involved in most instances had been on the market for several years and thus was likely to be in the mature stage of its product life cycle. In addition most of these spin-offs were not major contributors but did make more efficient use of facilities, generated new revenues, allowed the maintenance of in-house production or service facilities not otherwise affordable, and provided additional opportunities to test out management talent.[2]

As discussed earlier, packaging modifications or innovations can also enhance the marketability of a product. The flip-top can has enabled soft drink firms to expand the market for their products, making it easier for consumers to store or carry a soft drink product. It also provided extended opportunities in vending machines.

Pouch packs used for sauces, gravies, and instant potatoes also represent a modification that increased the use of the product by appealing to new market segments. In addition it provided the retailer with easier display opportunities and required less shelf space.

Future changes in packaging for certain products will continue to be a means of expanding market opportunities. Cans used for wine and boxes used for liquids are just a few innovations that we are likely to see in the near future. Management will however need to proceed cautiously with packaging changes and make sure that the change does satisfy a consumer need and is not simply a package that is more costly and difficult to use.

It is important for management to be prepared for possible product modifications and spin-offs. Since timing is so critical, management should not wait for the mature stage or for competition to get the jump on any of the possible strategies discussed here.

Following are suggestions that can be used to stimulate ideas and changes that can enhance future market share and profits for a mature product:[3]

1. Examine the major components of the product and determine possible options to reduce their materials or labor costs.
2. Conduct a marketing research study to determine what consumers are looking for in such a product.
3. Determine from the sales force what feature of the product is being attacked by the competition.
4. Evaluate appeals being used by competitors.
5. Have engineering evaluate the design to determine if the product actually "looks" like the price it carries.
6. Discuss the product category with the trade to determine what needs they have or perceive in their customers that are not being satisfied sufficiently.

Each of these suggestions can provide management with ideas and strategies for planning the destiny of mature products. Prior recognition and development of changes can lead to differentiation and increases in market share

[2]E. Patrick McGuire, *Spin-off Products and Services* (New York: The Conference Board, 1976), 3.
[3]Kenneth VanDyck, "New Products from Old: Short Cut to Profits," in *Product Planning*, 2nd ed., ed. A. Edward Spitz (New York: Petrocelli/Charter, 1977), 267–70.

and profits. It is also possible that these strategies will broaden the appeal of a product to a larger market, thus providing opportunities for market growth.

NEW USES FOR AN EXISTING PRODUCT

One of the most ideal strategies for a mature product is to find new uses for it. These new uses satisfy unmet consumer needs and can provide a significant boost in sales for a mature product. The case of Arm & Hammer baking soda is one of the most successful illustrations of finding new uses for a mature product.

For years Church & Dwight, the maker of Arm & Hammer baking soda, had promoted the fact that its product would absorb food odors. Approaching consumers with this issue was not a simple task, since no one wants to be identified as having a refrigerator-odor problem. However the firm was able to justify its product as insurance against ever encountering food odors. Thus the firm embarked on a campaign to promote Arm & Hammer baking soda as a product that, if left open in the refrigerator, could keep the refrigerator sweeter, cleaner, and fresher smelling.[4]

The results of the new use of this product were staggering within a very short time period. The success of this new use also stimulated the identification of other uses for this product. It is now also promoted as a low-cost replacement for chemicals used to keep swimming pools clean. In addition the firm has also introduced spin-offs: a personal deodorant and a laundry detergent.

Successful examples of finding new uses for a mature product are numerous. Johnson & Johnson found that many adults were using their baby shampoo, as well as other baby products, and utilized this information by initiating a campaign to promote the use of these products by adults. Since the baby market was declining, this opportunity to extend the product's life cycle was welcomed.

Food producers such as Kraft and Campbell have used new recipes through advertising as a means of extending the market for their products. Cookbooks and recipe promotions with other manufacturers are techniques used by Kraft for its cheeses and by Campbell for its soup lines.

Often these new uses either appeal to new market segments or require a minor product design or modification. These combinations have been successful for tea manufacturers (in the development of instant iced tea) and candy manufacturers (with Nestlé Ice Cream Bars and York Peppermint Bars). In these cases the product was modified (instant iced tea) to provide a new use or to appeal to a new market segment (ice cream novelties for children).

In the industrial market, Polaroid Corporation has found numerous new uses for its product when combined with other industrial products. For example Polaroid film is now used in conjunction with microscopes and X-ray machines. This equipment has been modified somewhat to be able to use the instant film. The advantages of having an instant picture of a test-sample slide

---

[4]Ernest Potischman, "How To Breathe New Life Into Your Old Brand," *Advertising Age*, 13 July 1973, 39–44.

under a microscope adds permanence to what might be a significant case. For X-ray machines, the film provides veterinarians with instant results of animal X-rays (e.g., horses) without having to leave the source to use a dark-room. These new uses are likely to continue to provide Polaroid with new opportunities for its instant photography process.

MARKETING MIX ADJUSTMENTS

In addition to finding new segments and uses, and redesigning or modifying products, a firm may consider revitalizing a product through specific changes in the marketing mix, pricing, promotion, and distribution. Table 20–1 identifies some of the general marketing strategies that may be employed for a mature product. In addition to price, promotion, and distribution, consideration of competition, research and development, and segmentation are also discussed in Table 20–1.

Discussion of product redesign, market segmentation, and competitive positioning were discussed earlier in this chapter. Although the use of the marketing mix may appear more subtle than other strategies, there are numerous examples of firms that have employed successful marketing mix adjustments to rejuvenate the mature product.

In the early 1970s Heinz Ketchup had reached a static market position. Management noted that more promotion emphasis had recently been di-

TABLE 20–1
Marketing strategies for the mature product

Competitive Strategy	Defense of brand position agsinst competing brands and the product category against other potential products.
Research and Development (Product Design)	Attention to product improvement and cost-cutting opportunities. Introduction of flanker products.
Market Segmentation	Analysis of market gaps representing unmet consumer needs. Concentration of market strategy by appealing to specific segments.
Price	Defensive pricing to maintain and protect market position. Use of incremental pricing opportunities, such as private labeling to build volume and reduce costs through added experience.
Promotion	Higher use of mass media directed to specific segments. Maintenance of dealer and trade loyalty through special dealer-oriented promotions.
Distribution	Intensive and extensive, and a strong emphasis on keeping dealer well supplied, but with a minimum inventory cost.

rected toward the trade rather than the consumer. Earlier successful growth of the product had resulted from extensive advertising that emphasized the qualities of the product to the consumer. With this information, management increased the consumer advertising budget and developed copy aimed at impressing consumers with the brand's benefits. The results were an increase in sales of 17 percent and an increase in market share of 12 percent.[5]

Pricing changes are often avoided by management if there is any likelihood of a price war, especially when there is little perceived consumer differentiation in the competitive offerings. However, in order to protect its market position, a firm may use temporary price changes to defend against an aggressive competitor. Special cents-off deals, 2-for-1 specials, or coupons may be used as temporary pricing changes for a mature product.

Ideally the firm would look for cost savings from research and develpment in order to actually reduce price yet maintain the profit margin. Alternatively research and development or product redesign may result in a new quality emphasis that would allow a price increase commensurate with the product quality.

Promotion strategy for the mature product could follow a similar pattern as that used by Heinz Ketchup. Generally the promotion mix is used to defend the market share by reminding previous buyers of the values obtained from the product and attracting market newcomers. Promotion may also encourage increased use of the product, which can enhance sales. Some of the promotion budget will also be needed to maintain interest and support among retailers. Premiums such as vacation trips and gifts based on volume can be offered to wholesalers and retailers.

Distribution strategy should include an analysis of the number of accounts, shelf space allotted, cooperative advertising programs, and the volume by account. The manufacturer will need to pay special attention to dealer service because the differentiation of alternative brands may be low. Channel logistics should be directed toward keeping the dealer inventory costs as low as possible.

Some selectivity at this point may also be warranted by the manufacturer. Certain customer accounts may no longer be profitable because of low volume and high support costs. Two alternatives may be possible for these accounts: 1) use manufacturer's reps, sharing the cost per sales call among other complementary manufacturers, 2) eliminate these accounts (no longer service them directly).

For all mature products the distribution strategy should be interfaced with the pricing and promotion elements in the marketing mix. Thus a change in one element, such as promotion (increase in dealer incentives), may affect pricing (lower price) or distribution (increase in intensity).

Firms have also used branding and packaging strategies to enhance market coverage. Added package sizes or new competing brands by the same manufacturer can often provide more shelf space and hence increase total sales. For example toothpastes are being introduced in many different packaging sizes and flavors to enhance shelf-space. However this strategy can also

---

[5]David S. Hopkins, *Business Strategies for Problem Products* (New York: The Conference Board, 1977), 23.

backfire. Too many competitive brands or packaging sizes can result in losses, even though sales for the category increase because of higher costs. Retailers and wholesalers also have limited space and often will not carry all the packaging sizes or flavors offered by one manufacturer.

PRODUCT SURVEILLANCE PROCEDURES

The mature product requires constant attention; otherwise it will move more rapidly to the decline stage and possibly become a cost drain in the company. Constant monitoring through the tracking of critical variables can be helpful to management during this period. During turbulent economic times, it is even more necessary for a firm to develop improved methods for tracking and reporting data.

Such surveillance techniques can take many forms. For example a metals supplier found that a more frequent reporting of economic indices in relationship to costs, pricing, sales and other market variables kept management better informed and resulted in better control over its mature products.

Use of special reports from the field sales force, supplemented by market research data of customer comments and complaints, can also enhance market awareness. Increased responsibility of group product managers regarding studies of the individual parts of their product lines can also help in identifying problems before they become too severe.

Another surveillance or monitoring device is a product review committee. This committee can exclude managers having day-to-day responsibilities for product sales in order to minimize bias. Any product identified by the committee as having serious problems can be referred to marketing research or another special group responsible for the task of developing corrective action.

The success of a company's strategy in maintaining sales market share and profits through a product's mature stage in the product life cycle is no simple task. Because many mature products are viewed by consumers as having little difference, concerted effort is needed to monitor and plan effective marketing strategy. As is often the case, the mature product represents the cash cow in the product mix. Longevity for this product is thus necessary to support other products and research and development to ensure the long-term success of the company. Without effective market strategy the product may quickly move into rapid decline and force management to consider its elimination.

SUMMARY

In earlier chapters we focused on the development process for a new product, planning (strategic and market/product planning), and marketing mix decisions prior to and during the early stages of the new product introduction. The previous chapter and this chapter were concerned with control of a new product and a mature product.

Since the mature stage tends to be the longest stage in the product life cycle, and since consumers usually perceive little difference in this stage, management must make a concerted effort to seek alternative market strategies to maintain market position and ensure a long and profitable maturity stage. Management may identify new market segments, modify or redesign the product, find new uses for the product, or adjust the marketing mix.

New market segments may involve international markets (Coke and Pepsi), marketing an industrial product in a consumer market (Janitor-in-a-Drum), or making minor modifications that appeal to a new segment (female market with cigarettes or adults with board games). Product modification may be part of a strategy to appeal to new market segments. Spin-offs or a redesign of an existing product are strategies used to expand the potential market for a mature product. Toothpastes, Cross pens, and Light n' Lively were examples discussed that achieved success by using spin-offs and product redesign to revitalize a mature product.

New uses of an existing product can also be used as an effective strategy in the mature stage. Arm & Hammer, Johnson & Johnson, Kraft, and Campbell were examples of firms that successfully stimulated sales of a mature product. Each of these strategies may not only impact sales, but could establish a broader appeal (new market segments) for the product.

Monitoring and tracking products through the later stages of their life cycle could provide management with a formal process to protect market positions from competition. A reluctance by management to track and monitor a product may prove to be risky and could hasten a product's decline and eventual elimination.

SELECTED READINGS

ABRAMS, BILL. "P & G May Give Crest A New Look After Failing To Brush Off Rivals." *Wall Street Journal*, 8 January 1981, 21.
> Describes the strategy introduced by P & G as a defense against two competitors. Strategy is typical of changes in the marketing mix necessitated by product maturity.

DAY, GEORGE S. "Product Life Cycle: Analysis and Applications Issues." *Journal of Marketing* 45 (Fall 1981): 60–67.
> Relates some of the strategy formulation of the various stages of the product life cycle. Discusses a product's evolution to maturity and decline.

ENIS, BEN M., LaGRACE, RAYMOND, and PRELL, ARTHUR E. "Extending the Product Life Cycle." *Business Horizons* 20, June 1977, 46–56.
> Discusses a revised model of the product life cycle to provide a better framework for marketing decision-making at the individual brand level. In addition to restructuring the stages of development, this new model incorporates cost and profit curves to the sales graph. Also discusses implications for marketing strategies and corporate planning/control.

HOPKINS, DAVID S. *Business Strategies for Problem Products*. New York. The Conference Board, 1977.
> Provides a procedure for diagnosing problem products and decision alternatives for these products as they reach the mature stage of the product life cycle. Also discusses preventive actions and strategies.

MASON, R. S. "Product Maturity and Marketing Strategy." *European Journal of Marketing* No. 1 (1976): 36–47.

Frequent review of product sales is essential as part of the planning and control process. Article describes the procedure of reviewing the status of mature products and modifying strategy.

McGUIRE, PATRICK E. *Spin-Off Products and Services.* New York: The Conference Board, 1976.

One possibility for extending the success of a mature product is to develop spin-off products that depend on the success of the mature product. Discusses some of the principles and practices of developing spin-off products. Also provides selected examples of spin-off ventures.

MONTGOMERY, DAVID G., and WEINBERG, CHARLES B. "Toward Strategic Intelligence Systems." *Journal of Marketing* 43 (Fall 1979): 41–52.

With the dramatic increase in strategic planning it has become necessary to develop a mechanism to provide quality information. Paper discusses an overview of a system to ensure quality information and explains how this information can be analyzed and processed.

INDEX